1983

CONDITIONS FOR LIFE

1983

Readings from
**SCIENTIFIC
AMERICAN**

CONDITIONS FOR LIFE

*With Introductions and
Additional Material by*
Aharon Gibor
*University of California
Santa Barbara*

W. H. Freeman and Company
San Francisco

Most of the SCIENTIFIC AMERICAN articles in
CONDITIONS FOR LIFE are available as separate
Offprints. For a complete list of more than 950
articles now available as Offprints, write to W. H.
Freeman and Company, 660 Market Street, San
Francisco, California 94104.

Library of Congress Cataloging in Publication Data
Main entry under title:

Conditions for life.

 Bibliography: p.
 Includes index
 1. Life (Biology) — Addresses, essays, lectures.
I. Gibor, Aharon, 1925-   II. Scientific
American.
QH501. C66   577   76-22196
ISBN 0-7167-0480-3
ISBN 0-7167-0479-X pbk.

Printed in the United States of America

9 8 7 6 5 4 3 2 1

In space there are countless constellations, suns, and
planets. We see only the suns because they give light;
the planets remain invisible, for they are small and dark.
There are also numberless earths circling around their
suns, no worse and no less inhabited than this globe of
ours. For no reasonable mind can assume that heavenly
bodies which may be far more magnificent than ours would
not bear upon them creatures similar or even superior
to those upon our human Earth.

— Giordano Bruno

Burned at the stake
for heresy; Rome, 1600

PREFACE

The title of this anthology suggests several interesting biological questions. One of these is the perennially intriguing question of the existence of life outside the earth. "Conditions for life" hints at the possibility of extraterrestrial environments where life similar to our own might exist. Indeed, increased understanding of the structure and evolution of the universe (astronomy and cosmology) and of the structure and function of the chemical components of living cells (molecular biology) has convinced many scientists that Earth is almost certainly not the one and only habitat of life in the universe. It is the belief of virtually all natural scientists that the living state represents an advanced level of organization of matter achieved by a long process of evolution. Any planet that resembles Earth in chemical composition and physical conditions is likely to contain some matter exhibiting certain properties that on earth we identify as life.

Another idea that the title suggests is that of the effects of modern human technology and lifestyle on the conditions of the biosphere. It is increasingly apparent that our casual disregard for the balance of nature that has evolved over the last three billion years could soon make the earth inhospitable not only to us but to all forms of life.

The pattern of subjects in this anthology follows that of an undergraduate course that I teach in the general physiology of the biosphere. In this course I discuss the physiological adaptations of living matter to the great variety of physical and chemical conditions that exist on the earth's surface.

Part I deals with the elemental composition of the biomass and its relation to the composition of the inanimate portions of the biosphere. How the vital elements circulate in the biosphere to assure the continuity of life is also discussed.

Part II illustrates the fact that the biomass is not just a conglomerate of specific elements and compounds in the correct proportions, but rather a complex and highly ordered arrangement of matter. The hierarchy of organization of the biomass and the forces that are responsible for it are considered. Starting with simple chemical bonds, we proceed through intermolecular forces to an example of the organization of a complex cellular structure—a consequence of these forces.

The remaining sections deal with the various environmental forces to which the biomass is subject. Interactions with radiant energy are discussed in Part III. Radiant energy is presented as a source of energy, as a sensing tool, and as a harmful agent for biological matter. Part IV deals with the temperature range in which biological activities occur. Part V deals with the effects of high pressures on living organisms and with the sensitivity of organisms to slight variations in external pressure. In addition, there is a discussion of gravity, another environmental force that organisms have learned to sense and utilize to their advantage.

A point that I tried to emphasize in choosing the articles for the last three sections is the great versatility and adaptability of living matter, which exists under almost all prevailing environmental conditions on earth. I have no doubt that the limits of this versatility and adaptability of living matter are not defined by the conditions that happen to exist on earth right now. It is widely believed that the original living organisms of this planet had to cope with and adapt to an environment quite different from that which exists at present.

Our environment continues to change, of course, but the imperceptibly slow rate of natural change has been drastically accelerated by the effects of human technology since the Industrial Revolution. Granted that we may be changing the average temperature of the atmosphere through industrial pollution (CO_2 tends to increase the average temperature whereas smoke and dust tend to decrease it) and that we may have begun to destroy the ozone layer of the upper atmosphere (which would allow more of the sun's harmful UV radiation to reach the earth's surface)—the question to ponder is whether these and other effects of our species will result in the rapid destruction of life on earth or whether they will merely accelerate the rate of biological evolution by imposing new environmental demands on the biomass.

As an optimistic final chapter, I added the stimulating article by Carl Sagan and Frank Drake, who present the arguments for the presence not just of life, but of *intelligent* life, in the universe. It is reassuring to think of the universe around us as a vast forest teeming with all kinds of living creatures, rather than as a great, sterile desert.

I wish to thank Fred Raab of W. H. Freeman and Company for his help in making my contributions to this book more succinct and coherent.

May 1976   *Aharon Gibor*

CONTENTS

V   PRESSURE, GRAVITY, AND LIFE

EPILOGUE

Note on cross-references to SCIENTIFIC AMERICAN *articles:* Articles included in this book are referred to by title and page number; articles not included in this book but available as Offprints are referred to by title and Offprint number; articles not included in this book and not available as Offprints are referred to by title and date of publication.

CONDITIONS FOR LIFE

I

ELEMENTAL COMPOSITION OF THE BIOMASS

ELEMENTAL COMPOSITION OF THE BIOMASS

I

INTRODUCTION

This anthology deals with the relations between living organisms and their environments. The planet Earth is the only place in the universe where we know that life exists. However, it is now a common belief among scientists that a similar state of matter also exists on many other planets. I say "belief" because we still await concrete evidence of life outside the earth.

Before embarking on a search for life we must define what we are looking for. What life is and how it originated are questions that philosophers have pondered from earliest recorded history. It was only in the last few decades, however, that scientists began to tackle these questions in earnest, with more than armchair tactics. Experiments were conducted in many laboratories in an attempt to simulate conditions that might be conducive to the spontaneous generation of life, or at least the basic chemical compounds of which living matter is composed. It was found that amino acids, the building blocks of proteins, are formed by reacting certain simple chemical compounds. For example, when gaseous mixtures of methane, water, and ammonia are mixed and energy supplied in the form of electric discharges, energetic radiation, shock waves, or heat, a spontaneous chemical reaction takes place that yields a mixture of amino acids and other organic compounds.

We think that similar gas mixtures and energy sources existed in the atmosphere during the early history of our planet. It is generally believed that very similar chemical conditions also exist or existed on many other planets in the vast universe. (See, e.g., the special issue of SCIENTIFIC AMERICAN of September 1975, devoted to recent research on the solar system.) Thus we are led to the inevitable conclusion that the molecular precursors to proteins, lipids, and nucleic acids accumulated on the earth during its early history, as well as on other planets. On earth these substances eventually evolved into living structures. Similar chemical evolution has probably taken place on other planets where organic substances accumulated.

Besides experiments in which we attempt to simulate presumed conditions on other planets, a more direct approach to the study of the chemistry of other worlds is the analysis of meteorites that occasionally strike the earth. Such analysis has revealed that these extraterrestrial rocks indeed contain amino acids and other organic compounds that are similar to the earth's biochemical compounds. We do not know the precise origin of the meteorites, but the intriguing findings reveal that the basic biochemical substances that characterize living matter as we know it are *not* unique to our planet.

As I write these words, we are about to land an instrument package on Mars to search for the presence of life there, a most exciting prospect. We will be searching for life, but how are we going to identify it? What are the basic characteristics of the living state of matter? An admission of our limited knowledge and imagination of what constitutes a life process is implicit in the phrase, "life as we know it," which almost invariably qualifies our statements concerning the search for extraterrestrial life. This qualifying phrase is interesting from a historical point of view. It may be a vestige of man's ancient, egocentric view of the universe.

It took the efforts of Copernicus, Galileo, and other astronomers to remove the earth from the center of the universe and place it in its humble orbit. Darwin is rightly credited with dislodging man from the unique pedestal upon which he had placed himself. But as far as the basic nature of life is concerned, we seem to be limited to the recognition of only those structures that possess a chemistry similar to our own as being "alive." Are we also overly egocentric in this regard? Only the future will tell.

At present, however, it is wise to limit our consideration to the forms of life with which we are familiar. At the outset we must distinguish two aspects of the question, "what is life?" One is the description of the unique properties of living organisms. The other is the quest for an explanation of the causes of these unique properties.

It is usually a simple task for anyone to distinguish intuitively between living and nonliving objects. However, every one of us has at one time or another encountered objects that necessitated closer study before we could decide whether they were animate or inanimate. Indeed, it is not easy to come up with a general definition that encompasses the properties common to all living organisms and to them only.

We usually list several properties, each of which alone might also be found among nonliving things, but which together serve well to characterize living organisms: (1) Living matter is composed of compounds of the element carbon of the type called *organic compounds*. (2) The living state is characterized by continuous exchange of matter and energy with the environment, a process called *metabolism*. (3) As a consequence of metabolism, living matter grows and produces more of itself, i.e., there is *growth* and *reproduction*. (4) Reproduction generally results in offspring that are essentially identical to their parents. Occasionally, however, significant modifications are found; such modified organisms are called *mutants*. The potential to mutate is a most significant property of living matter because it is this attribute that is responsible for evolution in the living world. (5) Living organisms are responsive to changing conditions of their environment; we refer to this property as *sensitivity*.

Using these criteria, we can recognize a living organism, but how are we to explain its existence? We will not dwell on the vitalistic view, which explains the living state as being due to mysterious vital forces with which such matter is endowed. A more fruitful view is the *mechanistic* view, which assumes that the living state represents only a vastly more complex state of matter than that of inanimate systems, but one that obeys all the same laws of chemistry and physics. It follows that with further research we will eventually be able to explain all the properties of living organisms.

Life processes occur within the framework of the inanimate environment, which includes all the chemical and physical conditions that stimulate or inhibit such processes. By regarding a life process as a complex set of simple chemical reactions and physical interactions, it is possible to examine the relations of a living system with its environment in terms of chemistry and physics. This is, in fact, the basic theme of this volume: the consideration of the intricate relations between living matter and its environment.

The environment should not be considered a passive abode for living organisms; on the contrary, it is in a dynamic and reciprocal interaction with all living matter. The environment provides the material—the atoms—from which living matter is built. It is the source of the energy that organisms utilize to synthesize their own molecules, and it provides the appropriate temperature for the biochemical reactions to proceed at the appropriate rates. The environment, in turn, is affected by life processes. End products of metabolic reactions are excreted into the environment and cause significant changes in it.

A dramatic example of the effects of life on the inanimate world is the chemical composition of the atmosphere. It is now generally accepted that the earth's atmosphere was without oxygen during the first half of its total age of 5 billion years. About 2.5 billion years ago biological evolution introduced an important chemical reaction in which radiant energy dissociated water molecules and released molecular oxygen. This photosynthetic reaction is responsible for the accumulation of oxygen in the atmosphere.

The consequences of this change in the chemistry of the atmosphere were far-reaching. Not only did the availability of oxygen impose special demands on the biochemistry of living organisms, which until then had lived anaerobically, it also caused the oxidation of elements in the rocks and the oceans. Furthermore, O_2 molecules absorb ultraviolet radiation. Therefore the transparency of the atmosphere to the sun's rays diminished, and less total radiant energy reached sea level. Significantly, since little of the biologically damaging UV radiation reached the surface of the earth, living organisms could emerge from the sea and occupy the land masses.

A less dramatic but more direct effect of living organisms on their environment is seen in the weathering of rocks and the production of soil. Roots of plants and various microorganisms directly affect the rate of weathering by the secretion of acids, which accelerate the dissolution of certain components of the rocks. Dissolution of rocks makes the mineral components available for use by living organisms, but this also erodes the land.

Evolution results from changes in living organisms that make them better adapted to cope with changes in the environment. Since living organisms, in turn, cause modification of their environment, we should regard evolution as a process of the biosphere as a whole rather than just of living matter.

Of all the important roles that the environment plays in regulating the living world, perhaps the most significant is that of reservoir of the atoms of which living matter is built. Living matter is composed of the same kinds of atoms of which the inanimate world is composed. The list of essential elements now includes about one-third of all the elements. During growth and reproduction, more living matter is produced by acquiring the necessary atoms and molecules from the environment. When an organism dies, its matter returns to become part of the inanimate portion of the biosphere. Only when all the required elements are present in their immediate environment are living organisms able to grow. Evolution has created organisms that differ somewhat in their needs for certain trace chemical elements. It is possible, therefore, that in a specific location where a certain chemical element is lacking, some organisms can grow while others cannot.

Some of the elements, such as carbon, oxygen, hydrogen, nitrogen, sulfur, and phosphorus, are essential to all living organisms. There are few habitats on earth where these elements are absent. There are, however, many localities where the abundance of one or more of these elements becomes the limiting factor in the growth of the living community. For example, the growth of corn in a fertile, moist soil on a sunny, still day is limited by the availability of carbon, which plants utilize in the form of carbon dioxide from the atmosphere. The concentration of this gas in the atmosphere is only 0.03% by volume, and the supply to an actively growing field of corn on a windless day is soon exhausted. As a matter of fact, some farmers add CO_2 to the air that they pass through greenhouses, a practice that increases the growth rate of their crops.

Fertilizer is added to soil to meet the need of intensively grown crops for some essential elements, notably nitrogen and phosphorus. This is necessary to counteract the continuous depletion of the soil as the crops are harvested.

An important consideration for the required elements, other than their actual presence in the environment, is their mobility. In order to maintain a biosphere for long periods, the chemical elements must remain available to living organisms as the elements move in an unending cycle between the animate and inanimate portions of the biosphere.

Uneven heating of the earth's surface creates winds and ocean currents. Wind carries and distributes sea salt and terrestrial dust over the land. Ocean currents and the upwelling of deep, nutrient-rich waters contribute to the redistribution and cycling of the elements. On a geological time scale, land masses erode and are carried to the depths of the ocean, but the ocean floor rises and forms new land masses.

Our primary consideration in this section is the elemental composition of living matter and the importance of the environment as the source of these elements. In his article, "The Chemical Elements of Life" (page 10), Earl Frieden reviews the accumulated information on the composition of the biomass and compares it to the composition of the earth's crust and the universe. He also presents recent findings on the role of the essential elements in the life of higher animals. The uniqueness of the chemistry of the biomass is apparent from the fact that 99.9% of it involves only 11 elements. The remaining 0.1% involves more than a dozen other elements, which are referred to, for obvious reasons, as trace elements.

Frieden's arguments concerning the chemical properties of carbon and silicon and the suitability of their compounds as the building blocks of a biochemistry are intriguing. He argues that carbon-carbon bonds are more stable than silicon-silicon bonds, whereas polymers built from silicon-oxygen-silicon backbones, the so-called silicones, are *too* stable. It is interesting to note in this regard that the backbone of the most stable biopolymer, DNA, is composed of the repeating units $(-C-C-C-O-P-O-)_n$, whereas that of a polypeptide (also a stable biopolymer) is composed of (-C-C-N-C-C-N-)$_n$. Obviously we are not dealing with pure chains of carbon atoms.

Silicon has been shown to be an essential element for chicks; it is also a major component of the cell walls of diatoms, and is found in the cell walls of many weeds. The biochemical pathways by which silicon is incorporated into these systems are not well known. The possibility that complexes of carbon and silicon compounds may serve on an alien planet as building blocks of a biochemistry different from our own cannot be dismissed outright. Such a silicon-rich biochemistry might be suitable at higher temperatures and pressures than those to which our biochemistry is adapted.

Studies such as those described by Frieden on the essential role of certain elements in chicks, mice, or other higher animals cannot be generalized to indicate that the same elements are required by *all* organisms. Qualitative differences exist between different organisms: simple forms of life might not require some elements that are essential to higher forms. Fluorine, for example, is known to be important for the healthy development of teeth and bones, but we do not know whether protozoa or even mammalian cell cultures require fluorine in order to live.

The differences might also be quantitative, some organisms requiring much less or much more of a specific element than other organisms. Silicon is a good example: diatom cultures require the addition of soluble silicon salts for active growth, whereas the culture media of other organisms do not require this element. If there is a need for silicon, the trace amounts that are present as contaminants or that dissolve from the glass containers are sufficient for most other algae.

The implications of these differences in the needs of different organisms are rather important. As we will see in the following articles, geographic areas that lack certain elements cannot sustain the life of some organisms, though a flora and fauna of different organisms exist in the very same locality.

The next article, "Mineral Cycles" (page 19), by Edward S. Deevey, Jr., emphasizes the significance of the cycling of the essential elements in the biosphere. I should point out the differences in the meaning of the term *biosphere* as used by Deevey and by myself and others. In his article, Deevey refers to the entire mass of living organisms and their immediate organic products as the biosphere. This is what I refer to as the *biomass* or sometimes as the animate portion of the biosphere. Biosphere in my usage refers to the biomass *and* that portion of the earth's crust with which it continuously exchanges matter, namely, the atmosphere, hydrosphere, and lithosphere. The important point to note in Deevey's article is the basic difference between those elements that have volatile forms and can thus cycle through the atmosphere and those, such as phosphorus, that are not volatile and are thus confined to the lithosphere/hydrosphere and the biomass.

Another path that nonvolatile substances can take in the recycling process is via volcanic smoke and ash, which sometimes cover large areas of the globe. Smoke and ashes from forest fires also distribute minerals over large areas.

Minerals also return from the ocean to the land via biological pathways. The droppings of birds that feed on fish form large guano deposits in some localities. Some marine fish migrate up rivers to lay their eggs; they then die and their decomposing bodies fertilize the river waters. Apparently there are no quantitative data to indicate the relative importance of these pathways.

Ocean spray is also a very important vehicle for mineral cycling, more so than previously suspected. Rainwater, even in inland locations, has been found to contain considerable amounts of minerals. The mineral composition of rainwater in New Zealand, for example, was equivalent to a solution of 0.5 milliliter of seawater in one liter of distilled water. Not only minerals but a considerable amount of organic substances of marine origin were found in the rainwater. [See also Ferren MacIntyre's article, "The Top Millimeter of the Ocean" (Offprint 913).]

Deevey refers briefly to the studies of F. Herbert Bormann and Gene E. Likens concerning a watershed called Hubbard Brook in New Hampshire. These authors tell in detail of their interesting studies in their article, "The Nutrient Cycles of an Ecosystem" (page 29).

Ecological studies on mineral cycling in a habitat were in the past mainly qualitative, with some rough quantitative estimates or guesses. A basic difficulty in attempting to measure the loss of minerals from a terrain is that losses via subsurface water flow cannot be measured. The unique terrain of the steep watershed studied by Bormann and Likens minimizes this problem and enabled them to measure lost minerals directly by analyzing stream waters flowing from the watershed. Careful measurements of the minerals brought in by rain and dust fall made possible a direct calculation of the quantitative cycling of important elements. These measurements show clearly the effects of the biomass on the mobility of some soluble elements. Deforestation caused a marked increase in the rate of leaching out of many elements from the watershed.

This article is a good example of the methods used by modern ecologists in attempting to understand the intricate relations between the components of the biosphere. A significant point brought out by the authors is the effect of the increased concentration of acids in the atmosphere (from industrial pollution) on the leaching of minerals from the soil. The acidic rainwater contains hydrogen ions, which readily exchange with calcium, magnesium, and potassium ions of the soil, thus causing their leaching out in the runoff water. An interesting finding is the fact that some of the elements that are tightly bound by the biomass have a positive annual balance, indicating an enrichment of the habitat with respect to them.

In the first three articles the essential role of various elements and how they recycle in the biosphere are considered. In the next article, "Trace-Element Deserts" (page 39), A. J. Anderson and E. J. Underwood describe situations in which the inhomogeneity of the biosphere with respect to its mineral composition is emphasized.

Not all the essential elements are equally distributed over the surface of the earth. Regions that are deficient in one or more elements become apparent, especially when demands for biological productivity of the specific region are increased. It is obvious that intense agricultural demands will readily show any deficiency of the soil. Apparently this situation caused severe problems to Australian farmers. The cooperation of soil scientists and experts in plant and animal nutrition produced results that were interesting and rewarding to biologists as well as the local farmers. Most significant was the demonstration that different organisms depend to different degrees on the availability of an ample supply of a specific element. Cobalt deficiency is an example: although good growth of the pasture plants could be maintained, the grazing sheep came down with deficiency symptoms. Copper deficiency was also much more noticeable in its effects on sheep than on cattle.

It is well known that soils that are especially rich in some elements will support a unique flora. Mineral prospectors often rely on such botanical indicators. Undoubtedly deficiencies of specific elements also enable the development of a characteristic biota that includes those organisms able to survive without the missing element. Such differences between the requirements for a given element suggest interesting studies on the comparative biochemistry of organisms that do require a trace nutrient and those that do not.

Because of geographic and climatic irregularities, the distribution of rainfall over the earth's surface is not uniform. The common term *desert* refers to areas in which water is a limiting factor for life. By analogy, we refer to areas that lack certain elements as *trace-element deserts*. An important ecological difference between a water desert and a trace-element desert relates to the time scale involved. In the latter, redistribution of trace mineral elements does occur, but only on a geological time scale. Water deserts, on the other hand, have short periods, annually or once every several years, when water becomes available. Organisms adapted to life in such deserts have evolved interesting strategies that enable them to survive under these conditions. These evolutionary adaptations of desert plants are the subject that Fritz Went considers in his article, "The Ecology of Desert Plants" (page 46).

Exobiologists who are searching for life on Mars are also considering the possibilities of biological activity over very short periods of time. Water is thought to be present in some areas of Mars, but it is probably frozen most of the time. It is likely, however, that at certain times in the Martian year and in some regions of the planet, liquid water becomes available. If there is any life there it might have adopted a strategy similar to that of the desert plants described by Went. The essence of this strategy is a rapid utilization of the available water so that the life cycle can be completed prior to the return of the drought.

1

The Chemical Elements of Life

by Earl Frieden
July 1972

Until recently it was believed that living matter incorporated 20 of the natural elements. Now it has been shown that a role is played by four others: fluorine, silicon, tin and vanadium

How many of the 90 naturally occurring elements are essential to life?. After more than a century of increasingly refined investigation, the question still cannot be answered with certainty. Only a year or so ago the best answer would have been 20. Since then four more elements have been shown to be essential for the growth of young animals: fluorine, silicon, tin and vanadium. Nickel may soon be added to the list. In many cases the exact role played by these and other trace elements remains unknown or unclear. These gaps in knowledge could be critical during a period when the biosphere is being increasingly contaminated by synthetic chemicals and subjected to a potentially harmful redistribution of salts and metal ions. In addition, new and exotic chemical forms of metals (such as methyl mercury) are being discovered, and a complex series of competitive and synergistic relations among mineral salts has been encountered. We are led to the realization that we are ignorant of many basic facts about how our chemical milieu affects our biological fate.

Biologists and chemists have long been fascinated by the way evolution has selected certain elements as the building blocks of living organisms and has ignored others. The composition of the earth and its atmosphere obviously sets a limit on what elements are available. The earth itself is hardly a chip off the universe. The solar system, like the universe, seems to be 99 percent hydrogen and helium. In the earth's crust helium is essentially nonexistent (except in a few rare deposits) and hydrogen atoms constitute only about .22 percent of the total. Eight elements provide more than 98 percent of the atoms in the earth's crust: oxygen (47 percent), silicon (28 percent), aluminum (7.9 percent), iron (4.5 percent), calcium (3.5 percent), sodium (2.5 percent), potassium (2.5 percent) and magnesium (2.2 percent). Of these eight elements only five are among the 11 that account for more than 99.9 percent of the atoms in the human body. Not surprisingly nine of the 11 are also the nine most abundant elements in seawater [*see illustration on page 12*].

Two elements, hydrogen and oxygen, account for 88.5 percent of the atoms in the human body; hydrogen supplies 63 percent of the total and oxygen 25.5 percent. Carbon accounts for another 9.5 percent and nitrogen 1.4 percent. The remaining 20 elements now thought to be essential for mammalian life account for less than .7 percent of the body's atoms.

The Background of Selection

Three characteristics of the biosphere or of the elements themselves appear to have played a major part in directing the chemistry of living forms. First and foremost there is the ubiquity of water, the solvent base of all life on the earth. Water is a unique compound; its stability and boiling point are both unusually high for a molecule of its simple composition. Many of the other compounds essential for life derive their usefulness from their response to water: whether they are soluble or insoluble, whether or not (if they are soluble) they carry an electric charge in solution and, not least, what effect they have on the viscosity of water.

The second directing force involves the chemical properties of carbon, which evolution selected over silicon as the central building block for constructing giant molecules. Silicon is 146 times more plentiful than carbon in the earth's crust and exhibits many of the same properties. Silicon is directly below carbon in the periodic table of the elements; like carbon, it has the capacity to gain four electrons and form four covalent bonds.

The crucial difference that led to the preference for carbon compounds over silicon compounds seems traceable to two chemical features: the unusual sta-

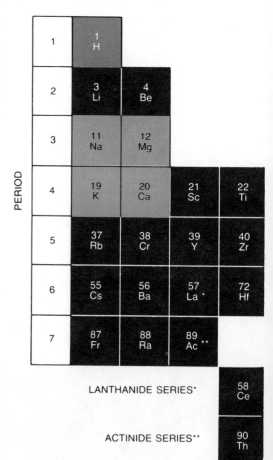

1	1 H				
2	3 Li	4 Be			
3	11 Na	12 Mg			
4	19 K	20 Ca	21 Sc	22 Ti	
5	37 Rb	38 Cr	39 Y	40 Zr	
6	55 Cs	56 Ba	57 La *	72 Hf	
7	87 Fr	88 Ra	89 Ac **		

PERIOD

LANTHANIDE SERIES*   58 Ce

ACTINIDE SERIES**   90 Th

ESSENTIAL LIFE ELEMENTS, 24 by the latest count, are clustered in the upper half of the periodic table. The elements are ar-

bility of carbon dioxide, which is readily soluble in water and always monomeric (it remains a single molecule), and the almost unique ability of carbon to form long chains and stable rings with five or six members. This versatility of the carbon atom is responsible for the millions of organic compounds found on the earth.

Silicon, in contrast, is insoluble in water and forms only relatively short chains with itself. It can enter into longer chains, however, by forming alternating bonds with oxygen, creating the compounds known as silicones (–Si–O–Si–O–Si–). Carbon-to-carbon bonds are more stable than silicon-to-silicon bonds, but not so stable as to be virtually immutable, as the silicon-oxygen polymers are. Nevertheless, silicon has recently been shown to be essential in a way as yet unknown for normal bone development and full growth in chicks.

The third force influencing the evolutionary selection of the elements essential for life is related to an atom's size and charge density. Obviously the heavy synthetic elements from neptunium (atomic number 93) to lawrencium (No. 103), along with two lighter synthetic elements, technetium (No. 43) and promethium (No. 61), were never available in nature. (The atomic number expresses the number of protons in the nucleus of an atom or the number of electrons around the nucleus.) The eight heavy elements in another group (Nos. 84 and 85 and Nos. 87 through 92) are too radioactive to be useful in living structures. Six more elements are inert gases with virtually no useful chemical reactivities: helium, neon, argon, krypton, xenon and radon. On various plausible grounds one can exclude another 24 elements, or a total of 38 natural elements, as being clearly unsatisfactory for incorporation in living organisms because of their relative unavailability (particularly the elements in the lanthanide and actinide series) or their high toxicity (for example mercury and lead). This leaves 52 of the 90 natural elements as being potentially useful.

Only three of the 24 elements known to be essential for animal life have an atomic number above 34. All three are needed only in trace amounts: molybdenum (No. 42), tin (No. 50) and iodine (No. 53). The four most abundant atoms in living organisms—hydrogen, carbon, oxygen and nitrogen—have atomic numbers of 1, 6, 7 and 8. Their preponderance seems attributable to their being the smallest and lightest elements that can achieve stable electronic configurations by adding one to four electrons. The ability to add electrons by sharing them with other atoms is the first step in forming chemical bonds leading to stable molecules. The seven next most abundant elements in living organisms all have atomic numbers below 21. In the order of their abundance in mammals they are calcium (No. 20), phosphorus (No. 15), potassium (No. 19), sulfur (No. 16), sodium (No. 11), magnesium (No. 12) and chlorine (No. 17). The remaining 10 elements known to be present in either plants or animals are needed only in traces. With the exception of fluorine (No. 9) and silicon (No. 14), the remaining eight occupy positions between No. 23 and No. 34 in the periodic table [see illustration below]. It is inter-

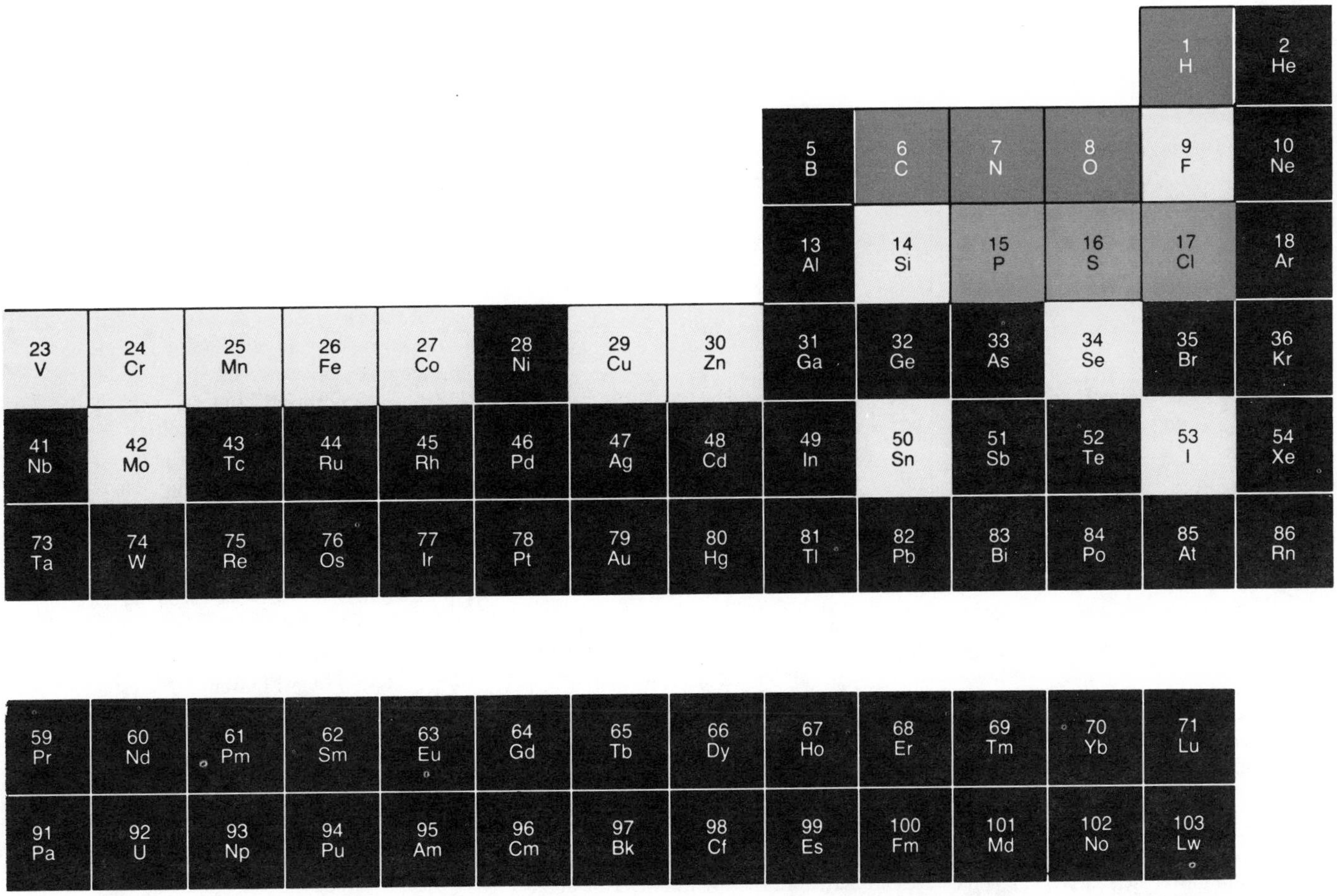

ranged according to their atomic number, which is equivalent to the number of protons in the atom's nucleus. The four most abundant elements that are found in living organisms (hydrogen, oxygen, carbon and nitrogen) are indicated by dark color. The seven next most common elements are in lighter color. The 13 elements that are shown in lightest color are needed only in traces.

esting that this interval embraces three elements for which evolution has evidently found no role: gallium, germanium and arsenic. None of the metals with properties similar to those of gallium (such as aluminum and indium) has proved to be useful to living organisms. On the other hand, since silicon and tin, two elements with chemical activities similar to those of germanium, have just joined the list of essential elements, it seems possible that germanium too, in spite of its rarity, will turn out to have an essential role. Arsenic, of course, is a well-known poison.

Functions of Essential Elements

Some useful generalizations can be made about the role of the various elements. Six elements—carbon, nitrogen, hydrogen, oxygen, phosphorus and sulfur—make up the molecular building blocks of living matter: amino acids, sugars, fatty acids, purines, pyrimidines and nucleotides. These molecules not only have independent biochemical roles but also are the respective constituents of the following large molecules: proteins, glycogen, starch, lipids and nucleic acids. Several of the 20 amino acids contain sulfur in addition to carbon, hydrogen and oxygen. Phosphorus plays an important role in the nucleotides such as adenosine triphosphate (ATP), which is central to the energetics of the cell. ATP includes components that are also one of the four nucleotides needed to form the double helix of deoxyribonucleic acid (DNA), which incorporates the genetic blueprint of all plants and animals. Both sulfur and phosphorus are present in many of the small accessory molecules called coenzymes. In bony animals phosphorus and calcium help to create strong supporting structures.

The electrochemical properties of living matter depend critically on elements or combinations of elements that either gain or lose electrons when they are dissolved in water, thus forming ions. The principal cations (electron-deficient, or positively charged, ions) are provided by four metals: sodium, potassium, calcium and magnesium. The principal anions (ions with a negative charge because they have surplus electrons) are provided by the chloride ion and by sulfur and phosphorus in the form of sulfate ions and phosphate ions. These seven ions maintain the electrical neutrality of body fluids and cells and also play a part in maintaining the proper liquid volume of the blood and other fluid systems. Whereas the cell membrane serves as a physical barrier to the exchange of large molecules, it allows small molecules to pass freely. The electrochemical functions of the anions and cations serve to maintain the appropriate relation of osmotic pressure and charge distribution on the two sides of the cell membrane.

One of the striking features of the ion distribution is the specificity of these different ions. Cells are rich in potassium and magnesium, and the surrounding plasma is rich in sodium and calcium. It seems likely that the distribution of ions in the plasma of higher animals reflects the oceanic origin of their evolutionary antecedents. One would like to know how primitive cells learned to exclude the sodium and calcium ions in which they were bathed and to develop an internal milieu enriched in potassium and magnesium.

The third and last group of essential elements consists of the trace elements. The fact that they are required in extremely minute quantities in no way diminishes their great importance. In this sense they are comparable to the vitamins. We now know that the great majority of the trace elements, represented by metallic ions, serve chiefly as key components of essential enzyme systems or of proteins with vital functions (such as hemoglobin and myoglobin, which respectively transports oxygen in the blood and stores oxygen in muscle). The heaviest essential element, iodine, is an essential constituent of the thyroid hormones thyroxine and triiodothyronine, although its precise role in hormonal activity is still not understood.

The Trace Elements

To demonstrate that a particular element is essential to life becomes increasingly difficult as one lowers the threshold of the amount of a substance recognizable as a "trace." It has been known for more than 100 years, for example, that iron and iodine are essential to man. In a rapidly developing period of biochemistry between 1928 and 1935 four more elements, all metals, were shown to be

COMPOSITION OF UNIVERSE		COMPOSITION OF EARTH'S CRUST		COMPOSITION OF SEAWATER		COMPOSITION OF HUMAN BODY	
PERCENT OF TOTAL NUMBER OF ATOMS							
H	91	O	47	H	66	H	63
He	9.1	Si	28	O	33	O	25.5
O	.057	Al	7.9	Cl	.33	C	9.5
N	.042	Fe	4.5	Na	.28	N	1.4
C	.021	Ca	3.5	Mg	.033	Ca	.31
Si	.003	Na	2.5	S	.017	P	.22
Ne	.003	K	2.5	Ca	.006	Cl	.03
Mg	.002	Mg	2.2	K	.006	K	.06
Fe	.002	Ti	.46	C	.0014	S	.05
S	.001	H	.22	Br	.0005	Na	.03
		C	.19			Mg	.01
ALL OTHERS <.01		ALL OTHERS <.1		ALL OTHERS <.1		ALL OTHERS <.01	

Al ALUMINUM   C CARBON   Fe IRON   O OXYGEN   S SULFUR
B BORON   Cl CHLORINE   Mg MAGNESIUM   K POTASSIUM   Ti TITANIUM
Br BROMINE   He HELIUM   Ne NEON   Si SILICON
Ca CALCIUM   H HYDROGEN   N NITROGEN   Na SODIUM

CHEMICAL SELECTIVITY OF EVOLUTION can be demonstrated by comparing the composition of the human body with the approximate composition of seawater, the earth's crust and the universe at large. The percentages are based on the total number of atoms in each case; because of rounding the totals do not exactly equal 100. Elements in the colored boxes in the last column appear in one or more columns at the left. Thus one sees that phosphorus, the sixth most plentiful element in the body, is a rare element in inanimate nature. Carbon, the third most plentiful element, is also very scarce elsewhere.

ELEMENT	SYMBOL	ATOMIC NUMBER	COMMENTS
HYDROGEN	H	1	Required for water and organic compounds.
HELIUM	He	2	Inert and unused.
LITHIUM	Li	3	Probably unused.
BERYLLIUM	Be	4	Probably unused; toxic.
BORON	B	5	Essential in some plants; function unknown.
CARBON	C	6	Required for organic compounds.
NITROGEN	N	7	Required for many organic compounds.
OXYGEN	O	8	Required for water and organic compounds.
FLUORINE	F	9	Growth factor in rats; possible constituent of teeth and bone.
NEON	Ne	10	Inert and unused.
SODIUM	Na	11	Principal extracellular cation.
MAGNESIUM	Mg	12	Required for activity of many enzymes; in chlorophyll.
ALUMINUM	Al	13	Essentiality under study.
SILICON	Si	14	Possible structural unit of diatoms; recently shown to be essential in chicks.
PHOSPHORUS	P	15	Essential for biochemical synthesis and energy transfer.
SULFUR	S	16	Required for proteins and other biological compounds.
CHLORINE	Cl	17	Principal cellular and extracellular anion.
ARGON	A	18	Inert and unused.
POTASSIUM	K	19	Principal cellular cation.
CALCIUM	Ca	20	Major component of bone; required for some enzymes.
SCANDIUM	Sc	21	Probably unused.
TITANIUM	Ti	22	Probably unused.
VANADIUM	V	23	Essential in lower plants, certain marine animals and rats.
CHROMIUM	Cr	24	Essential in higher animals; related to action of insulin.
MANGANESE	Mn	25	Required for activity of several enzymes.
IRON	Fe	26	Most important transition metal ion; essential for hemoglobin and many enzymes.
COBALT	Co	27	Required for activity of several enzymes; in vitamin B_{12}.
NICKEL	Ni	28	Essentiality under study.
COPPER	Cu	29	Essential in oxidative and other enzymes and hemocyanin.
ZINC	Zn	30	Required for activity of many enzymes.
GALLIUM	Ga	31	Probably unused.
GERMANIUM	Ge	32	Probably unused.
ARSENIC	As	33	Probably unused; toxic.
SELENIUM	Se	34	Essential for liver function.
MOLYBDENUM	Mo	42	Required for activity of several enzymes.
TIN	Sn	50	Essential in rats; function unknown.
IODINE	I	53	Essential constituent of the thyroid hormones.

SOME TWO-THIRDS OF LIGHTEST ELEMENTS, or 21 out of the first 34 elements in the periodic table, are now known to be essential for animal life. These 21 plus molybdenum (No. 42), tin (No. 50) and iodine (No. 53) constitute the total list of the 24 essential elements, which are here enclosed in colored boxes. It is possible that still other light elements will turn out to be essential. The most likely candidates are aluminum, nickel and germanium. The element boron already appears to be essential for some plants.

essential: copper, manganese, zinc and cobalt. The demonstration can be credited chiefly to a group of investigators at the University of Wisconsin led by C. A. Elvehjem, E. B. Hart and W. R. Todd. At that time it seemed that these four metals might be the last of the essential trace elements. In the next 30 years, however, three more elements were shown to be essential: chromium, selenium and molybdenum. Fluorine, silicon, tin and vanadium have been added since 1970.

The essentiality of five of these last seven elements was discovered through the careful, painstaking efforts of Klaus Schwarz and his associates, initially located at the National Institutes of Health and now based at the Veterans Administration Hospital in Long Beach, Calif. For the past 15 years Schwarz's group has made a systematic study of the trace-element requirements of rats and other small animals. The animals are maintained from birth in a completely isolated sterile environment [see illustration on page 16].

The apparatus is constructed entirely of plastics to eliminate the stray contaminants contained in metal, glass and rubber. Although even plastics may contain some trace elements, they are so tightly bound in the structural lattice of the material that they cannot be leached out or be picked up by an animal even through contact. A typical isolator system houses 32 animals in individual acrylic cages. Highly efficient air filters remove all trace substances that might be present in the dust in the air. Thus the animals' only access to essential nutrients is through their diet. They receive chemically pure amino acids instead of natural proteins, and all other dietary ingredients are screened for metal contaminants.

Since the standards of purity employed in these experiments far exceed those for reagents normally regarded as analytically pure, Schwarz and his coworkers have had to develop many new analytical chemical methods. The most difficult problem turned out to be the purification of salt mixtures. Even the purest commercial reagents were contaminated with traces of metal ions. It was also found that trace elements could be passed from mothers to their offspring. To minimize this source of contamination animals are weaned as quickly as possible, usually from 18 to 20 days after birth.

With these precautions Schwarz and his colleagues have within the past several years been able to produce a new deficiency disease in rats. The animals grow poorly, lose hair and muscle tone, develop shaggy fur and exhibit other detrimental changes [see illustration on page 17]. When standard laboratory food is given these animals, they regain their normal appearance. At first it was thought that all the symptoms were caused by the lack of one particular trace element. Eventually four different elements had to be supplied to complete the highly purified diets the animals had been receiving. The four elements proved to be fluorine, silicon, tin and vanadium. A convenient source of these elements is yeast ash or liver preparations from a healthy animal. The animals on the deficiency diet grew less than half as fast as those on a normal or supplemented diet. Growth alone, however, may not tell the entire story. There is some evidence that even the addition of the four elements may not reverse the loss of hair and skin changes resulting from the deficiency diet.

Functions of Trace Elements

The addition of tin and vanadium to the list of essential trace metals brings

Ala	ALANINE	His	HISTIDINE	Phe	PHENYLALANINE
Cys	CYSTEINE	Ile	ISOLEUCINE	Pro	PROLINE
Gln	GLUTAMINE	Lys	LYSINE	Thr	THREONINE
Gly	GLYCINE	Met	METHIONINE		

THE METALLOENZYME CYTOCHROME c is typical of metal-protein complexes in which trace metals play a crucial role. Cytochrome c belongs to a family of enzymes that extract energy from food molecules. It consists of a protein chain of 104 amino acid units attached to a heme group (color), a rosette of atoms with an atom of iron at the center. This simplified molecular diagram shows only the heme group and several of the amino acid units closest to it. The iron atom has six coordination sites enabling it to form six bonds with neighboring atoms. Four bonds connect to nitrogen atoms in the heme group itself, and the remaining two bonds link up with amino acid units in the protein chain (histidine at site No. 18 and methionine at site No. 80). The illustration is based on the work of Richard E. Dickerson of the California Institute of Technology, in whose laboratory the complete structure of horse-heart cytochrome c was recently determined.

to 10 the total number of trace metals needed by animals and plants. What role do these metals play? For six of the eight trace metals recognized from earlier studies (that is, for iron, zinc, copper, cobalt, manganese and molybdenum) we are reasonably sure of the answer. The six are constituents of a wide range of enzymes that participate in a variety of metabolic processes [see illustration at right].

In addition to its role in hemoglobin and myoglobin, iron appears in succinate dehydrogenase, one of the enzymes needed for the utilization of energy from sugars and starches. Enzymes incorporating zinc help to control the formation of carbon dioxide and the digestion of proteins. Copper is present in more than a dozen enzymes, whose roles range from the utilization of iron to the pigmentation of the skin. Cobalt appears in enzymes involved in the synthesis of DNA and the metabolism of amino acids. Enzymes incorporating manganese are involved in the formation of urea and the metabolism of pyruvate. Enzymes incorporating molybdenum participate in purine metabolism and the utilization of nitrogen.

These six metals belong to a group known as transition elements. They owe their uniqueness to their ability to form strong complexes with ligands, or molecular groups, of the type present in the side chains of proteins. Enzymes in which transition metals are tightly incorporated are called metalloenzymes, since the metal is usually embedded deep inside the structure of the protein. If the metal atom is removed, the protein usually loses its capacity to function as an enzyme. There is also a group of enzymes in which the metal ion is more loosely associated with the protein but is nonetheless essential for the enzyme's activity. Enzymes in this group are known as metal-ion-activated enzymes. In either group the role of the metal ion may be to maintain the proper conformation of the protein, to bind the substrate (the molecule acted on) to the protein or to donate or accept electrons in reactions where the substrate is reduced or oxidized.

In 1968 the complete three-dimensional structure of the first metalloenzyme, cytochrome c, was published [see "The Structure and History of an Ancient Protein," by Richard E. Dickerson; SCIENTIFIC AMERICAN Offprint 1245]. Cytochrome c, a red enzyme containing iron, is universally present in plants and animals. It is one of a series of enzymes, all called cytochromes, that extract en-

METAL	ENZYME	BIOLOGICAL FUNCTION
IRON	FERREDOXIN	Photosynthesis
	SUCCINATE DEHYDROGENASE	Aerobic oxidation of carbohydrates
IRON IN HEME	ALDEHYDE OXIDASE	Aldehyde oxidation
	CYTOCHROMES	Electron transfer
	CATALASE	Protection against hydrogen peroxide
	[HEMOGLOBIN]	Oxygen transport
COPPER	CERULOPLASMIN	Iron utilization
	CYTOCHROME OXIDASE	Principal terminal oxidase
	LYSINE OXIDASE	Elasticity of aortic walls
	TYROSINASE	Skin pigmentation
	PLASTOCYANIN	Photosynthesis
	[HEMOCYANIN]	Oxygen transport in invertebrates
ZINC	CARBONIC ANHYDRASE	CO_2 formation; regulation of acidity
	CARBOXYPEPTIDASE	Protein digestion
	ALCOHOL DEHYDROGENASE	Alcohol metabolism
MANGANESE	ARGINASE	Urea formation
	PYRUVATE CARBOXYLASE	Pyruvate metabolism
COBALT	RIBONUCLEOTIDE REDUCTASE	DNA biosynthesis
	GLUTAMATE MUTASE	Amino acid metabolism
MOLYBDENUM	XANTHINE OXIDASE	Purine metabolism
	NITRATE REDUCTASE	Nitrate utilization
CALCIUM	LIPASES	Lipid digestion
MAGNESIUM	HEXOKINASE	Phosphate transfer

WIDE VARIETY OF METALLOENZYMES is required for the successful functioning of living organisms. Some of the most important are given in this list. The giant oxygen-transporting molecules hemoglobin and hemocyanin are included in the list (in brackets) even though they are not strictly enzymes, that is, they do not act as biological catalysts.

ergy from food molecules by the stepwise addition of oxygen.

The complete amino acid sequence of cytochrome c obtained from the human heart was determined some 10 years ago by a group led by Emil L. Smith of the University of California at Los Angeles and by Emanuel Margoliash of Northwestern University. The iron atom is partially complexed with an intricate organic molecule, protoporphyrin, to form a heme group similar to that in hemoglobin. Of the iron atom's six coordination sites, four are attached to the heme group through nitrogen atoms. The other two sites form bonds with the protein chain; one bond is through a nitrogen atom in the side chain of a histidine unit at site No. 18 in the protein sequence and the other bond is through a sulfur atom in the side chain of a methionine unit at site No. 80 [see illustration on opposite page].

Although the cytochrome c molecule is complicated, it is one of the simplest of the metalloenzymes. Cytochrome oxidase, probably the single most important enzyme in most cells, since it is responsible for transferring electrons to oxygen to form water, is far more complicated. Each molecule contains about 12 times as many atoms as cytochrome c, including two copper atoms and two heme groups, both of which participate in transferring the electrons.

More complicated yet is cysteamine oxygenase, which catalyzes the addition of oxygen to a molecule of cysteamine; it contains one atom each of three different metals: iron, copper and zinc. There are many other combinations of metal ions and unique molecular assemblies. An extreme example is xanthine oxidase, which contains eight iron atoms, two molybdenum atoms and two molecules incorporating riboflavin (one of the B vitamins) in a giant molecule more than 25 times the size of cytochrome c.

The metal-containing proteins of another group, the metalloproteins, closely resemble the metalloenzymes except that they lack an obvious catalytic function. Hemoglobin itself is an example. Others are hemocyanin, the copper-containing blue protein that carries oxygen in many invertebrates, metallothionein, a protein involved in the absorption and storage of zinc, and transferrin, a protein that transports iron in the bloodstream. There may be many more such compounds still unrecognized because their function has escaped detection.

The Newest Essential Elements

Much remains to be learned about the specific biochemical role of the most recently discovered essential elements. In 1957 Schwarz and Calvin M. Foltz, working at the National Institutes of Health, showed that selenium helped to prevent several serious deficiency diseases in different animals, including liver necrosis and muscular dystrophy. Rats were protected against death from liver necrosis by a diet containing one-tenth of a part per million of selenium. Comparably low doses reversed the white muscle disease observed in cattle and sheep that happen to graze in areas where selenium is scarce.

In April a group at the University of Wisconsin under J. T. Rotruck reported a direct biochemical role for selenium. Oxidative damage to red blood cells was detected in rats kept on a selenium-deficient diet. This damage was related to reduced activity of an enzyme, glutathione peroxidase, that helps to protect hemoglobin against the injurious oxidative effects of hydrogen peroxide. The enzyme uses hydrogen peroxide to catalyze the oxidation of glutathione, thus keeping hydrogen peroxide from oxidizing the reduced state of iron in hemoglobin. Oxidized glutathione can readily be converted to reduced glutathione by a variety of intracellular mechanisms. There is some reason to believe glutathione peroxidase may even contain some form of selenium acting as an integral part of the functional enzyme molecule.

The physiological importance of chromium was established in 1959 by Schwarz and Walter Mertz. They found that chromium deficiency is characterized by impaired growth and reduced life-span, corneal lesions and a defect in sugar metabolism. When the diet is deficient in chromium, glucose is removed from the bloodstream only half as fast as it is normally. In rats the deficiency is relieved by a single administration of 20 micrograms of certain trivalent chromic salts. It now appears that the chromium ion works in conjunction with insulin, and that in at least some cases diabetes may reflect faulty chromium metabolism.

After developing the all-plastic trace-element isolator described above, Schwarz, David B. Milne and Elizabeth Vineyard discovered that tin, not previously suspected as being essential, was necessary for normal growth. Without one or two parts per million of tin in their diet, rats grow at only about two-thirds the normal rate.

The next element shown to be essential in mammals by the Schwarz group was vanadium, an element that had been detected earlier in certain marine invertebrates but whose essentiality had not been demonstrated. On a diet in which vanadium is totally excluded rats suffer a retardation of about 30 percent in growth rate. Schwarz and Milne found that normal growth is restored by adding one-tenth of a part per million of vanadium to the diet. At higher concentrations vanadium is known to have several biological effects, but its essential role in trace amounts remains to be established. A high dose of vanadium blocks the synthesis of cholesterol and reduces the amount of phospholipid and cholesterol in the blood. Vanadium also promotes the mineralization of teeth and is effective as a catalyst in the oxidation of many biological substances.

The third element most recently iden-

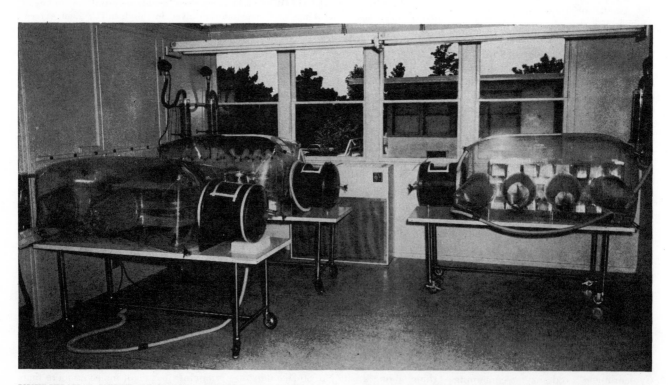

NUTRITIONAL NEEDS OF SMALL ANIMALS are studied in a trace-element isolator, a modification of the apparatus originally conceived to maintain animals in a germ-free environment. To prevent unwanted introduction of trace elements the isolator is built completely of plastics. It holds 32 animals in separate cages, individually supplied with food of precisely known composition. The system was designed by Klaus Schwarz and J. Cecil Smith of the Veterans Administration Hospital in Long Beach, Calif.

tified as being essential is fluorine. Even with tin and vanadium added to highly purified diets containing all other elements known to be essential, the animals in Schwarz's plastic cages still failed to grow at a normal rate. When up to half a part per million of potassium fluoride was added to the diet, the animals showed a 20 to 30 percent weight gain in four weeks. Although it had appeared that a trace amount of fluorine was essential for building sound teeth, Schwarz's study showed that fluorine's biochemical role was more fundamental than that. In any case fluoridated water provides more than enough fluorine to maintain a normal growth rate.

Although there were earlier clues that silicon might be an essential life element, firm proof of its essentiality, at least in young chicks, was reported only three months ago. Edith M. Carlisle of the School of Public Health at the University of California at Los Angeles finds that chicks kept on a silicon-free diet for only one or two weeks exhibit poor development of feathers and skeleton, including markedly thin leg bones. The addition of 30 parts per million of silicon to the diet increases the chicks' growth more than 35 percent and makes possible normal feathering and skeletal development. Considering that silicon is not only the second most abundant element in the earth's crust but is also similar to carbon in many of its chemical properties, it is hard to see how evolution could have totally excluded it from an essential biochemical role.

Nickel, nearly always associated with iron in natural substances, is another element receiving close attention. Also a transition element, it is particularly difficult to remove from the food used in special diets. Nickel seems to influence the growth of wing and tail feathers in chicks but more consistent data are needed to establish its essentiality. One incidental result of Schwarz's work has been the discovery of a previously unrecognized organic compound, which will undoubtedly prove to be a new vitamin.

Synergism and Antagonism

The interaction of the various essential metals can be extremely complicated. The absence of one metal in the diet can profoundly influence, either positively or negatively, the utilization of another metal that may be present. For example, it has been known for nearly 50 years that copper is essential for the proper metabolism of iron. An animal deprived of copper but not iron develops anemia because the biosynthetic machinery fails to incorporate iron in hemoglobin molecules. It has only recently been found in our laboratories at Florida State University that ceruloplasmin, the copper-containing protein of the blood, is a direct molecular link between the two metals. Ceruloplasmin promotes the release of iron from animal liver so that the iron-binding protein of the serum, transferrin, can complex with iron and transfer it to the developing red blood cells for direct utilization in the biosynthesis of hemoglobin. This represents a synergistic relation between copper and iron.

As an example of antagonism between elements one can cite the instance of copper and zinc. The ability of sheep or cattle to absorb copper is greatly reduced if too much zinc or molybdenum is present in their diet. Evidently either of the two metals can displace copper in an absorption process that probably involves competition for sites on a metal-binding protein in the intestines and liver.

The recent discoveries present many fresh challenges to biochemists. One can expect the discovery of previously unsuspected metalloenzymes containing vanadium, tin, chromium and selenium. New compounds or enzyme systems requiring fluorine and silicon may also be uncovered. The multiple and complex interdependencies of the elements suggest many hitherto unrecognized and important facts about the role and interrelations of metal ions in nutrition and in health and disease.

TRACE-ELEMENT DEFICIENCY developed when the rat at the top of this photograph was kept in the trace-element isolator for 20 days and fed a diet from which fluorine, tin and vanadium had been carefully excluded. The healthy animal at the bottom was fed the same diet but was kept under ordinary conditions. It was evidently able to obtain the necessary trace amounts of fluorine, tin and vanadium from dust and other contaminants.

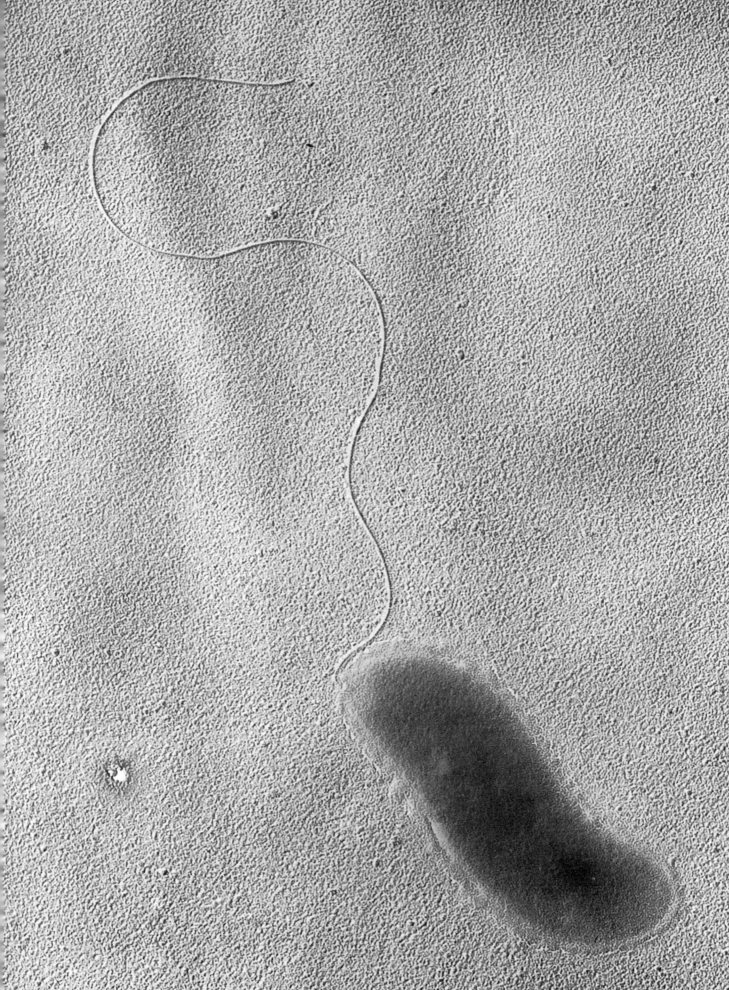

Mineral Cycles

by Edward S. Deevey, Jr.
September 1970

Although the biosphere is mainly composed of hydrogen, carbon, nitrogen and oxygen, other elements are essential constituents of living matter. Notable among them are phosphorus and sulfur

The periodic table lists more than 100 chemical elements. Yet ecologists have defined the biosphere as the locus of interaction of only four of them: hydrogen, carbon, nitrogen and oxygen. In the periodic table these four are numbered 1, 6, 7 and 8. This definition, although it deals handsomely with much of the chemistry of life, turns out to be a little too restrictive. But when we enlarge it to include phosphorus and sulfur, as we do here, we have gone no farther up the table than element No. 16. From this it should be apparent that no element lighter than sulfur can be ignored, either by ecologists or by anyone else. The fact is that most human problems—all environmental ones, anyway—arise from the exceptional reactivity of six of the 16 lightest elements.

Because our definition of the biosphere is based more on reactivity than on atomic number, it is a minimum definition. It is not intended to exclude heavier elements that react with the primary six. As a matter of empirical fact it is known that no element lighter than iron and cobalt, elements No. 26 and No. 27, is unimportant to the biosphere. Beyond copper, No. 29, there are a few conspicuously reactive elements such as the heavy halogens bromine and iodine. Most of the heavies are metals, such as gold, mercury and lead (Nos. 79, 80 and 82), however, and their main effect on the lightweight biosphere is to depress it. Toward the end of the periodic table are some famously overweight metals whose tendency to lighten themselves has disastrous effects on any light substances that get in the way.

In order to understand how it is that many elements interact with the essential six, one must briefly reflect on the biosphere as a whole. Because the biosphere is so reactive, its influence on the hydrosphere, the lithosphere and the atmosphere is inversely proportional to its mass. This mass is very small. An average square centimeter of the earth's surface supports a tiny amount of biosphere: 580 milligrams, less than the weight of two aspirin tablets. A roughly equivalent mass is found in the same area of hydrosphere a single centimeter deep, or in a paper-thin slice of lithosphere. Still, from a worm's-eye view the biosphere has real substance, particularly on land, where it amounts to 200 oven-dry tons on an average hectare.

A glance at a partial list of the elements that compose the biosphere shows why hydrogen, oxygen, carbon and nitrogen dominate conceptions of biosphere chemistry. Together these elements constitute all but a tiny fraction of the average terrestrial vegetation, which in turn constitutes more than 99 percent of the world's standing crop. The quantities are shown in the chart on the next page, based on a splendid compilation by L. E. Rodin and N. I. Basilevich. What I have done is to weight their chemical analyses in proportion to the kinds of land area they represent. The weighting factors, for desert, forest, tundra and so on, are the same ones I used to calculate the earth's production of carbon in an earlier article ["The Human Population," by Edward S. Deevey, Jr.; SCIENTIFIC AMERICAN Offprint 608]. Incidentally, on the basis of this new calculation terrestrial carbon production comes out at 65×10^9 tons of carbon per year, about 15 percent more than the figure I computed before.

What chemical compounds do these elements form? The standard way to determine the chemical composition of an organic substance is to burn it and collect the products. The list of components that results from this destructive procedure expresses some obvious facts, such as the familiar one that the biosphere is mainly carbon dioxide and water. Nitrogen, a major constituent of protein, seems surprisingly scarce (about five parts per 1,000 by weight) until we remember that the biosphere is chiefly wood, that is, not protein but the carbohydrate cellulose.

The destructive procedure would also leave a smudge, about 12 parts per 1,000 of the total, loosely called ash. Its dominant elements calcium, potassium, silicon and magnesium have important biochemical functions. One atom of magnesium, for instance, lies at the center of every molecule of chlorophyll, and silicon, the stuff of sand, is obviously useful for building hard structures. Iron and manganese also play central roles in the biosphere, a fact that could not be guessed from their position in our chart. In biochemistry as in geochemistry the importance of these elements is in governing oxidation-reduction reactions, but the masses involved are small. As for the major cations—ions of such elements as calcium, potassium, magnesium and sodium—new insights have just begun to flood in with their discovery in rainwater.

There are many other metallic elements that appear in trace amounts. Not all of them are listed in the chart because some could be accidental con-

SULFUR-FIXING BACTERIUM shown in the electron micrograph on the opposite page is one of five species that make sulfur available to the biosphere. This bacterium, *Desulfovibrio salexigens*, metabolizes the sulfates in seawater and releases the sulfur as hydrogen sulfide. This sulfur enters the atmosphere and is used by other forms of life. The micrograph enlarges the bacterium 31,000 diameters. It was made by Judith A. Murphy of the University of Illinois.

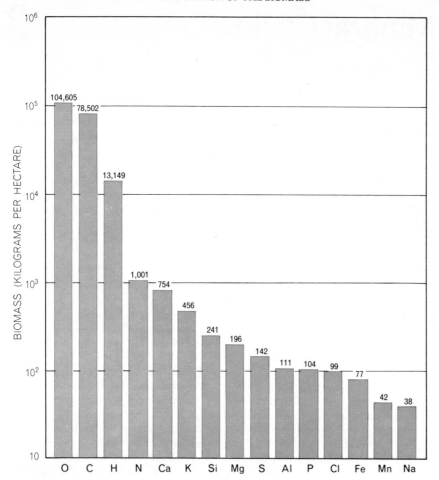

COMPOSITION OF THE BIOSPHERE is dominated by oxygen, carbon and hydrogen, as is indicated by the bars in this logarithmic chart. The units are kilograms per hectare of land surface. Key to the symbols for the chemical elements is at the bottom of the page.

charts on the opposite page can serve as a substitute, we can cast a Holmesian eye over the list. Our thinking will be more productive if we compare the composition of the biosphere with the composition of the lithosphere, the hydrosphere and the atmosphere. For this comparison all four of the "spheres" in the chart are converted from parts by weight to atoms per 100 atoms. (The masses of the four spheres being very different, these percentages will give no idea of the earth's mean or total composition.)

At first glance the four spheres do not seem to belong in the same universe. Not surprisingly, the lithosphere turns out to be a slightly metallic aluminum silicate. ("Here is no water but only rock/Rock and no water and the sandy road," as T. S. Eliot put it in "The Waste Land.") The biosphere, in sharp contrast, is both wet and carbonaceous. A single class of compounds, formaldehyde (CH_2O) and its polymers, including cellulose, could make up more than 98 percent of the total (by weight). Still, even when it is dried in an oven at 110 degrees Celsius, life is mainly hydrogen and oxygen, in close approximation to the proportions known as water. In other words, the biosphere is notably carboxylated: it is both more hydrated and chemically more reduced (hydrogenated) than is the lithosphere from which, in some sense, it came. Among the 10 most abundant elements of the lithosphere there is no obvious source for life's carbon. Hydrogen is also fairly far down the list for rock (and would be farther down if I had not copied some old figures from Frank W. Clarke's *The Data of Geochemistry*, which overweight the acidic rocks of continents).

Even the elementary Dr. Watson might conclude that life's hydrogen comes from some inorganic hydrate—water, for instance—and indeed the hydrosphere provides an ample and ready supply. This will not work for carbon, though, and in trying to account for carboxylation we can make a deduction that is truly elementary in the Holmesian, or nonobvious, sense. We begin

taminants. There remain two, sulfur and phosphorus, each amounting to more than 10 percent of the nitrogen, that do not look like contaminants. To ignore these elements as "traces" or even to think of them as "ash" or "inorganic" elements is to misconstrue the chemical architecture of the biosphere.

A listing of elements and compounds does not reveal that architecture. There is a big difference between a finished house and a pile of building materials. Nevertheless, a list is a useful point of departure. If it is made with care, it can protect ecologists from the kind of mistake that architects sometimes make, such as forgetting the plumbing.

When a list contains as much information as a shopping list—when it shows amounts as well as kinds of materials—some conclusions can be drawn from the relative proportions. (As a former bureaucrat I have learned that a "laundry list" contains even more ambiguous information than a "shopping list"; good bureaucrats keep both.) If a housewife's shopping list showed a pound of coffee, four pork chops and 100 pounds of

sugar, for example, we would know that madame is either hoarding or running a private business. If she also wants a ton of flour, she is evidently baking, not distilling. The inclusion of two dozen light bulbs would suggest that she works mainly at night, but the listing of 10 dozen light bulbs would point to a faulty generator.

As it happens, this kind of semiquantitative ratiocination was applied to ash, and to biogeochemistry, by the master of nonobvious deduction, Sherlock Holmes. Unfortunately no copy of his analytical results (the monograph on cigar ash, cited in Chapter 4 of "A Study in Scarlet") has yet come to light. If the

Al	ALUMINUM	Cl	CHLORINE	Mn	MANGANESE	P	PHOSPHORUS
Ar	ARGON	Fe	IRON	N	NITROGEN	S	SULFUR
B	BORON	H	HYDROGEN	Na	SODIUM	Si	SILICON
C	CARBON	K	POTASSIUM	Ne	NEON	Ti	TITANIUM
Ca	CALCIUM	Mg	MAGNESIUM	O	OXYGEN		

RELATIVE AMOUNTS OF ELEMENTS in the biosphere, the lithosphere, the hydrosphere and the atmosphere are presented in the charts on the opposite page. Here, however, amounts are given not as kilograms per hectare but as atoms per 100 atoms. Here again scale is logarithmic to show less abundant elements, which otherwise could not be compared.

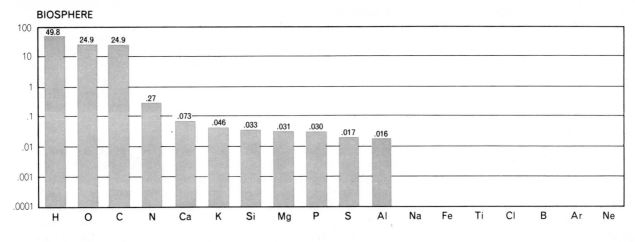

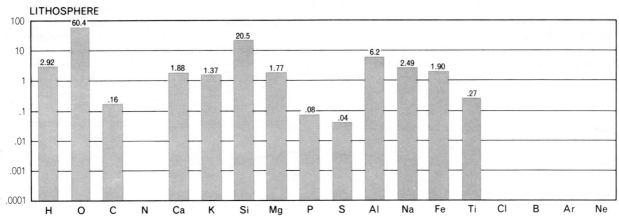

ATOMIC COMPOSITION (PERCENT)

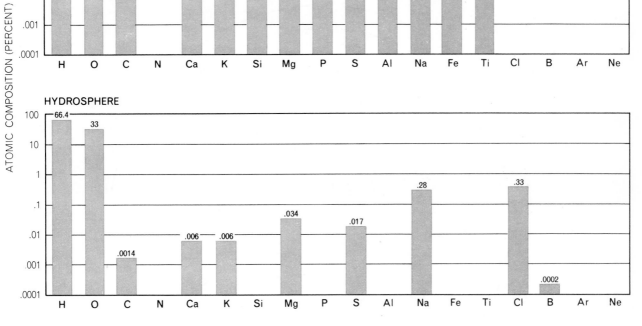

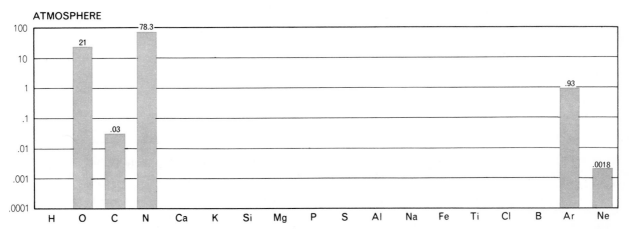

again by noting that life is mainly aqueous, and also that it concentrates carbon in proportions far greater than those in any accessible source. Is it possible that these facts are related? If they are, what do we know about water that throws any light on this relation and on the behavior of carbon? (At this point a lesser detective might reach for the carbonated water and pause for a reply.)

Instead of guessing, Holmes would proceed with his review of the evidence. Water, of course, is continuously recycled near the earth's surface, by runoff, evaporation and condensation. That is, it flows in rivers from the lithosphere to the hydrosphere, and it returns to rewash the land by way of the atmosphere. Any water-soluble elements are certain to track this cycle at least partway, from land to sea, although they may find the sea to be a sink, as boron does. If they

are to get out, they can reach the land as part of an uplifted sea bottom, but that is a chancy mechanism. Recycling is both faster and surer if the element is volatile as well as soluble, so that one of its compounds can move landward through the atmosphere as water does.

In the biosphere there are at least three elements besides those of water—carbon, nitrogen and sulfur—that fall in this doubly mobile class. Among their airborne compounds are carbon dioxide (CO_2), methane (CH_4), free nitrogen (N_2), ammonia (NH_3), hydrogen sulfide (H_2S) and sulfur dioxide (SO_2). It is interesting that when carbon, nitrogen and sulfur are recycled, their valence changes. It may not be an accident that all three are more reduced in the biosphere than they are in the external world. Be that as it may, they all seem

to belong to the biosphere, which is otherwise mainly water. Hence all three must be recycled together, *along with the water* (said Holmes with an air of quiet triumph), if the earth is to sustain its most unusual hydrate. ("And what is that?" I asked. "Why, *carbohydrate*, of course," said Holmes.)

I call this deduction nonobvious, because in an obvious variant it has become so familiar as to inhibit thought. The outlines of the carbon cycle, in organisms at any rate, have been evident since Joseph Priestley's day. The critical step, "obviously," is the photosynthetic reduction of carbon dioxide. That reaction is a hydrogenation, yielding formaldehyde. Its source of hydrogen is the dehydrogenation of water, with the liberation of oxygen. The chemical energy thus captured, by a process unique to green plants, becomes available, inside

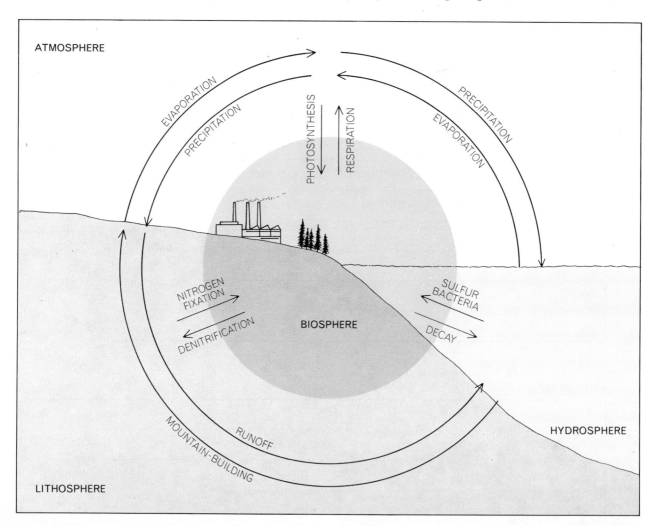

CARBOXYLATION CYCLE supplies the biosphere with carbon, oxygen, hydrogen, nitrogen and sulfur by carrying them from the lithosphere, the hydrosphere and the atmosphere. Curved arrows at upper right and left represent any or all of these five elements that travel from the atmosphere to the lithosphere or to the hydrosphere by precipitation, or back to the atmosphere by evaporation. Curved arrows at bottom indicate direct routes between the lithosphere and the hydrosphere such as runoff, mountain-building and the hydration of minerals. Biosphere (*color*) captures these elements by providing alternative routes. Top pair of straight arrows show exchange between the biosphere and the atmosphere, carbon, for example, being exchanged by photosynthesis and respiration. Pair of straight arrows at right show exchange between the biosphere and the hydrosphere, that of sulfur being mediated by bacteria. Pair of arrows at left indicate soil-biosphere exchanges including nitrogen fixation and denitrification by microorganisms.

the cell, for all other vital reactions (does it not?). After its utilization, which includes consumption by animals, the re-oxidized carbon dioxide can rejoin any geochemical cycles it likes.

All other vital reactions? Well, not quite all. The chemical reduction of nitrogen is one hydrogenation essential to green plants that they cannot perform for themselves. As one result, even elementary textbooks admit, the carbon and nitrogen cycles are necessarily interdependent. Without microorganisms that take nitrogen from the air and hydrogenate it (they can use carbon dioxide as a carbon source), all the nitrogen in the biosphere would soon appear in the atmosphere in stable, oxidized form. (The textbooks concede this point somewhat grudgingly, because much of the biological nitrogen cycle operates below the oxidation state of free nitro-

gen by the reversible reduction of nitrate and nitrite to amino acids and ammonia.)

If, as it turns out, sulfur too is recycled by way of the hydrologic cycle but independently of green plants, it becomes necessary to look beyond carbon and water for the clue to carboxylation. In other words, some biologists are not unlike architects who forget about the plumbing. In their preoccupation with carbon dioxide reduction as the starting point for cell biochemistry they tend to forget two other hydrogenations, those of sulfur and nitrogen, that are just as important.

A check is needed here, to be sure that these two elements are really intrinsic to the biosphere. In the case of sulfur the figures show it to be very scarce, and if it is a contaminant, the whole

argument might be superfluous. Sulfur, however, is no contaminant; no protein can be made without it. In fact, sulfur is the "stiffening" in protein. A protein cannot perform its function unless it is folded and shaped in a particular way. This three-dimensional structure is maintained by bonds between sulfur atoms that link one segment of a protein molecule to another. Without these sulfur bonds a protein would coil randomly, like a carelessly dropped rope.

The reason for the apparent scarcity of sulfur is the low protein content of woody tissue; any animal body contains much more. Cod-meal protein, for example, with 2.26 percent of the sulfurous amino acid methionine, has the empirical formula $H_{555}C_{265}O_{174}N_{83}S$. Although other proteins differ in the proportions, the substance of the biosphere must always contain these five elements.

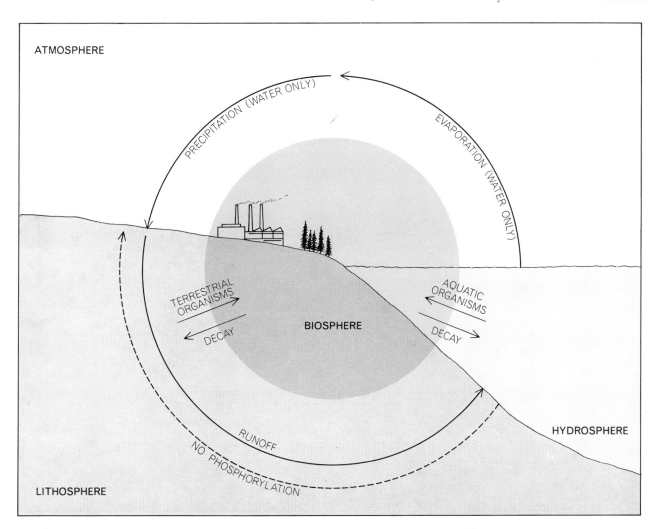

SOLUBLE-ELEMENT CYCLE is followed by minerals such as phosphorus that dissolve in water but are not volatile, that is, they are not carried into the air by evaporation (*curved arrow at right*). The curved arrow at bottom shows that phosphorus is washed from the lithosphere into the hydrosphere by runoff from rainfall (*curved arrow at top left*). The broken curved arrow at bottom indicates that phosphorus in the hydrosphere does not normally return to the lithosphere and that therefore the ocean would be-

come a phosphorus sink. The upper straight arrows at right and left, however, show that the organisms of the biosphere impede this development by absorbing some phosphorus. The straight arrows pointing from the biosphere to the lithosphere and to the hydrosphere indicate the decay of organic matter. On land the soluble-element cycle is continued when decay returns phosphorus to the lithosphere. Without an atmospheric link from ocean to land, however, the cycle is actually a one-way flow with interruptions.

It has been known for many years that sulfur is recycled from the sea back to the land by way of the atmosphere. Calculations confirming this fact by Erik Eriksson of the International Meteorological Institution show that the world's rocks contain too little sulfur, by a factor of about three, to account for the sulfate delivered annually by the world's rivers. About three-quarters of the total budget (in 1940) is therefore inferred to have come from the atmosphere. Of this amount about a third, or a quarter of the total, can have come from industrial sources—better known these days as "sulfur dioxide pollution." The other two-thirds, or half the total budget as of 1940, must take some more natural route from the hydrosphere.

When Eriksson wrote, in 1959, the question was still open, whether the cycled sulfur reaches the land as an aerosol from sea spray or as hydrogen sulfide (H_2S). If the principal volatile compound is a sulfide, it must be made by sulfate-reducing bacteria, because no other "room temperature" source of sulfide is known. M. LeRoy Jensen and Noboyuki Nakai, then working at Yale University, settled this question in favor of the bacteria, by showing that atmospheric sulfur, although it falls in rain as sulfate, contains less of the heavy isotope sulfur 34 than seawater sulfate does. What the natural isotopic label shows is that the sulfate in rain entered the atmosphere not as sea spray but as sulfide, there to be oxidized to sulfur dioxide. After dissolution in rainwater, sulfate (and sulfuric acid) are formed.

The principle of the Jensen-Nakai demonstration is worth noticing, because it applies to the cycling of carbon as well as of sulfur, and barring some technical difficulties it could also apply to nitrogen. The route followed through oxidation-reduction reactions by ordinary sulfur (sulfur 32) is analogous to the route followed by ordinary carbon (carbon 12) in photosynthesis. These lighter, more mobile isotopes appear preferentially in reduced compounds such as hydrogen sulfide, methane and formaldehyde. At equilibrium in a closed system the oxidation products (carbon dioxide or sulfate) have correspondingly more of the heavier isotopes carbon 13 and sulfur 34 without change in the total mass. If, however, a reduced and isotopically light product escapes, as hydrogen sulfide does from the hydrosphere, equilibrium is not attained, and if the gaseous product is trapped and reoxidized in a separate system, the oxide (sulfur dioxide in this case) remains light.

Exactly where within the hydrosphere

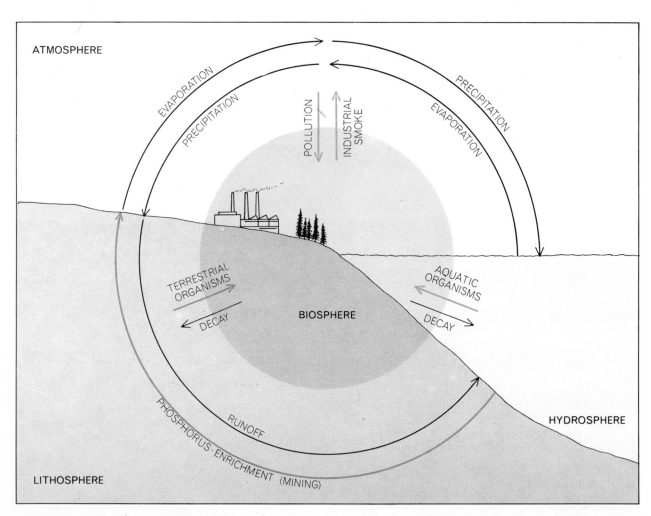

EUTROPHICATION OF THE BIOSPHERE is the intensive cycling of phosphorus, nitrogen and sulfur. Colored curved arrow at bottom represents beginning of the process: the human use of phosphorus as fertilizer, which returns phosphorus to the lithosphere, thereby reversing the phosphorus cycle. Colored straight arrows at left and right indicate that phosphorus added to the lithosphere (and to phosphorus already present) is then taken up by phytoplankton and other organisms as well as by crops. Other straight arrows at right and left show that phosphorus and other elements return to the lithosphere and hydrosphere by decay. Once phosphorus is plentiful, scarcity of nitrogen and sulfur may limit eutrophication. Arrows at top represent carbon dioxide, nitrate and sulfate from industrial activity rising into atmosphere and falling in rain. They may promote eutrophication of dry land since vegetation may reabsorb them from air and soil. Curved arrows indicate routes followed by elements that are both soluble and volatile.

most sulfate-reducers reside is still unclear. The known ones are obligate anaerobes, and their habitat is mud. Swamps, marshes and the floor of eutrophic lakes must all be important, and they may be quantitatively more important than the blue mud of estuaries and continental shelves. The sulfur metabolism of such large systems is not easy to study, even with isotopic tools. Minze Stuiver, now at the University of Washington, injected radioactively labeled sulfate ions into one eutrophic lake, Linsley Pond in Connecticut. Sulfate reduction proved to be intense, as had been expected. This lake, however, has quite a bit of ferrous iron in its deeper waters, and more in the mud itself. In the presence of the ferrous iron all the labeled sulfide was firmly held in the mud as ferrous sulfide, and no hydrogen sulfide escaped. At least for the duration of the radioactive label, with its half-life of 89 days, this mass of mud was not a source of atmospheric sulfur but a sink.

It follows from all of this that the cycling of sulfur in nature is no less relevant to carboxylation than the cycling of carbon and nitrogen. Without downgrading photosynthesis, we can say that carbon fixation is only one of at least three critical steps in the global synthesis of protein. All three are hydrogenations, achieved with the aid of enzymes, which are themselves proteins, and therefore occur only in the biosphere. Of the three reductions, however, only the reduction of carbon calls for green plants and sunlight. The other two, the reduction of nitrogen and of sulfur, are accomplished anaerobically, by microbes. Thus the locus of the nitrogen and sulfur reductions is, broadly speaking, oxygen-deficient soil and mud. Both loci are separated spatially from that airy, sunlit world where green plants (addicted, like human societies, to the external disposal of wastes) are thoughtlessly liberating oxygen.

With three critical steps for five elements, moving through four "spheres" of abstract space, one feels the need for a picture—a "systems model"—just to keep track of the relations. The two-dimensional analogue on page 6 is simple but adequate. Although it fails to specify fluxes, or any chemical quantities, it provides a mental framework for the movement of five elements: hydrogen, oxygen, carbon, nitrogen and sulfur, either alone or in combinations such as water, nitrate, the dioxides of carbon and sulfur, and carbohydrate. The synthetic output is the biosphere, with the

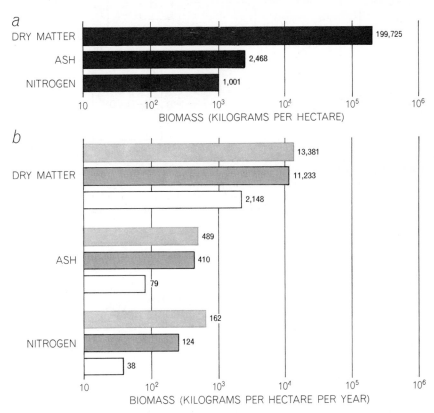

	INPUT (KILOGRAMS PER HECTARE)	OUTPUT (KILOGRAMS PER HECTARE)	NET EXPORT (KILOGRAMS PER HECTARE)
CALCIUM (Ca^{++})	2.8	3.0	+0.2
MAGNESIUM (Mg^{++})	1.1	1.8	+0.7
SODIUM (Na^+)	2.1	4.2	+2.1
POTASSIUM (K^+)	1.8	1.1	−0.7
CHLORIDE (Cl^-)	2.8	4.1	+1.3
SULFATE (SO_4^{--})	30.0	29.4	−0.6
AMMONIUM (NH_4^+)	2.1	0.3	−1.8
NITRATE (NO_3^-)	6.7	4.8	−1.9
SILICON DIOXIDE (SiO_2)	1.9	20.9	+19.0
ALUMINUM (Al^{+++})	−−	1.4	+
BICARBONATE (HCO_3^-)	0	0.7	+0.7
TOTAL	51.4	71.7	+20.3

EUTROPHICATION OF DRY LAND is indicated by the imbalance between the quantity of certain ions falling from the atmosphere on the forest at Watershed No. 6 at Hubbard Brook in New Hampshire and the output of these ions in the brook itself. Input (*smaller arrows at left*) of some elements such as calcium, magnesium and sodium is smaller than the output (*larger arrows at right*). The input of potassium, ammonium, sulfate and nitrate, however, is larger (*larger arrows at left*) than output of these substances (*smaller arrows*). The excess of input indicates that the forest is utilizing these four substances as it grows.

OTHER EVIDENCE FOR EUTROPHICATION is provided by studies of the earth's standing crop on dry land. In *a* biomass, or weighable dry matter that includes ash and nitrogen, totals about 200,000 kilograms per hectare. In *b* the first set of bars shows that the dry matter increases in an average year by 13,381 kilograms per hectare (*color*). About 11,000 kilograms is lost in the form of litter fall (*gray*) such as fallen leaves and branches, giving a "mean increment" of 2,148 kilograms per hectare (*open bar*). The second set of bars shows that ash increases by 489 kilograms per hectare (*color*) but is reduced by litter fall (*gray*) to 79 kilograms (*open bar*). The third set of bars shows that nitrogen increases by 162 kilograms (*color*), a gain that is reduced by litter fall (*gray*) to 38 kilograms per hectare.

empirical composition of protein. The central or regulatory position of the biosphere in this model follows from the fact that for all five elements it is both a source and a temporary sink. For any element that might be tempted to cycle around the edges of the model, the biosphere provides several high-energy alternatives. The most interesting of these are the reductions of carbon, nitrogen and sulfur, each concentrated at a different interface, two being out of immediate contact with air. Water, although it is able in principle to cycle independently, is the source of the hydrogen that energizes the biosphere, and cannot long avoid the biospheric loop as long as the biosphere functions.

It is easy to be bemused by so fascinating a model. Its function, however, is to clarify thought. If further thought disrupts the model, nothing is lost but a few lines on paper. More or less instantly, by reference to the table of biospheric composition, we can see that the model is incomplete. Phosphorus has been left out, along with calcium, potassium, silicon and magnesium, four elements that are commoner in the biosphere than sulfur is. Will any or all of these cycle tamely through the model, or will they disrupt it beyond repair?

For phosphorus, but not yet for the others, the answer is clear: With one significant modification, the model can accommodate phosphorus. First, let us be sure, as we made sure for sulfur, that phosphorus is necessary to the biosphere. It is not a constituent of protein, but no protein can be made without it. The "high-energy phosphate bond," reversibly moving between adenosine diphosphate (ADP) and adenosine triphosphate (ATP), is the universal fuel for all biochemical work within the cell. The photosynthetic fixation of carbon would be a fruitless tour de force if it were not followed by the phosphorylation of the sugar produced. Thus although neither ADP nor ATP contains much phosphorus, one phosphorus atom per molecule of adenosine is absolutely essential. No life (including microbial life) is possible without it.

UNIVERSAL FUEL of living matter is adenosine triphosphate (ATP). High-energy phosphate bonds of ATP ($\sim$) each store 12,000 calories and release 7,500 calories when broken.

PRODUCTION OF ATP, shown in generalized form, consists of two stages. The first stage begins as aldehyde reacts with an inorganic phosphate to produce hydrogen and acid phosphate. In second stage (*bottom*) acid phosphate (*shading*) reacts with ADP (adenosine diphosphate) to make an organic acid and ATP (*color*). R stands for radicals, or side groups.

Our provisional model of the biosphere has been constructed on two explicit assumptions: (1) the biosphere necessarily contains the five elements of protein and (2) all five are both soluble and volatile. If we now add phosphorus as a sixth necessary element, we can safely assume its solubility in water, and the crucial question concerns its volatility. Except as sea spray in coastal

SYNTHESIS OF SUCROSE is an example of a reaction for which ATP (*color*) supplies energy. The reaction begins at upper left as the ATP molecule combines with glucose molecule, releasing 7,500 calories. The reaction produces ADP and glucose-1-phosphate. In a second stage of the reaction (*bottom*) glucose-1-phosphate combines with fructose, yielding sucrose and inorganic phosphate.

BASIC FUNCTION OF SULFUR in living matter appears to be to provide a linkage between the polypeptide chains in a protein molecule. These linkages help the protein maintain its three-dimensional shape so that it can perform its function. In this segment of a bovine insulin molecule disulfide bonds (*color*) are formed between sulfur atoms, which are present in the amino acid cystine. Cystine is a subunit of both polypeptide chains. Because the molecule is displayed in two dimensions it is flattened. Therefore the bond between the top and bottom cystine groups on the upper chain appears broken. In the normal three-dimensional state, however, this chain is twisted and folded because of the disulfide bond in a way indicated by the colored line that joins the two sulfur atoms. (The other bond is one of two that links the chains.) The shape of the insulin molecule maintained by these bonds enables it to control the metabolism of sugar. The other amino acids in this molecular segment, whose side chains are indicated by the letter *R*, are glutamic acid (GLU), alanine (ALA), serine (SER), valine (VAL), histidine (HIS), leucine (LEU) and glycine (GLY).

zones, or as dust in the vicinity of exposed phosphate rock, phosphorus is unknown in the atmosphere; none of its ordinary compounds has any appreciable vapor pressure. It therefore tracks the hydrologic cycle only partway, from the lithosphere to the hydrosphere, and in a world uncomplicated by a biosphere the ocean would be its only sink. In terms of my model this amounts to uncoupling the atmospheric reservoir (except for water), omitting the half-arrow showing the return of phosphorus from the hydrosphere to the lithosphere and leaving the biosphere's phosphorus as a feedback loop, diverting some of the one-way flow from rock to ocean. Geometrically at least, the model is general enough to accommodate these changes, some version of which will be needed if any permanent sinks are discovered in the system.

For any soluble but nonvolatile element a closed natural cycle is possible only through the biosphere. The model hints at the reason why many elements—vanadium, cobalt, nickel and molybdenum among them—are best known in aquatic organisms and cycle mainly within the hydrosphere. Now, however, the model, nonquantitative though it is, suggests something else. If the biosphere demands such an element as phosphorus (and the cases of iron and manganese should be similar), two alternative inferences are permissible, depending on the magnitudes of reservoirs and fluxes. If the lithosphere contains an ample supply of phosphorus, or if the flux to the hydrospheric sink is large, the biosphere can take off what it needs and waste the rest. It is commonly believed such elements as sodium and calcium are thus wasted by terrestrial vegetation, although ecologists are beginning to doubt it. On the other hand, if the quantity is scanty or the flux small, the element will be in critically short supply. And if short supply is chronic, the output of the entire system could be expected to be adjusted to the rate of exploitation of one critical element, much as the performance of a bureaucracy is closely geared to the supply of paper clips.

In undisturbed nature the chronic shortage of phosphorus is notorious; that is what most people mean by "soil infertility." In the lithosphere phosphorus is scarcer than carbon, and in the hydrosphere, because phosphorus falls in the parts-per-billion range, it fails to show up at all in the chart of constituents on page 3. Apart from its natural scarcity, phosphorus is freely soluble only in acid solution or under reducing conditions. On the surface of an alkaline and oxidized earth it tends to be immobilized as calcium phosphate or ferric phosphate. In lake waters, where the output of carbohydrate is thriftily attuned to phosphorus concentrations of the order of 50 micro-

grams per liter, doubling the phosphorus input commonly doubles the standing crop of plankton and pondweeds.

Under these conditions the situation in a lake changes drastically. If phosphorus is plentiful, nitrate may become the critically short nutrient for a crop that needs about 15 atoms of nitrogen for one of phosphorus. Blue-green algae may then take over the plankton because by reducing atmospheric nitrogen they escape the dependence that other algae have on nitrate. Meanwhile, judging from much recent experience, the phosphorus input will probably have doubled again—but the subject under discussion is no longer undisturbed nature. What started as "cottage eutrophication," by seepage from a few septic tanks, has been escalated into a noisome mess by "treated" sewage and polyphosphate detergents. To conserve biogeochemical parity the atmosphere has begun to deliver into lakes nitrate and sulfate from the combustion of fossil fuels.

It would be wrong to read too much into a systems model. "Conserving biogeochemical parity" is just a figure of speech, technically hyperbole, and ironical at that. After all, the pollution of the air by nitrate and sulfate is quite independent—technologically, spatially and politically—of the pollution of water by phosphorus. If the accelerated phosphorus cycle in lakes takes advantage of these added inputs, we dare not say that nitrate and sulfate have been drawn into the biosphere from the atmosphere, as U.S. power was drawn to a Canadian circuit breaker in the Northeast black-out of 1965. What the model tells us is that matters can look that way from the standpoint of the biosphere. If the lake segment of the phosphorus cycle is accelerated to the point where nitrogen and sulfur are as critical as phosphorus used to be, and if there is a new source of nitrate and sulfate in the atmosphere, the atmosphere is adequately coupled to all other subsystems to ensure the success of the newly accelerated loop. The loop, known as eutrophication, is thus amplified from a lacustrine nuisance to a systems problem, and around such lakes as Lake Erie it threatens to become a cancer in the global ecosystem.

The trouble started, of course, when the world's one-way phosphorus cycle was first reversed and then accelerated by human activity. Since bird guano was discovered on desert islands, later to be supplemented in fertilizers by phosphate rock, marine phosphate has been restored to the lithosphere in ever increasing amounts. As a device for growing people in ever increasing numbers the practice cannot be faulted, but if people are to continue to flourish in the biosphere, they will have to pay more attention to scarce resources. Phosphorus is much too valuable to be thoughtlessly shared with blue-green algae.

The term eutrophication, which means enrichment, usually inadvertent, is not ordinarily applied to forests and deserts. I dare to extend it to the terrestrial biosphere because two new lines of evidence have suddenly appeared to suggest that the known pollution of air by nitrate and sulfate also encourages the bloom on dry land. The first line of evidence comes from Hubbard Brook, N.H., where F. Herbert Bormann of the Yale School of Forestry and Gene E. Likens of Cornell University have had six forested watersheds under close study since 1963. What interests us here is the difference, per hectare of ecosystem, between the input of ions in rainfall (plus dry fallout, if any) and the output as measured at a dam at the foot of each drainage basin.

Among the common ions entering and leaving Watershed No. 6 at Hubbard Brook, chloride and three positive ions—calcium, magnesium and sodium—show an excess of output over input, pre-sumably derived from the local rocks and soil. These four ions conform, if only barely, to the idea that the biosphere wastes excess salts on their way to the sea. In contrast, potassium and ammonium (NH_4) and the two major negative ions, sulfate and nitrate, are avidly held by this segment of the biosphere, as is indicated by the fact that their input exceeds their output. In the case of potassium all but 700 grams per hectare is captured. Collectively the "nonvolatile" minerals, including silica, that fall from the clear New Hampshire sky amounted to some 13 kilograms per hectare in a typical year. With sulfate, nitrate and ammonium added, the total reached 51.4 kilograms per hectare.

The second line of evidence indicates that the biosphere as a whole is becoming larger. Ecologists expect to find growth in secondary forests, but climax vegetation should be in a steady state, with annual gains balancing losses. According to figures I have recompiled from Rodin and Basilevich, the mean world vegetation is not yet at climax. After the known quantity of dead leaves, branches and other litter is subtracted from the net production of new tissue, the difference is always positive, at an average 2,148 kilograms of new biomass per hectare of land per year. With ash making up 1.2 percent of this biomass, about 26 kilograms of ash is annually withdrawn from an average hectare to sustain the increment of carbohydrate. The input of airborne elements at Hubbard Brook could provide this ash twice over, with no contribution from the local lithosphere.

This comparison is impressionistic, and it may be misleading. Apart from industrial sulfate, which (as sulfuric acid) is perhaps as likely to corrode the biosphere as to nourish it, the world's vegetation may be in no danger of instant eutrophication. (If the biosphere is really becoming larger, the input of industrial carbon dioxide may constitute another major nutrient.) The modes of recycling discovered at Hubbard Brook are nonetheless astonishing. Added to what we know or can safely infer about other volatile elements, such studies underscore the necessity of a global view of biochemistry. What can be said with assurance is that there is a unique and nearly ubiquitous compound, with the empirical formula $H_{2960}O_{1480}C_{1480}N_{16}P_{1.8}S$, called living matter. Its synthesis, on an oxidized and uncarboxylated earth, is the most intricate feat of chemical engineering ever performed—and the most delicate operation that people have ever tampered with.

GUANO-COVERED ISLAND off the coast of Peru is a source of phosphate and nitrate for fertilizer. Guano has been deposited during many millenniums by generations of birds.

3

The Nutrient Cycles of an Ecosystem

by F. Herbert Bormann and Gene E. Likens
October 1970

When all vegetation was cut in a 38-acre watershed in an experimental forest in New Hampshire, the output of water and nutrients increased. The experiment illustrates ecological principles of forest management

An ecological system has a richly detailed budget of inputs and outputs. One of the reasons it is difficult to assess the impact of human activities on the biosphere is the lack of precise information about these inputs and outputs and about the delicate adjustments that maintain a balance. As a result the people planning a project such as the logging of a forest or the building of a power plant often cannot take into account or even foresee the full range of consequences the project will have. Even if they could, the traditional practice in the management of land resources has been to emphasize strategies that maximize the output of some product or service and give little or no thought to the secondary effects of the strategies. As a result one sees such ecological maladjustments as the export of food surpluses while natural food chains become increasingly contaminated with pesticides and runoff waters carry increasing burdens of pollutants from fertilizers and farm wastes; the cutting of forests with inadequate perception of the effects on regional water supplies, wildlife, recreation and aesthetic values, and the conversion of wetlands to commercial use with little concern over important hydrologic, biological, aesthetic and commercial values lost in the conversion.

It seems evident that a new conceptual approach to the management of resources would be desirable. One approach that has been suggested is to consider entire ecological systems. In an experimental forest in New Hampshire we and the U.S. Forest Service have been conducting a large-scale investigation aimed at supplying the kind of information that is usually lacking about ecosystems. The investigation represents a multidisciplinary collaboration between workers at a number of universities and in several government agencies.

The investigation has involved two major operations. First we attempted to determine the inputs and outputs affecting the forest under normal circumstances. Then we inflicted a serious disruption on an ecosystem by cutting down an entire section of the forest. Over a period of years we have been measuring the results of that drastic action on the ecosystem's budget of nutrients. We believe our findings help to show how a natural ecosystem works and what happens when man perturbs it. Moreover, our data provide a conceptual basis for taking into account the ecological factors that in the long run determine whether or not a new technology or a particular economic policy is wise.

An ecosystem, as we use the term, is a basic functional unit of nature comprising both organisms and their nonliving environment, intimately linked by a variety of biological, chemical and physical processes. The living and nonliving components interact among themselves and with each other; they influence each other's properties, and both are essential for the maintenance and development of the system. An ecosystem, then, can be visualized as a grouping of components—living organisms, organic debris, available nutrients, primary and secondary minerals and atmospheric gases—linked by food webs, flows of nutrients and flows of energy.

A typical forest ecosystem might be visualized as a 1,000-hectare stand of mature deciduous forest. (A hectare is 2.47 acres.) The lateral boundaries of the system can be either an edge of the stand or an arbitrarily determined line. The upper boundary is the treetop level, and the lower one is the deepest level of soil where significant biological activity takes place.

Our ecosystem is in the Hubbard Brook Experimental Forest, which is maintained in the White Mountain National Forest by the Forest Service. We have focused on the movement of nutrients across the boundary of the system (both in and out) and their circulation within it. Nutrients are found in four basic compartments of the ecosystem that are intimately linked by an array of natural processes. The organic compartment consists of the living organisms and their debris. (There are probably more than 2,500 species of plants and animals in a 1,000-hectare system.) The available-nutrient compartment is composed of nutrients held on the surface of particles of the clay-humus complex of the soil or in solution in the soil. Roots as they grow produce positively charged hydrogen ions that exchange with the nutrient ions (of calcium, magnesium and so on) held on the negatively charged particles, and the nutrients are then taken up by the roots. The third compartment consists of soil and rocks containing nutrients in forms temporarily unavailable to living organisms. The atmospheric compartment is made up of gases, which can be found not only in the air but also in the ground.

Nutrients can flow between these compartments along a variety of pathways. In most cases the flow is powered directly or indirectly by solar energy. Available nutrients are taken up and assimilated by vegetation and microor-

DRAINAGE OF WATERSHED is measured by means of a weir at the base of the cutover watershed in the White Mountain National Forest. The chemical content of the water in the stream feeding into the weir that is shown in the photograph on page 32 is also ascertained by means of water samples taken from the stream periodically.

ganisms. They also circulate in complex food webs within the organic compartment, subsequently being made available again through decomposition or leaching. Minerals in soil and rock are decomposed by weathering, so that nutrients are made available to organisms. Sometimes available nutrients are returned to the soil-and-rock compartment through the formation of new minerals such as clay. Nutrients tend to cycle between the organic, available-nutrient and soil-and-rock compartments, forming an intrasystem cycle. Nutrients in gaseous form are continually being transferred to and from other compartments by inorganic chemical reactions such as oxidation and reduction and by organic reactions related to such processes as photosynthesis, respiration and the fixation and volatilization of nitrogen.

An ecosystem is connected to the surrounding biosphere by its system of inputs and outputs. They arrive or leave in such forms as radiant energy, gases, inorganic chemicals and organic substances. Inputs and outputs can be transported across ecosystem boundaries by meteorological forces such as precipitation and wind, geological forces such as running water and gravity and biological vectors involving the movement of animals in and out of the system.

Ordinarily it is difficult to measure the input-output relations of an ecosystem, particularly those involving nutrients. The nutrient cycle is closely connected to the water cycle: precipitation brings nutrients in, water leaches them from rocks and soil and stream flow carries them away. Hence one cannot measure the input and output of nutrients

without simultaneously measuring the input and output of water. The problem usually is that subsurface flows of water, which can be a significant fraction of the hydrologic cycle, are almost impossible to measure.

Several years ago it occurred to us that under certain circumstances the interaction of the nutrient cycle and the hydrologic cycle could be turned to good advantage in the study of an ecosystem. The requirements are that the ecosystem be a watershed underlain by tight bedrock or some other impermeable base. In that case the only inputs would be meteorological and biological; geological input need not be considered because there would be no transfer between adjacent watersheds. In humid areas where surface wind is a minor factor losses from the system would be only

SCENE OF EXPERIMENT is the Hubbard Brook Experimental Forest in the White Mountain National Forest. After the normal inputs and outputs of precipitation and nutrients had been ascer- tained in six contiguous watersheds in the forest, all vegetation in one of the watersheds was cut and dropped in place, and regrowth was inhibited by herbicide. The purpose of the treatment was to

geological and biological. Given an impermeable base, all the geological output would inevitably turn up in the streams draining the watershed. If the watershed is part of a larger and fairly homogeneous biotic unit, the biological output tends to balance the biological input because animals move randomly in and out of the watershed, randomly acquiring or discharging nutrients. Thus one need measure only the meteorological input and the geological output of nutrients in order to arrive at the net gain or loss of a given nutrient in the ecosystem.

This is the approach we use at Hubbard Brook, where we have been studying six contiguous watersheds ranging in size from 12 to 43 hectares [see top illustration on page 34]. They are

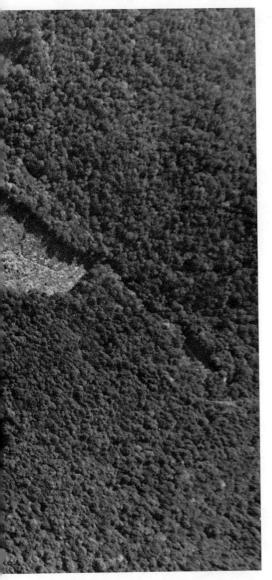

ascertain the effect on the outputs of the ecological system. The drainage of the cutover watershed is to south, which is at right.

all tributary to Hubbard Brook and forested with a well-developed, second-growth stand of sugar maple, beech and yellow birch. The forest has been undisturbed by cutting or fire since 1919, when much of the first growth was removed in lumbering operations.

We measure the meteorological inputs to these watersheds by means of a network of gauging stations. We measure the geological outputs by means of a weir built at the foot of each watershed, that is, at the point where the principal stream leaves the watershed [see illustration on page 32]. With the weir, which also includes a ponding basin, one can both measure the water that is leaving the watershed and, by combining these data with frequent chemical measurements, ascertain the quantities of chemical substances that are leaving the watershed.

Inasmuch as the impermeable base prohibits deep seepage in these watersheds, the loss of water by evaporation and by transpiration through leaves is calculated by subtracting the hydrologic output from the hydrologic input. Water budgets for the six watersheds from 1955 to 1968 indicate an average annual precipitation of 123 centimeters and a runoff of 72 centimeters, with evapotranspiration therefore averaging 51 centimeters. Precipitation is distributed rather evenly throughout the year, but runoff is uneven. Most of the runoff (57 percent) occurs during the snow-melt period of March, April and May; indeed, 35 percent of the total runoff occurs in April. In contrast, only .7 percent of the yearly runoff takes place in August.

We accomplish chemical measurements by taking weekly samples of the water output (stream water at the weir) and collecting the total weekly water input (rain and snow) and analyzing them for calcium, magnesium, potassium, sodium, aluminum, ammonium, nitrate, sulfate, chloride, bicarbonate, hydrogen ion and silicate. The concentrations of these elements in precipitation and in stream water are entered in a computing system, where weekly concentrations are multiplied by the weekly volume of water entering and leaving the ecosystem. In this way the input and output of chemicals is computed in terms of kilograms of an element per hectare of watershed.

Knowing the input and output of chemicals, we have made nutrient budgets for nine elements. Considering four of the major ones, we find the following annual averages in kilograms per

hectare entering the system and being flushed out of it: calcium, 2.6 and 11.8; sodium, 1.5 and 6.9; magnesium, .7 and 2.9, and potassium, 1.1 and 1.7. These inputs and outputs represent connections of the undisturbed forest ecosystem with worldwide biogeochemical systems. The data also provide a comparative basis for judging the effects of managerial practices on biogeochemical cycles.

Net losses of calcium, sodium and magnesium were recorded each year even though the period of measurement included wet and dry years as well as years of average precipitation. Potassium, a major component of the bedrock, showed net gains in two years and a smaller average net loss than was recorded for the other elements. Evidently potassium is accumulating in the ecosystem with respect to other elements. One reason may be that it is retained in the structure of illitic clays developing in the ecosystem. Perhaps also potassium is retained in proportionately larger amounts than other elements are in the slowly increasing biomass of the system.

Highly predictable relations appear between the concentrations of dissolved chemicals in stream water and the discharge rates of the stream. For example, the concentrations of sodium and silica are inversely related to discharge rates, whereas the concentrations of aluminum, hydrogen ion and nitrate increase as discharge rates increase. Magnesium, calcium, sulfate, chloride and potassium are relatively independent of discharge rate.

The magnitudes of change of concentration, however, are fairly small. The concentrations of potassium, calcium and magnesium hardly change at all, and the concentration of sodium decreases by only three times as the discharge rate increases by four orders of magnitude. These results were unexpected: we had thought that during the spring melt period there would be considerable dilution, making the concentrations of elements in stream water relatively low. All these relations show how strongly stream-water chemistry is under the control of processes inherent in the forest ecosystem.

Because of the comparative constancy of chemical concentrations, the total output of elements is strongly dependent on the volume of stream flow. Hence it is now possible, knowing only the hydrologic output, to predict with fair accuracy both the output and the concentration of chemicals in the stream water draining from our mature, forested eco-

system. This relation would seem to have considerable value for regional planners concerned with water quality.

A particularly interesting finding is that almost the entire loss of cations (positively charged nutrient ions) from the undisturbed forest ecosystem is balanced by the input of positively charged hydrogen ions in precipitation. The proportion of hydrogen ions is related to the amount of sulfate in precipitation. It is estimated that 50 percent of the sulfate in precipitation results from industrial activities that put sulfur dioxide and other sulfur products into the air. These sulfur compounds may ultimately form ionized sulfuric acid, which consists of hydrogen ions and sulfate ions. When the precipitation enters the ecosystem, the hydrogen ions replace the nutrient cations on the negatively charged exchange sites in the soil, and the cations are washed out of the system in stream water. Thus air pollution is directly related to a small but continuous loss of fertility from the land area of the ecosystem and a small but continuous chemical enrichment of streams and lakes. The relation appears to represent an important hidden cost of air pollution, made apparent through the analysis of ecosystems.

So far we have mentioned only chemical losses appearing as dissolved substances. Losses also arise when chemicals locked up in particulate matter such as rock or soil particles and in organic matter such as leaves and twigs are washed out of the ecosystem by the stream. We have measured these outputs and developed equations expressing the loss of particulate matter as a function of the discharge rate of the stream. The loss is highly dependent on the discharge rate.

Losses of dissolved substances account for the great bulk of the chemical loss from our undisturbed ecosystem. Whereas they are largely independent of the discharge rate, losses of particulate matter are highly dependent on it. This point is of particular interest since forest-management practices can either increase or decrease stream-discharge rates and thereby shift the balance between the loss of dissolved substances and the loss of particulate matter.

Weathering, or the release of elements bound in primary minerals, is another factor that must be considered in an ecosystem, since the elements thus released are made available as nutrients to the vegetation and animals. Based on net losses of elements from our ecosystem, a relatively uniform geology in the region

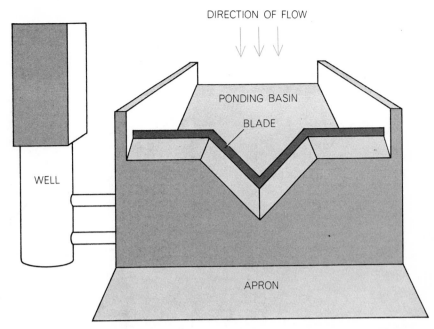

DIRECTION OF FLOW

PONDING BASIN

BLADE

WELL

APRON

STRUCTURE OF WEIR is designed to collect all the water draining from a watershed and to release it in a measurable way over the *V*-shaped blade. Well is used to gauge water level, which is continuously recorded in well house. Each Hubbard Brook watershed has one weir.

and a knowledge of the bulk chemistry of the rock and soil, we estimate that the nutrients contained in some 800 kilograms per hectare of rock and soil are made available each year by weathering.

We now have for the undisturbed Northern hardwood ecosystem of Hubbard Brook estimates of chemical input in precipitation and output in stream water and the rates of generation of ions by the weathering of minerals within the system. To complete the picture of nutrient cycling it is necessary to measure the nutrient content of the four compartments and the flow rates among them arising from uptake, decomposition and leaching and the formation of new minerals. A typical set of relations, using calcium as an example, is shown in the illustration on page 36.

The annual net loss of calcium from the ecosystem is 9.2 kilograms per hectare. This loss represents only about .3 percent of the calcium in the available-nutrient and organic compartments of the system and only 1.3 percent of what is in the available-nutrient compartment alone. The data suggest that Northern hardwood forests have a remarkable ability to hold and circulate nutrients.

It was against this background that we and the Forest Service embarked on the experiment of cutting down everything that was growing in one watershed. One of the objectives of this severe treatment was to block a major pathway of the ecosystem—the uptake of nutrients

by higher plants—while the pathway of ultimate decomposition continued to function. We questioned whether or not the ecosystem had the capacity under these circumstances to hold the nutrients accumulating in the available-nutrient compartment. We also wanted to determine the effect of deforestation on stream flow, to examine some of the fundamental chemical relations of the forest ecosystem and to evaluate the effects of forest manipulation on nutrient relations and the eutrophication of stream water.

The experiment was begun in the winter of 1965–1966 when the forest of Watershed No. 2, covering 15.6 hectares, was completely leveled by the Forest Service. All trees, saplings and shrubs were cut and dropped in place; their limbs were removed so that no slash was more than 1.5 meters above the ground. No products were removed from the forest, and great care was taken to prevent disturbance of the surface of the soil that might promote erosion. The following summer regrowth of vegetation was inhibited by an aerial application of the herbicide Bromacil at a rate of 28 kilograms per hectare.

Deforestation had a pronounced effect on runoff. Beginning in May, 1966, runoff from the cut watershed began to increase over the levels that would have been expected if there had been no cutting. The cumulative runoff for 1966 exceeded the expected amount by 40 percent. The largest difference was re-

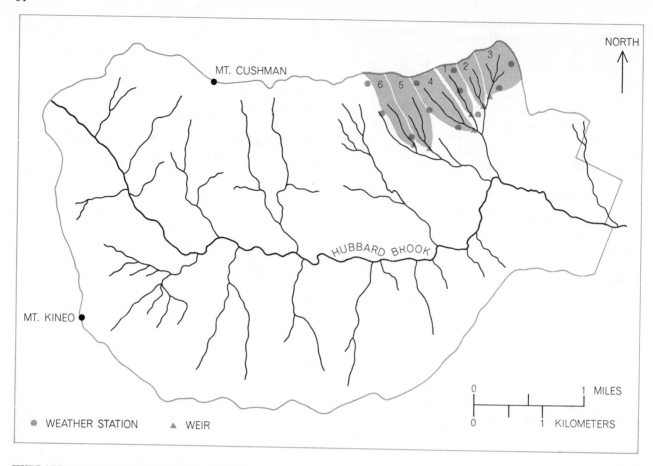

NORTH

MT. CUSHMAN

6 5 4 1 2 3

HUBBARD BROOK

MT. KINEO

0   1 MILES

0   1 KILOMETERS

● WEATHER STATION   ▲ WEIR

HUBBARD BROOK EXPERIMENTAL FOREST is in central New Hampshire, about 30 miles north of Laconia. The six water-sheds figuring in experiment are in the northeast corner (*color*). Vegetation was cut and regrowth repressed in Watershed No. 2.

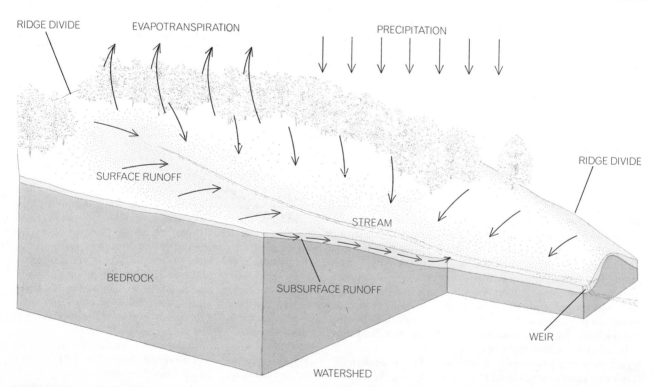

RIDGE DIVIDE   EVAPOTRANSPIRATION   PRECIPITATION

SURFACE RUNOFF

RIDGE DIVIDE

STREAM

BEDROCK

SUBSURFACE RUNOFF

WEIR

WATERSHED

UNDISTURBED WATERSHED receives inputs from precipitation and discharges outputs through its principal stream. In many watersheds there also would be outputs arising from deep seepage, which is almost impossible to measure, but watersheds of Hubbard Brook ecosystem are underlain by impermeable bedrock, so that all liquid output is by stream. Another aspect of the ecosystem is the intrasystem cycle, involving the release of nutrients into the soil by weathering of rocks, the uptake of nutrients by vegetation and their return to the soil by decomposition and leaching. Destruction of vegetation blocks a major pathway, nutrient uptake.

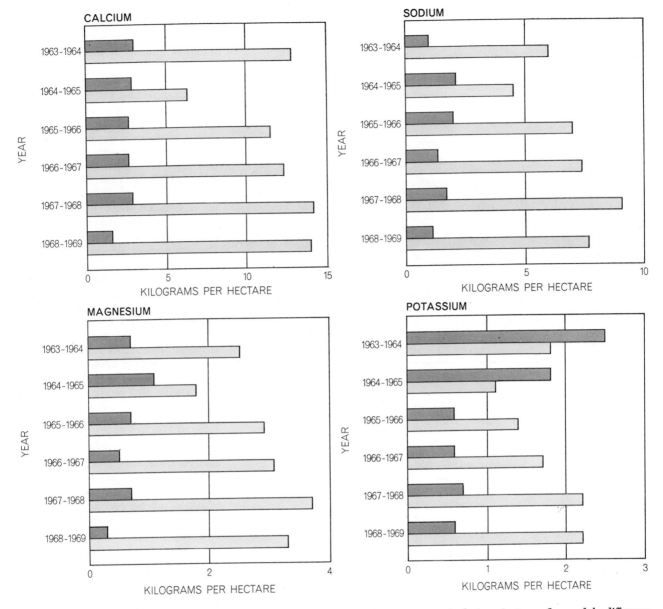

INPUT AND OUTPUT of four major nutrients were recorded for the undisturbed watersheds at Hubbard Brook. Input (*gray*) was in precipitation; output (*color*) was in stream flow, and the difference represents the net loss or gain from the ecosystem in a given year.

corded during the four months from June through September, when the runoff was 418 percent higher than the expected amount. This difference is directly attributable to changes in the hydrologic cycle resulting from the removal of the transpiring surface. Accelerated runoff has continued through the succeeding summers.

Our treatment also resulted in a fundamental alteration of the nitrogen cycle, which in turn caused extraordinary losses of soil fertility. In an undisturbed ecosystem nitrogen incorporated in organic compounds is ultimately decomposed in a number of steps to ammonium nitrogen (NH_4^+), a positively charged ion that can be held fairly tightly in the soil on the negatively charged exchange

sites. Ammonium ions can be taken up directly by green plants and used in the fabrication of nitrogen-containing organic compounds. Ammonium ions can also be used as the substrate for the process of nitrification. In this process two genera of soil bacteria, *Nitrosomonas* and *Nitrobacter,* oxidize ammonium to nitrate (NO_3^-). Two hydrogen ions are produced for every ion of ammonium oxidized to nitrate. As we have already mentioned, hydrogen ions can play a key role in the release of nutrient cations from the soil. Nitrate, being negatively charged, is highly leachable. If it is not taken up by higher plants, it can easily be removed from the ecosystem in drainage water.

The available evidence indicates that

nitrification is of minor importance in undisturbed forests such as ours, underlain by acid podzol soil. The nitrate drained by our streams (invariably in low concentration) can be largely accounted for by its input in precipitation. In fact, our budgetary analyses show that undisturbed ecosystems are accumulating nitrogen at a rate of about two kilograms per hectare per year.

The concentration of nitrate in stream water from undisturbed forests shows a seasonal cycle, being higher from November through April than it is from May through October. The decline in May and the low concentration in the summer are correlated with heavy demand for nutrients by the vegetation

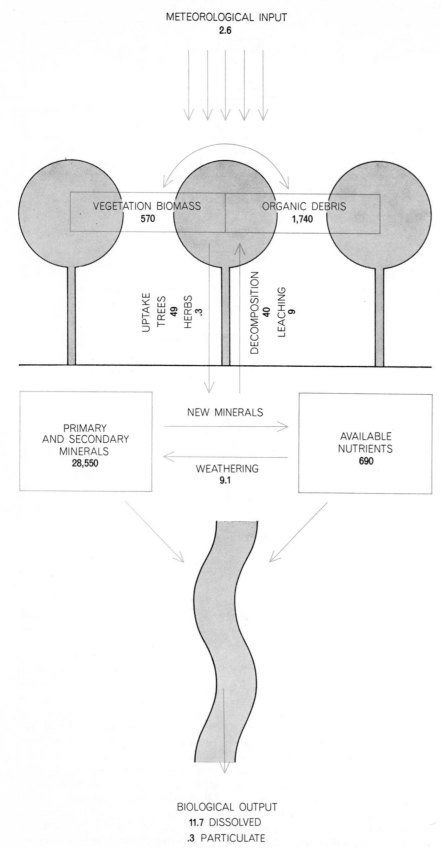

METEOROLOGICAL INPUT
2.6

VEGETATION BIOMASS
570

ORGANIC DEBRIS
1,740

UPTAKE
TREES 49
HERBS .3

DECOMPOSITION 40
LEACHING 9

PRIMARY
AND SECONDARY
MINERALS
28,550

NEW MINERALS

WEATHERING
9.1

AVAILABLE
NUTRIENTS
690

BIOLOGICAL OUTPUT
11.7 DISSOLVED
.3 PARTICULATE

CALCIUM CYCLE is depicted for an undisturbed forest ecosystem. Numerals represent the average number of kilograms per hectare per year. Thus the meteorological input to the ecosystem in precipitation and dust is 2.6 kilograms per hectare annually. A substantial amount of calcium is in soil and rock; 9.1 kilograms is released annually by weathering. Vegetation takes up 49.3 kilograms; 49 kilograms is returned to the soil by decomposition and leaching. Gross loss in stream drainage is 12 kilograms, so the net loss is 9.4 kilograms.

and generally increased biological activity associated with warming of the soil.

Beginning in June, 1966, the concentration of nitrate in the deforested watershed rose sharply. At the same time the undisturbed ecosystem showed the normal spring decline [see *illustration on next page*]. The high concentration has continued in the deforested watershed during the succeeding years. Average net losses of nitrate nitrogen were 120 kilograms per hectare per year from 1966 through 1968. We estimate that the annual turnover of nitrogen in our undisturbed forests is about 60 kilograms per hectare. Therefore an amount of elemental nitrogen equivalent to double the amount normally taken up by the forest has been lost from the deforested watershed each year since cutting. The magnitude of nitrate loss is a clear indication of the acceleration of nitrification in the watershed. There is no doubt that the cutting drastically altered the conditions controlling the nitrification process.

Another factor of interest about nitrification is the body of evidence from other regions that certain types of vegetation can inhibit nitrification chemically. Presumably the effect is to inhibit production of the highly leachable nitrate ion. At the same time the positively charged ammonium ions may be held within the system on the negatively charged exchange sites in the soil. If this inhibition process goes on at Hubbard Brook, cutting the vegetation would promote nitrification. The effect may account for much of the nitrate loss from the deforested ecosystem.

The export of nitrate to the small stream draining the cut watershed has resulted in nitrate concentrations exceeding the levels established by the U.S. Public Health Service for drinking water. In general the deforestation has led to eutrophication of the stream and to the development of algal "blooms." This finding indicates that in some circumstances forest-management practices can contribute significantly to the eutrophication of streams.

Since nitrification produces hydrogen ions that replace metallic cations on the exchange surfaces in the soil, one would expect a loss of metallic nutrients from the deforested watershed. We have recorded substantial losses of this kind. The concentrations of calcium, magnesium, sodium and potassium in the stream water increased almost simultaneously with the increase in nitrate. About a month later the concentration of aluminum rose sharply.

Net losses of potassium were 21 times higher than those in an undisturbed watershed; of calcium, 10 times; of aluminum, nine; of magnesium, seven, and of sodium, three. These figures represent a substantial loss of nutrients from the ecosystem. The finding suggests that commercial forestry should focus more on the effect of harvesting practices on the loss of nutrients, giving more consideration to such corrective procedures as selective cutting and the promotion of regrowth on cut areas.

Our results indicate that the capacity of the ecosystem to retain nutrients is dependent on the maintenance of nutrient cycling within the system. When the cycle is broken, as by the destruction of vegetation, the loss of nutrients is greatly accelerated. This effect is related both to the cessation of nutrient uptake by plants and to the larger quantities of drainage water passing through the system. The loss may also be related to increased rates of decomposition resulting from such changes in the physical environment as higher soil temperature and moister soil.

We also found a basic change in the pattern of loss of particulate matter in the deforested watershed. The data for three years indicate an increase of some ninefold over a comparable undisturbed ecosystem. After an initial surge the loss of particulate organic matter has declined as a result of the virtual elimination of the production of primary organic matter in the ecosystem. In contrast, the loss of inorganic material from the stream bed has accelerated because of the greater erosive capacity of the augmented stream flow and also because several biological barriers to the erosion of surface soil and stream banks have been greatly diminished. The continuous layer of litter that once protected the soil surface is now discontinuous. The extensive network of fine roots that tended to stabilize the stream bank is now dead, and the dead leaves that tended to plaster over exposed banks are gone. The erosive trend can be expected to rise exponentially as long as regrowth of vegetation is inhibited in the watershed.

Our study clearly shows that the stability of an ecosystem is linked to the orderly flow of nutrients between the living and the nonliving components of the system and the production and decomposition of biomass. These processes, integrated with the seasonal changes in climate, result in relatively tight nutrient cycles within the system, a minimum output of nutrients and wa-

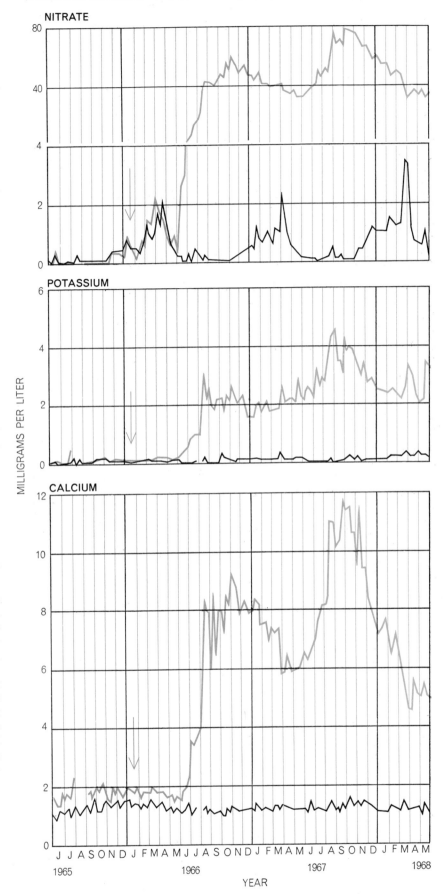

RESULTS OF DEFORESTATION shown by the output of three nutrients in stream water appear in a comparison of Watershed No. 6 (*black*), which was undisturbed, with Watershed No. 2 (*color*), which was deforested. The arrows indicate the time of the deforestation.

ter and good resistance to erosion. Destruction of the vegetation sets off a chain of interactions. Their net effect is an increase in the amount and flow rate of water and the breakdown of biological barriers to erosion and transportation, coupled with an increase in the export of nutrient capital and inorganic particulate matter.

Three points, which are inherent in the ecosystem concept and are emphasized by the Hubbard Brook study, should be recognized as being basic to any wise scheme for managing the use of land. First, the ecosystem is a highly complex natural unit composed of or-

ganisms (plants and animals, including man) and their inorganic environment (air, water, soil and rock). Second, all parts of an ecosystem are intimately linked by natural processes that are part of the ecosystem, such as the uptake of nutrients, the fixation of energy, the movement of nutrients and energy through food webs, the release of nutrients by the decomposition of organic matter, the weathering of rock and soil minerals to release nutrients and the formation of new minerals. Third, individual ecosystems are linked to surrounding land and water ecosystems and to the biosphere in general by connections with

food webs and the worldwide circulation of air and water.

Failures in environmental management often result from such factors as failure to appreciate the complexity of nature, the assumption that it is possible to manage one part of nature alone and the belief that somehow nature will absorb all types of manipulation. Good management of the use of land—good from the viewpoint of society at large—requires that managerial practices be imposed only after a careful analysis and evaluation of all the ramifications. A focus for this type of analysis and evaluation is the ecosystem concept.

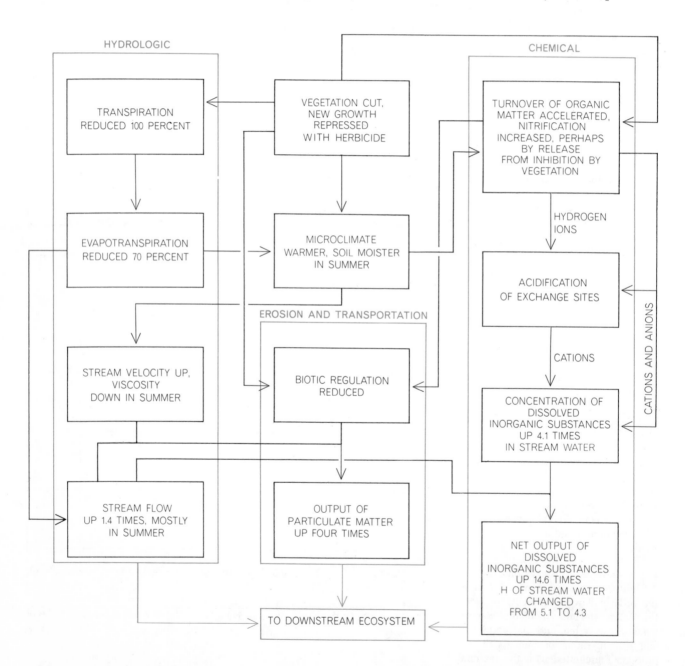

IMPACT OF DEFORESTATION on the ecosystem relations of Watershed No. 2 is portrayed in terms of three types of effect (*color*). Cations are positively charged nutrient ions; anions are negatively charged. Normally cations are held fairly tightly within the ecosystem, but with deforestation they tended to leak away. The accelerated loss was related to the intensified nitrification.

4

Trace-Element Deserts

by A. J. Anderson and E. J. Underwood
January 1959

*Throughout the world potential farmland goes to waste for
the lack of elements required in traces by plants and
animals. By rectifying such deficiencies, Australia hopes to
reclaim some 300 million acres*

Until recently South Australia's Ninety Mile Desert was a scrubby wasteland of heath and eucalyptus thickets. Today its six million once worthless acres are being swiftly transformed into bounteous pastureland. The "Desert," which formerly supported one sheep per 20 acres, can now sustain 40 times that number. Indeed, the Desert—in name, at any rate—is no more: its prosperous new residents have rechristened it Coonalpyn Downs.

The new fertility of Coonalpyn Downs was not obtained by expensive irrigation or clearance projects. Like much of the world's unproductive land, it suffered from an ailment subtler than lack of water. Recent investigations have made it possible to diagnose a large group of deficiencies of soil, plants and animals. The infertility of the soil at Coonalpyn Downs was found to arise from the absence of an almost infinitesimal sprinkling of zinc and copper.

In the context of biology, zinc and copper are numbered among the "trace elements," that is, elements that com-

RECLAIMED DESERT in eastern Australia was once infertile from lack of molybdenum, which plants require in order to fix nitrogen from the air. Shown here are sheep being mustered for shearing on the reclaimed Southern Tablelands near Canberra.

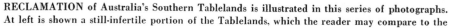

RECLAMATION of Australia's Southern Tablelands is illustrated in this series of photographs. At left is shown a still-infertile portion of the Tablelands, which the reader may compare to the reclaimed area illustrated on the preceding page. The middle photograph shows

prise so small a part of the substance of an animal or plant that chemists of an earlier time, unable to measure them exactly, could state only that they had found a trace of them. Trace amounts of 20 to 30 elements occur in living matter. Some of the elements may be present by accident Others—we cannot yet be sure how many—are indispensable to higher organisms. Most of them apparently contribute to the catalytic activity of particular enzymes in the chains of molecular events that constitute the processes of life. The trace elements, along with the vitamins, are often called "micronutrients."

The study of trace elements promises rich rewards for the agriculture of the future. The world's food-producing capacity is already sorely taxed by the explosive multiplication of the human species. Yet it seems likely that hundreds of millions of acres are now kept from productivity by nothing more than the easily remedied lack of trace elements. The reclamation of Coonalpyn Downs is today being repeated in many parts of Australia; tomorrow the same methods may be ameliorating trace-element deficiencies in other underdeveloped regions of our planet.

The trace elements now recognized as essential to plant life are seven in number: iron, manganese, copper, zinc, molybdenum, boron and chlorine. High-

er forms of animal life also need seven: the first five listed above plus cobalt and iodine. As research proceeds, these lists will doubtless grow. There is already suggestive, if not conclusive, evidence that plants need cobalt, sodium and vanadium, and that animals require selenium, bromine, barium and strontium.

The need of the human body for certain trace elements has been known for some 100 years. Nineteenth-century French physicians found that iron therapy remedied the "green sickness" (anemia) of adolescent girls; the French investigator Eusebe Gris ascertained that plants grown in a medium free of iron were yellow and stunted. The careful observations of A. Chatin, another French worker, disclosed the connection between goiter and lack of environmental iodine. Thus iron and iodine were the first trace elements to be identified. Many years passed, however, before investigators traced these elements to the metabolic system: oxidative enzymes in the case of iron, and the thyroid hormone thyroxine in the case of iodine.

Not until this century was it known that organisms require trace elements other than iron and iodine. As early as 1860 the German workers W. Knop and J. von Sachs were able to raise green plants in artificial media without soil or organic matter of any kind. By supply-

ing measured amounts of apparently pure mineral salts and water, they and their successors sought to determine the exact chemical requirements of plant life. The need for trace elements went unnoticed, because they were unwittingly supplied in the form of unobserved contaminants of the salts.

Experiments of this kind were conducted for more than half a century before the "pure" mineral salts were observed to contain trace elements. Without these hidden elements the plants could not have lived. While the trace elements had escaped detection by chemists, they were easily found and utilized by the plants growing in culture solution.

Animal experiments similarly overlooked the trace elements, which contaminated not only the mineral content of "pure" diets, but were also present in the crude vitamin supplements which the animals needed for growth. After it was recognized that trace elements might be essential to life, new techniques were developed which made it possible to control their content in both plant and animal nutrients. The present list of vital elements is the work of the past 35 years. New methods of purifying the diet of experimental animals, developed particularly at the University of Wisconsin, showed that mammals require copper, manganese and zinc. The

the continuing infertility of this land when sown to clover and treated with fertilizers (but no trace elements). At right the same land has received molybdenum as well as fertilizers. An ounce of molybdenum per acre amply insures the clover's health and nitrogen-fixing ability.

latest addition to the list of elements needed by plants is chlorine, which is acquired largely from the air—a discovery made possible by a new technique, developed at the University of California, for eliminating trace elements from experimental atmospheres.

In our country—Australia—more than 400 million acres of adequately watered land lie undeveloped, much of it for lack of the trace elements required by farm crops and animals. To appreciate the significance of this fact for Australia, one must compare the enormous figure of 400 million acres with our 22 million acres of cropland, 27 million acres of forest and 28 million acres of improved pasture. Suppose that we eliminate 25 per cent of the 400 million acres as unsuitable for agriculture or needed for urban development. That still leaves 300 million acres of potentially useful land—more than four times the present acreage! This area may ultimately be of enormous importance not only to Australia but to a world hard-pressed for food. Of course most of the land needs the benefit of routine agricultural measures, such as treatment with superphosphate and the planting of nitrogen-accumulating legumes. But the fact remains that much of it would have to remain forever in the category of "inherently infertile" land unless it is

treated with a tiny but vital dose of trace elements.

Copper, zinc and molybdenum are the elements that cropland and pasture most often lack. Copper and cobalt are those most generally needed for the health and productivity of sheep and cattle. At Coonalpyn Downs the sowing of seven pounds per acre of zinc sulfate and seven pounds of copper sulfate made the difference between infertility and fertility. More recently workers of the Western Australia Department of Agriculture have found that traces of zinc and copper, together with superphosphate, provide the key to the development of the three-million-acre Esperance Plain, near the southwest corner of the continent. This area now bears nothing but harsh native scrub; agriculturists have always thought it worthless despite its adequate rainfall. Very likely the entire region will soon contribute to Australia's food-producing capacity.

While copper and zinc deficiencies occur in southern and western Australia, the eastern part of the continent suffers mostly from a lack of molybdenum. About a third of the so-called podsolic soils in eastern Australia is more or less deficient in the metal. The podsolic soils stretch for 1,000 miles along the east coast, reaching more than 150 miles inland into Victoria, New South Wales and Queensland. They also cover much

of the island of Tasmania. The whole of the podsolic belt receives more than 20 inches of rainfall per year, and most of it can be sown to pasture if treated with a molybdenum-superphosphate fertilizer. This is rapidly being done.

The lack of molybdenum in Australian soils was discovered only in 1942. It works in an unusual way on the affected plants. Symptoms of molybdenum deficiency were observed in plants in the laboratory before any attempt was made to diagnose the infertility of podsolic soils. But the laboratory symptoms were quite unlike those observed in the plants of podsolic pastures. The first clue to what was wrong with the pastures was gained when they were found to respond well to treatment with wood-ash, lime or other alkaline matter. At first it was thought that the lime aided pasture growth by counteracting phosphate fixation in the soil. Experiments disproved this theory, however, and the search continued. Finally it was observed that clover grown on the podsolic soils responded spectacularly to molybdenum. Now it has been discovered that plants of many species all over the world grow faster in soils to which molybdenum has been added.

But why did the molybdenum-deficient pasture plants not resemble those grown in the laboratory? The reason is that the lack of molybdenum in the soil interfered with the ability of the pasture plants to fix nitrogen from the air, but was not severe enough to induce the symptoms obtained in the laboratory. The laboratory plants, on the other hand, received nitrogen in their well-balanced diet, and did not exhibit the symptoms of nitrogen deficiency. Thus the pasture plants were stunted by lack of nitrogen, and did not show the other symptoms of molybdenum deficiency. The laboratory plants were afflicted only with the other symptoms. To appreciate the importance of this fact, one must realize that clover is especially valued for its capacity to enrich the soil with nitrogen. A small dose of molybdenum restored its dark-green color and normal growth, just as if the pasture had been heavily dressed with a nitrogen fertilizer.

Remarkably little molybdenum is needed to correct its deficiency in the soil. On some soils even one sixteenth of an ounce per acre is sufficient to effect a

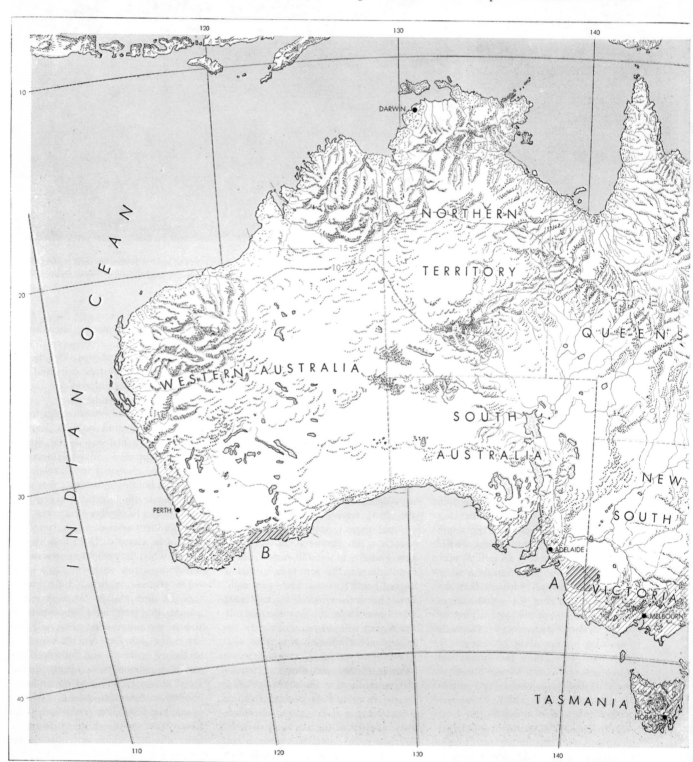

COMMONWEALTH OF AUSTRALIA has been scene of pioneer efforts to amend trace-element deficiencies, which occur in the areas hatched in color. Areas A and B (*black hatching*) are, respectively, the zinc- and copper-deficient Ninety Mile Desert (now Coonalpyn Downs) and a similar area, the Esperance Plain. The broken colored lines define areas of equal average rainfall per

clear-cut response in clover plants. The normal commercial application for pasture is of the order of one ounce per acre. This homeopathic dose, by far the smallest in trace-element therapy, is known to be effective for at least 10 years.

Some trace-element deficiencies affect plants, some affect animals and some

affect both. Even where pasture growth is poor from lack of zinc or manganese, the sparse herbage will contain enough of these elements to meet the needs of livestock grazing on it. Naturally the livestock will benefit greatly from treatment of their pasturage with zinc and manganese. But the benefit to them will lie in the increase of herbage, not in the increase of zinc and manganese. It is otherwise in the case of cobalt or copper. Pasture plants can apparently thrive with no cobalt at all, yet a lack of this element is deadly to sheep and cattle. Both plants and animals require copper.

When sheep suffer a deficiency of copper, their wool often loses its crimp and takes on a stringy or steely appearance. Such sheep are poor wool producers, and their product brings much less in the market than normal wool. Extensive areas exist where wool production and quality are thus affected. Fortunately the condition is easily remedied, either by feeding the land five pounds per acre of copper (in the form of "copperized" superphosphate) or by providing the stock with copper-treated salt licks.

In some parts of Australia, notably the south and west, the copper deficiency is more serious. Here the breeding performance of ewes as well as the quality of wool is likely to suffer. Lambs may be stillborn, or they may be born with a curious wobbly gait and die soon afterward. It has been found that lambs which are dropped by copper-deficient ewes have malformed brains and spinal cords. The condition is completely prevented by the addition of copper to the pasturage, by salt licks or by drenching the pregnant ewes with a solution of a copper salt.

Calves born on such land rarely suffer from symptoms as severe as those observed in lambs. But cows often have a strange malady, locally known as "falling disease" because the afflicted animals suddenly drop dead in the fields. These cows have died of heart failure. Prolonged copper deficiency has so weakened their heart muscles that any momentary stress can kill them. This disease, too, can be prevented. We should add that in all cases the copper not only prevents the specific ailment but markedly improves the over-all health and productivity of the flock or herd.

Some parts of southern Australia have suffered from a deficiency of both copper and cobalt so severe that sheep and cattle could not survive unless they were regularly transferred to a healthier area. This problem has also been solved, with the result that several hundred thousand acres of well-nigh valueless land have

been transformed into thriving communities. It has recently been found that the cobalt requirements of a ruminant stem from a unique symbiosis of the animal and the vitamin-producing microorganisms in its gut. For this reason the cobalt deficiency and its cure merit special attention.

The fact that cobalt is essential to life was first discovered in 1935 by Australian workers. Their observation emerged from the study of a peculiar and highly localized disease. In some parts of southern and western Australia, animals from seemingly healthy pastures weakened and died, apparently of starvation. New Zealand also suffered from this disease: workers there attributed it to deficiency of iron, having observed that the malady did not affect animals dosed with crude iron compounds.

This explanation did not satisfy students of the problem in western Australia. The effect of various iron compounds seemed unrelated to their iron content; moreover, the doses needed were suspiciously large. Perhaps, these workers thought, the effectiveness of the compounds was due not to iron but to some other element. Accordingly they fractionated one of the compounds and tested its constituent elements separately. It at once became clear that cobalt was the effective substance. Soil from the sick pastures and from the livers of diseased animals were found to be abnormally low in cobalt. An exceedingly small amount of cobalt is needed by ruminants: less than a tenth of their copper requirement. A sheep needs only a tenth of a milligram of cobalt per day; cattle must have five to 10 times as much. A single ounce of cobalt will sustain nearly 800 sheep or 80 cattle for a year!

While this investigation was proceeding, investigators in southern Australia, working on a similar disease of sheep, arrived independently at the same discovery. These workers found that cobalt is ineffective if injected with a needle; it must be taken by mouth. To obviate the need for repeated oral doses, they invented an ingenious cobalt "bullet"— four to five grams of cobalt oxide mixed with clay and baked into a small, heavy slug. Placed in the sheep's throat with a special gun, the bullet lodges in the upper alimentary canal, where it yields a steady supply of cobalt. One bullet lasts for months, and sometimes for years. It is possible that this novel technique can be used with other trace elements.

New interest in cobalt was aroused 10 years ago by the discovery in England

year. Deficient areas total about 400 million acres, three fourths of which (three times the now productive area) may be reclaimed.

COBALT-DEFICIENT SHEEP (*number 38*) is compared to healthy sheep (*number 37*) in an experiment performed by H. A. Keener at the University of New Hampshire. Cobalt deficiency, identified in the U. S., Britain and Australia, can be remedied by dosage (for sheep) of one tenth of a milligram per day. The cobalt is needed for synthesis of vitamin B-12 by microorganisms in gut.

and the U. S. that the element appears in the molecule of vitamin B-12. Within three years workers at Cornell University and in Australia had demonstrated that vitamin B-12 injections swiftly secured the remission of cobalt deficiency symptoms. Ruminants, they found, derive their natural supply of the vitamin entirely from the microorganisms in their gut. The microorganisms must have a steady supply of cobalt for their synthesis of the vitamin. It would be more accurate to call cobalt deficiency a nutritional disease of the microorganisms, rather than of their animal host.

Many chemical reactions in plants and animals are now known to require the presence of a trace element. But it is difficult to connect these reactions in cells and tissues with the outward symptoms of trace-element deficiency. Boron deficiency, for example, profoundly inhibits the growth of plants. Yet we know next to nothing of the biochemical role of boron. It is possible the element takes part in the transport of sugars by forming ionizable sugar-borate complexes, but evidence for this is meager.

Encouraging progress is being made in the study of trace elements as components of enzymes. Within the past few years several enzymes containing molybdenum have been isolated from living tissues. One of these is nitrate reductase, a plant enzyme which abets the synthesis of proteins by converting nitrate nitrogen from the soil into nitrite nitrogen. When this enzyme is deprived of its molybdenum, it ceases to function; when the molybdenum is restored, the enzyme is reactivated. In the nitrate reductase molecule the molybdenum serves as an electron carrier, alternately undergoing oxidation and reduction. It is not surprising that some plants lacking in molybdenum contain high levels of raw nitrate nitrogen. Of course this does not apply to clover, which fixes nitrogen from the air, or to plants that obtain their nitrogen from the soil as ammonia.

Zinc also plays a part in enzymes, notably the carbonic anhydrase of animals, which breaks down carbonic acid into carbon dioxide and water. Carbonic anhydrase must play an important role in respiration, for it helps convert carbon dioxide flushed from body tissues into blood-borne bicarbonate, and to convert bicarbonate back into carbon dioxide in the lungs. Though the role of this and several other zinc-containing enzymes has been well established, no one so far has been able to connect any deficiency-disease symptom to a reduction of their activity, even in gravely zinc-deficient animals. The symptoms of zinc deficiency—failure of growth and appetite, skin lesions and, in birds, poorly formed bones and feathers—remain unexplained.

More is known about the complex role of copper in enzymes. Recently investigators at the University of Utah and the University of London have traced certain of the signs caused by copper starvation in rats and pigs to a lack of the enzyme cytochrome oxidase, which is essential for the respiration of cells. While there is no copper in cytochrome oxidase, this enzyme has an iron component, called heme *a*, the synthesis of which is catalyzed by copper. Without heme *a*, cytochrome oxidase cannot be formed. It has been known for some time that copper deficiency plays a similar role in anemia; copper-starved bone marrow fails to mobilize the available iron and incorporate it in the red blood cells which the marrow manufactures. In addition it now seems that the red blood cells themselves utilize copper. Radioactive-tracer studies conducted at the University of Utah suggest that the cells require copper in order to complete their normal life span.

Investigators at the University of London have established a suggestive connection between lack of copper and ataxia, or poor coordination, in newborn lambs. The ataxia of the lambs is caused by underdevelopment of the myelin, or fatty outer coating of the nerve fibers, in their brains and spinal cords, and this in turn is due to a lack of the phospholipids of which myelin is largely composed. The lambs' tissues require copper to catalyze a crucial stage in the synthesis of the phospholipids.

Many recent studies suggest that trace-element disorders tend to involve pairs or triads of elements. A striking instance is the interplay of copper, molybdenum and inorganic sulfate in a disorder afflicting some Australian sheep. This disease develops on pastures rich in copper but poor in molybdenum. Sheep on such land will succumb to copper poisoning unless their molybdenum intake is increased. But molybdenum will not counteract the effect of copper except in the presence of a third substance: inorganic sulfate. The chemical reactions involved are not simple, and there are indications that still other substances are involved.

What of trace elements in man's own diet? The indications are that civilized man is not likely to suffer trace-element deficiencies. Unlike range-fed cattle or sheep, modern man derives his diet from many localities and soil types. The inadequacies of one food source are made up from another. After 100 years of study, iron and iodine are still the only trace elements that human populations are known to lack. While iodine deficiency is a genuinely regional disease, iron deficiency stems from loss of blood or poor choice of foods rather than from a local lack of iron in the soil.

So trace-element deficiencies have little direct effect upon man. Their indirect effects, on the other hand, are profoundly important. How we husband the chemical health of our crops, our stock and our land may have immense import for the future of man. As population pressure mounts, we will have to evolve ever faster-growing and higher-yielding strains of domestic plants and animals. Will the quality of our food keep pace with its quantity? That may well depend upon how carefully we monitor its content of the trace elements.

5

The Ecology of Desert Plants

by Frits W. Went
April 1955

What selects the plants in an extreme environment? Is it a sort of war? The study of plants in the desert and in the laboratory indicates otherwise, which points a moral for the human species

The laws of human behavior are very much in dispute, largely because there are no obvious experimental approaches to them. But animal and plant behavior can be studied both in nature and in the laboratory, and the science of their ecology should eventually be helpful in the understanding of human relationships, for the basic laws which govern the interrelations among organisms in general must also underlie human behavior. Ecology is an extremely complex study. For a relatively uncomplicated case, from which we may be able to extract some generalizations about behavior, I want to take you with me to the desert to survey its plant life.

The desert is an ideal area for research. It is usually unspoiled by the encroachment of civilization. Its plant life is sparse enough to be studied conveniently in detail, and it shows clearly and primitively the effects of the physical factors at play in the environment. Most important of all, the desert climate is violent: winds sweep over it unchecked, and its temperature and rainfall swing between wide extremes. Rainfall may vary fivefold from year to year. There are so few rainstorms that the effects of individual rains can be measured. The desert's sharply contrasting conditions can be reproduced in the laboratory for convenient experimental investigation of the germination and growth of plants. And the desert has an unending lure for the botanist; in the spring it is a delightful place.

The most extreme desert in the U. S. is Death Valley. Screened off from the nearest source of water vapor—the Pacific Ocean—by the tall Sierra Nevada, the valley bottom has an average annual rainfall of only 1.35 inches. It has almost no surface water—only a few springs bringing up the scanty runoff

from the dry surrounding mountains. Since it is sunk below sea level, Death Valley has no drainage. As a basin which holds and collects all the material that may be washed into it from the mountain canyons, it has accumulated salts in its central part. Seen from above, this salt bed glistens like a lake, but a traveler on foot finds it a dry, rough surface, studded by sharp salt pinnacles which crackle and tinkle as they expand or contract in the heat of the day and the cold of the night.

In the salt plain no green plants can grow: there are only bare rocks, gravel and salt. But on the fringes of the plain plant life begins. Here and there are patches of a lush green shrub—the mesquite. With their tender green leaflets, which suggest plenty of water, the plants seem completely out of place. Actually they do have a considerable source of water, but it is well underground. The mesquite has roots from 30 to 100 feet long, with which it is able to reach and tap underground lenses of fresh water fed by rain percolating down from the mountains.

The mesquite is the only shrub that can reach the water table here with its roots. But a mesquite seedling must send its roots down 30 feet or more through dry sand before it reaches this water. How, then, does it get established? This is one of the unsolved mysteries of the desert. Most of the mesquite shrubs in Death Valley are probably hundreds of years old. Some are all but buried by dunes of sand, piled around them over the years by the winds that sometimes blow with great force through the valley. There are places where dozens or hundreds of stems protrude from a dune, all probably the offshoots of a single ancient shrub rooted beneath the dune.

Another Death Valley plant endowed with a remarkable root system is the evergreen creosote bush. It has wide-reaching roots which can extract water from a large volume of soil. The creosote bush is spread with amazingly even spacing over the desert; this is especially obvious from an airplane. The spacing apparently is due to the fact that the roots of the bush excrete toxic substances which kill any seedlings that start near it. The distance of spacing is correlated with rainfall: the less rainfall, the wider the spacing. This probably means that rain leaches the poisons from the soil so that they do not contaminate as wide an area. We commonly find young creosote bushes along roads in the desert, where the road builders have torn up the old bushes.

During prolonged periods of drought creosote bushes lose their olive-green leaves and retain only small brownish-green leaves. Eventually these also may drop off, and the bush then dies unless rain comes soon afterward. However, it takes a really long drought to kill off all the creosote bushes in an area. They have suffered severely in some areas of the southern California deserts during the drought of the past five years. Because a killing drought tends to remove them wholesale, there are usually only a few age classes of creosote bushes in an area; each group springs up after a drought or during a period of unusual rainfall.

There are other shrubs that master the harsh conditions of the desert, among them the lush green *Peucephyllum*, which seems to be able to live without water, and the white-leaved desert holly, which grows in fairly salty soil.

Two prime factors control the abundance and distribution of plants: the number of seeds that germinate, and

JOSHUA TREES (*Yucca brevifolia*) grow among the fantastic rock formations of the Joshua Tree National Monument in south-ern California east of Los Angeles. These plants are characteristic of high steppe-desert vegetation in the southwestern U. S.

MESQUITE (*Prosopis*) grows in a large clump, indicating a source of water not more than 30 feet down. In the foreground are salt-bushes, reflecting salty soil near the surface. The absence of annuals among the shrubs shows that there has been no rain for months.

CREOSOTE BUSHES (*Larrea*) dot the desert landscape in the middle distance. After a series of dry years the creosote bushes in this area died; their stems are visible in the foreground. Here the ground is covered with annual plants because of recent rains.

the growing conditions the seedlings encounter while they seek to establish themselves. In the case of the desert shrubs the main controlling factor is the growing conditions rather than germination, for though many seedlings may come forth in a rainy season, few survive long enough to become established. The story is entirely different for the annual plants in the desert.

There are years when the desert floor in Death Valley blooms with a magic carpet of color. In the spring of 1939 and again in 1947 the nonsalty portion of the valley was covered with millions of fragrant, golden-yellow desert sunflowers, spotted here and there with white evening primroses and pink desert five-spots. The bursts of flowering are not necessarily correlated with the year's rainfall. For instance, the wettest year in Death Valley was 1941, when 4.2 inches of rain fell, but there was no mass flowering that year or the following spring. If Death Valley is to bloom in the spring, the rain must come at a certain time—during the preceding November or December. There will be a mass display of spring flowers if November or December has a precipitation of well over one inch: in December of 1938 and in November of 1946 the rainfall was 1.4 inches. Rain of this magnitude in August, September, January or February seems ineffective.

Let us consider these annual plants in greater detail. Probably their most remarkable feature is that they are perfectly normal plants, with no special adaptations to withstand drought. Yet they are not found outside the desert areas. The reason lies in the peculiar cautiousness of their seeds. In dry years the seeds lie dormant. This itself is not at all amazing; what is remarkable is that they refuse to germinate even after a rain unless the rainfall is at least half an inch, and preferably an inch or two. Since the upper inch of soil, where all the viable seeds lie, is as wet after a rain of a tenth of an inch as after one or two inches, their discrimination seems hard to explain. How can a completely dormant seed measure the rainfall? That it actually does so can easily be verified in the laboratory. If seed-containing desert soil is spread on pure sand and wet with a rain sprinkler, the seeds will not germinate until the equivalent of one inch of rain has fallen on them. Furthermore, the water must come from above; no germination takes place in a container where water only soaks up from below.

Of course this sounds highly implausible—how can the direction from which the water molecules approach make any difference to the seed? The answer seems to be that water leaching down through the soil dissolves seed inhibitors. Many seeds have water-soluble germination inhibitors in their covering. They cannot germinate until the inhibitors are removed. This can be done by leaching them in a slow stream of water percolating through the soil, which is what happens during a rainstorm. Water soaking up in the soil from below of course has no leaching action. Some seeds refuse to germinate when the soil contains any appreciable amount of salt. A heavy rain, leaching out the salts, permits them to sprout. Other seeds, including those of many grasses, delay germination for a few days after a rain and then sprout if the soil is still moist—which means that the rain probably was fairly heavy. Still other seeds have inhibitors that can be removed only by the action of bacteria, which requires prolonged moisture. Many seeds preserve their dormancy until they have been wet by a series of rains.

In the washes (dry rivers) of the desert we find a completely different vegetation with different germination requirements. The seeds of many shrubs that grow exclusively in washes (paloverde, ironwood, the smoke tree) have coats so hard that only a strong force can crack them. Seeds of the paloverde can be left in water for a year without a sign of germination; but the embryo grows out within a day if the seed coat is opened mechanically. In nature such seeds are opened by the grinding action of sand and gravel. A few days after a cloudburst has dragged mud and gravel over the bottom of a wash,

the bottom is covered with seedlings. It is easy to show that this germination is due to the grinding action of the mudflow: for instance, seedlings of the smoke tree spring up not under the parent shrub itself but about 150 to 300 feet downstream. That seems to be the critical distance: seeds deposited closer to the shrub have not been ground enough to open, and those farther downstream have been pulverized. Smoke-tree seedlings form about three leaves, then stop their above-ground growth until their roots have penetrated deep enough to provide an adequate supply of moisture for the plant. Thereafter the roots grow about five times as fast as the shoots. Few of these seedlings die of drought, but a flood will destroy most of them; only the oldest and biggest shrubs resist the terrific onslaught of rocks, gravel, sand and mud streaming down the wash.

The ability of the smoke tree to make the most of the available moisture was demonstrated by the following experiment. Cracked smoke-tree seeds were sown on top of an eight-foot-high cylinder containing sand moistened with a nutrient solution. Rain water was then sprinkled on them for a short time. Six seeds germinated, and five of the plants survived and have grown for 18 months in a high temperature with only a single watering midway in that period. Indeed, they have grown better than seedlings which were watered daily!

We have studied the control of germination in great detail in our laboratory at the California Institute of Technology. We have learned, for instance, that two successive rains of three tenths of an inch will cause germination provided

DESERT WASH is traced by ironwood (*Olneya*) and paloverde (*Cercidium*) plants. The seeds of these species were made to germinate when they were abraded by a violent flow of mud.

they are given not longer than 48 hours apart. Rain in darkness has a different effect from rain during the day. Most amazing is the seeds' specific responses to temperature. When a mixture of rain-treated seeds of various annuals is kept in a warm greenhouse, only the summer-germinating plants sprout; the seeds of the winter annuals remain dormant. When the same seed mixture is kept in a cool place, only the winter annuals germinate. From this it is obvious that the annuals will not germinate unless they can survive the temperatures following their germination—and unless there has been enough rain to allow them to complete their life cycle. Since these desert plants cannot depend on "follow-up" rains in nature, they germinate only if they have enough rain beforehand to give them a reasonable chance for survival.

A very small percentage of seeds (less than 1 per cent) germinate after an insufficient rain. Such seedlings almost invariably perish before reaching the flowering stage. On the other hand, more than 50 per cent of all seedlings that have sprouted after a heavy rain survive, flower and set seed. And here we find a remarkable fact: even though the seedlings come up so thickly that there are several thousand per square yard, a majority of them grow to maturity. Though crowded and competing for water, nutrients and light, they do not kill one another off but merely fail to grow to normal size. In one case 3,000 mature plants were found where an estimated 5,000 seedlings had originally germinated. The 3,000 belonged to 10 different species. All had remained small, but each had at least one flower and produced at least one seed. This phenomenon is not peculiar to desert plants. In fields of wheat, rice and sugar cane, at spots where seeds happen to have been sown too thickly, all the seedlings grow up together; they may be spindly but they do not die. It is true that in gardens weeds often crowd out some of the desirable plants, but usually this happens only because these plants have been sown or planted out of season or in the wrong climate. Under those conditions they cannot compete with the plants fully adapted to the local growing conditions—plants which we usually call weeds.

We must conclude, then, that all we have read about the ruthless struggle for existence and the "survival of the fittest" in nature is not necessarily true. Among many plants, especially annuals, there is no struggle between individuals for

precedence or survival. Once an annual has germinated, it matures and fulfills its destiny of forming new seed. In other words, after successful germination annual plants are less subject to the process of "natural selection." Very likely this accounts for the fact that so few of the desert annuals seem to show adaptations to the desert environment. This does not mean that the plants have avoided evolution, but the evolution has operated on their seeds and methods of germination rather than on the characteristics of the grown plants. Selection on the basis of germination has endowed the plants with a remarkable variety of mechanisms for germinating, and at the same time it has made them slow to germinate except under conditions insuring their later survival. The opposite is true of the cultivated plants that man has developed: his selection has favored the plants that germinate most easily and quickly. This has given us the wrong perspective on the significance of germination in plant survival.

We return now to our original theme: Can the ecology of plants in the desert teach us anything about human ecology or human relations? At least one moral stands out. In the desert, where want and hunger for water are the normal burden of all plants, we find no fierce competition for existence, with the strong crowding out the weak. On the contrary, the available possessions—space, light, water and food—are shared and shared alike by all. If there is not enough for all to grow tall and strong, then all remain smaller. This factual picture is very different from the time-honored notion that nature's way is cutthroat competition among individuals.

Actually competition or warfare as the human species has developed it is rare in nature. Seldom do we find war between groups of individuals of the same species. There are predators, but almost always they prey on a different species; they do not practice cannibalism. The strangler fig in the tropical jungle, which kills other trees to reach the light, is a rare type [see "Strangler Trees," by Theodosius Dobzhansky and João Murça-Pires; SCIENTIFIC AMERICAN, January, 1954]. Even in the dense forest there is little killing of the small and weak. The forest giants among the trees do not kill the small fry under them. They hold back their development, and they prevent further germination. In a mountain forest in Java it was observed that the small trees living in the shade of the forest giants had not grown after 40 years, but they were still alive.

Hundreds of different species of trees, large and small, grow in a tropical jungle. This diversity of vegetation is one of the jungle's most typical characteristics. Some trees grow faster, taller or wider than others, but these growing characteristics, which we have always considered as useful adaptations in the struggle for existence, do not really control the trees' survival. If they did, we would find very few species of trees in a jungle, and there would be an evolutionary tendency for these trees to become taller and taller. Actually the tallest trees are found not in jungles but in more open forests in temperate climates; remarkably enough, tropical jungles often have no particularly high or large trees. All this shows that selection does not work on the basis of growth potential. It works on the ability of plants to grow and survive with very little light.

In our minds the struggle for existence is usually associated with a ruthless extermination of the less well adapted by those better adapted—a sort of continuous cold war. There is no cold war or even aggression in the desert or jungle. Most plants are not equipped with mechanisms to combat others. All plants grow up together and share whatever light or water or nutrients are available. It is only when the supply of one of these factors becomes critical that competition starts. But it appears likely that in the jungle, as in the desert, survival is taken care of by the control of germination. Competition and selection occur during germination, and we can speak of germination control of the plant community—comparable to birth control in human society.

Apparently evolution has already eliminated most of the plant types that are unable to compete successfully. Fast-growing, slow-growing or tall plants all have the same chances once they have germinated. The struggle for existence is not waged among the well-established plant forms but tends to eliminate new types which germinate at inopportune times, have a decreased ability to photosynthesize or are less frost-resistant. This explains why so few plants die in the desert from drought or in the jungle from lack of light or in cold climates from frost.

As a general moral we conclude that war as man wages it finds no counterpart in nature, and it has no justification on the basis of evolution or natural selection. If we want to describe the process of control of the plant population in human terms, we should talk about birth control.

II

THE STRUCTURE OF LIVING MATTER

THE STRUCTURE
OF LIVING MATTER

II

INTRODUCTION

Elemental composition is only one of the important factors that determine the suitability of an environment for supporting life. Living matter is not just a conglomerate of atoms. The atoms are bonded in unique combinations to form molecules. Molecules are also not randomly distributed in living matter. Many of them combine to form polymers. For example, proteins are polymers of various amino acids, and starch is polymerized glucose. These biopolymers are organized, in turn, to form structures that serve as functional components of cellular organelles. Organelles are the active parts of living cells, and cells are the building blocks of living organisms. This entire hierarchy of organization of matter results from forces that attract atoms to each other, forming bonds of varying degrees of stability.

To understand the properties of any kind of matter we must understand the nature of the forces that hold it together. The nature of chemical bonds and their associated energies are perhaps the most important concepts for understanding chemical reactions between substances. In a chemical reaction a change in the distribution of the electrons of the reacting molecules takes place, causing the breaking of some bonds between atoms of the reacting molecules and the formation of different bonds, those of the product molecules.

While considering these intramolecular forces it is pertinent to bear in mind how they might be affected by environmental forces. The article, "High-Energy Reactions of Carbon" (page 58), by Richard M. Lemmon and Wallace R. Erwin addresses itself in part to this theme. The discussion of high-energy reactions of carbon is also relevant to our thoughts on exobiology.

In their introductory remarks, Lemmon and Erwin consider the basics of the nature of bonds between atoms. They go on to consider the unique properties of carbon, which make it so suitable as the backbone of biochemistry. Organic chemical reactions, whether they occur in a living cell or a test tube, are due to what the authors call low-energy interactions. Temperature fluctuations and visible light quanta are sources of energy that can activate molecules to a degree sufficient to cause alterations in the bonds between some atoms. This is, of course, the reason for heating the reaction flask when we want to initiate many chemical reactions.

The major theme of the article, however, is the unusual chemistry of carbon when the atoms are energized to much higher levels. By means of specialized physical tools, highly energized carbon atoms can be made to undergo new and unexpected types of reactions. Novel products are formed by what to the organic chemists appear to be new mechanisms of molecular interactions. The importance of these reactions is that such highly energized carbon atoms are not the unique creation of man—they are abundant in the universe.

Such revelations of new and unexpected organic reactions under these conditions lead us to speculate on the sorts of compounds that might form on a planetary surface under very high atmospheric pressures, exposed to energetic cosmic rays, or in a temperature range different from that to which we are accustomed. Lemmon and Erwin suggest that some of the unusual organic compounds produced by the high-energy reactions of carbon might have played a role in the origin of life on earth. Similar reactions can also explain the synthesis of the many carbon compounds that have recently been detected in interstellar space by radio astronomers.

Many chemical reactions require large amounts of energy for their initiation. Many changes that we observe in the world around us, however, are due to changes in forces other than those that bond atoms into molecules. Our common experiences are often the result of gross changes in the binding of molecules to each other, rather than of intramolecular changes.

Consider, for example, the transformation of water from its liquid state to a solid ice crystal or a vapor. The water molecules themselves do not change in such transformations of their physical state. Solid, liquid, and gaseous water molecules all have the composition H_2O. The differences in the physical states of water are due to the forces that hold the molecules together. These intermolecular forces are affected by the kinetic energy that the molecules acquire from their environment. The kinetic energy of a molecule is manifested in three kinds of motion: (1) vibration of the atoms with respect to each other; (2) rotation of the molecule itself; and (3) translation, or motion of the molecule with respect to neighboring molecules.

As we raise the temperature of a body of liquid water, the molecules acquire kinetic energy, especially of translation; at 100°C this finally liberates the molecules from each other, and the water evaporates by boiling. Molecules liberated from bonds to their neighbors are characteristic of the gaseous state.

When the temperature drops, the molecular motion (kinetic energy) decreases and the molecules find themselves at decreasing distances from each other. Below 100°C the attraction between the water molecules restrains their free mobility, and they condense to the liquid state. Further cooling results in slightly closer packing of the molecules and stronger bonding between them. Eventually the translational motion is arrested: the molecules freeze into a crystal lattice of ice.

What binds the molecules together? In water an important force acting between the molecules is the *dipolar* electric charge of each molecule. Owing to unequal attraction of the hydrogen and oxygen atoms for the valence electrons, the oxygen atom is on the average more negatively charged, and the hydrogen atoms remain on the average more positively charged. When such dipolar molecules approach each other, the negative end of one molecule strongly attracts the positive ends of neighboring molecules. These electrostatic forces between dipolar molecules are much weaker than the covalent bonds that hold the atoms of the molecules together. Covalent bond energies are of the order of 100 kcal/mole, whereas dipolar attraction energies are only about 5 kcal/mole.

Other forces even weaker than those between dipolar molecules are the so-called London dispersion forces. When two nonpolar molecules, such as oxygen or nitrogen, approach each other very closely (to within about 10 angstroms), a synchronously fluctuating polarity is induced in the molecules. This induced polarity causes attractions between closely packed molecules. The magnitude of the attractive force increases as the intermolecular distance decreases. Collectively, the dipole-dipole forces and London dispersion forces are known as van der Waals forces.

A common manifestation of these weak intermolecular forces is the phenomenon of adhesion. Very smooth, clean surfaces, when pressed hard against each other, tend to adhere because of the proximity of the molecules of both surfaces. (See "The Action of Adhesives" by Norman A. de Bruyne, April 1962.)

A direct way to estimate the amounts of energy associated with intermolecular forces is to measure the latent heat of vaporization of a liquid. This is the amount of energy that must be added to a liquid at its boiling point to convert it to a vapor at the same temperature. For example, to change water at 100°C to steam at the same temperature, we must add about 10 kcal/mole (18 g) of water. This added energy is required to overcome the van der Waals forces between the molecules; thus it serves as an estimate of the strength of those forces.

The relatively high latent heat of vaporization of water (those of methane and ammonia are approximately 2 kcal/mole and 5.6 kcal/mole, respectively) indicates that in the liquid state water molecules are held together rather tightly compared to other substances of similar molecular weight. The reason for this high latent heat is that the water molecules are held not only by their van der Waals forces, but by yet another type of weak bonding, the *hydrogen bond*.

Hydrogen bonding occurs when a hydrogen atom that is covalently bonded to a highly electronegative atom such as oxygen comes near another such atom, say, nitrogen. The hydrogen atom, being attracted to both electronegative atoms, forms a weak bridge between them. Hydrogen bonds are weak, with a typical energy of about 4 kcal/mole. However, because each water molecule has two hydrogen atoms, these form a network of hydrogen bonds connecting many molecules and contributing greatly to the cohesive forces that hold water together.

Intermolecular attractive forces reach their maximal values when a substance is in its solid state. The primary characteristic of this state is that the molecules are firmly positioned with respect to their neighboring molecules, and have virtually no mobility.

In his article, "The Force between Molecules" (page 67), Boris V. Derjaguin, a distinguished Russian physical chemist, discusses intermolecular forces in general, especially in the solid state. His theoretical considerations of the subject are followed by a detailed description of a direct experimental procedure that he devised for measuring the attractive forces between two solid surfaces. Note that, although the solid objects that Derjaguin and his coworkers studied are simple, homogeneous substances, the interpretation of the results is rather complicated.

Interactions among different molecules or atoms in heterogeneous solids are much more complicated and sometimes have unexpected consequences. A clear manifestation of the changes that occur among the intermolecular bonds in heterogeneous solids is the new properties that metals acquire when they are blended with other metals to form alloys. Metal alloys are just slightly more complex than pure metals, but their physical properties are quite different. Imagine, therefore, the complications arising from the interactions of many different kinds of large, complex molecules that are mixed in a liquid environment.

An interesting example of the effects of intermolecular forces in liquid solutions is found in the properties of some water-soluble polymers. If dilute aqueous solutions of gum arabic, a natural biopolymer, and polyethylene glycol, a synthetic polymer, are mixed, the mixture gradually separates into two phases. Each phase contains a specific mixture of the two polymers. The two aqueous phases obviously differ in their properties—so much so that, if a low-molecular-weight substance is added to the system, it becomes unequally distributed between the two phases. An explanation in terms of the intermolecular forces that cause the difference between the two aqueous phases awaits further research.

Most biochemical processes occur among substances in aqueous solutions. Water is therefore a most important component of living matter. The importance of water in the biosphere is associated with many of its unusual properties. Water serves as a thermal buffer in the biosphere because of its high heat capacity and the ability of water molecules to absorb infrared radiation. Water also functions as raw material for carbon reduction in photosynthesis, and participates as an intermediate in many other biochemical reactions.

In their article, "Water" (page 75), Arthur M. Buswell and Worth H. Rodebush discuss many of the interesting properties of this substance. It appears that the most important ones are attributable to the extensive intermolecular hydrogen bonding. The size and shape of the water molecule and its ability to form several H-bonds permit water molecules to form several types of packing arrangements. The stability of these supramolecular structures depends not only on the temperature but also on the presence of solutes. The properties of these dissolved molecules, such as size, electric charge, and polarity, determine how effective they are in "organizing" the water molecules in their vicinity.

Some nonpolar molecules cause water to assume cage-like structures called hydrates (or clathrates). These structures resemble tiny ice crystals and give such solutions unusual properties. For example, Linus Pauling has pointed out a direct correlation between the ability of certain molecules to form hydrate microcrystals and their ability to induce anesthesia in animals. This correlation implies that water organization plays a role in the normal function of nerve cells. It is a good example of the significance of intermolecular bonding and supramolecular structures in living processes.

So far we have dealt with relatively simple systems: molecules of one species interacting with each other or solute molecules interacting with molecules of a solvent. We noted the increasing complexity of these interactions as the molecules increase in size and asymmetry. The situation becomes immensely more complex when a large number of different molecular species are interacting with each other. The presence of ions in the vicinity of macromolecules that possess charges on their own surfaces has strong effects on the attractive forces among the macromolecules. The net charge on a protein molecule, for example, is directly related to the concentration of hydrogen ions in the solution. These effects are responsible for phenomena such as protein coagulation and precipitation in acid solutions or in high salt concentrations.

I have attempted to stress the complexity of the intermolecular relations as molecules mingle with other large and small molecules. At this point we should again consider living organisms and their relations to the environment, interpreting these relations in terms of intramolecular and intermolecular forces.

Living matter is made up of hundreds of different kinds of giant molecules. These molecules form a hierarchy of organized structures, the complexity of which has only recently become apparent. It appears that the forces responsible for the organization of the structures within a living cell are primarily those that we classified as weak intermolecular bonds. These bonds can be readily modified by changes in temperature, ionic composition, and hydrogen ion concentration.

A good example of the complexity of a biological structure is the cell membrane. In "The Structure of Cell Membranes" (page 83), C. Fred Fox describes a modern view of the organization of biological membranes. These structures are composed of a semi-solid lipid phase in which are embedded numerous kinds of large protein aggregates. Both sides of the membrane are in contact with an aqueous solution that contains a large number of solutes. Proteins are giant molecules; some parts of their surfaces are highly charged (ionic) and polar, whereas other parts are neutral (lipoidal) and nonpolar. The degree of immersion of such molecules in the lipid layer depends on the relative sizes of these two types of surfaces.

The entire membrane structure is in an active, dynamic state. Proteins change their shape and thus their position within the lipid layer. This occurs because of interactions between the proteins and various solutes in the surrounding solution. Proteins also change their shape when they absorb light energy, and pigmented proteins have been shown to change their position in the membrane after they absorb light.

The effects of temperature on the fluidity of the lipid layer are an important point to remember. It is thought that one of the major effects of abnormal temperature on a living cell is to either harden or soften the lipid layer of the membrane, thus interfering with its normal function. It appears, therefore, that the delicate structures of a living cell, such as membranes, are readily affected by environmental factors like light, temperature, or various solutes.

6

High-Energy Reactions of Carbon

by Richard M. Lemmon and Wallace R. Erwin
January 1975

Much of the carbon chemistry in the universe takes place at high energies. Experiments with accelerated carbon ions point to processes that may have played a role in the emergence of life

No element is more basic to life than carbon. If one disregards water, living cells are almost entirely an assemblage of carbon compounds. In the chemical processes of life the reactions of carbon take place at modest temperatures, which is another way of saying that they take place at modest energies. In living cells these reactions proceed at temperatures near 38 degrees Celsius (100 degrees Fahrenheit). Even in laboratory studies of carbon compounds a chemist rarely employs temperatures above 300 degrees C. Although these low-energy carbon reactions are obviously of great importance to us, the recent discovery of many complex carbon compounds in interstellar space makes it clear that the carbon chemistry of the universe is almost entirely a high-energy chemistry. The earth is constantly bombarded by high-energy carbon atoms in cosmic rays and in the "wind" flowing outward from the sun's corona. It is entirely possible that the emergence of life on our planet depended on some key organic reactions promoted by these high-energy carbon atoms. In order to gain any understanding of this possibility it is necessary to investigate high-energy carbon reactions in the laboratory.

Over the past few years we and our colleagues at the Lawrence Berkeley Laboratory of the University of California have been producing such reactions by means of a chemical accelerator. Carbon ions are accelerated in an electrostatic field and aimed at selected target molecules. By analyzing the products of the reactions we are beginning to gain some knowledge of the mechanisms by which high-energy carbon atoms form new carbon compounds.

There is much evidence for the pervasiveness of high-energy carbon chemistry in the universe. Radio astronomers have now found some 20 carbon compounds in interstellar space. Some of these compounds may well have been created in collisions involving high-energy carbon atoms such as those in cosmic rays and the solar wind. It is clear that the carbon atoms of the solar wind have played an important role in the chemistry of the moon's surface [see "The Carbon Chemistry of the Moon," by Geoffrey Eglinton, James R. Maxwell and Colin T. Pillinger; SCIENTIFIC AMERICAN, October, 1972]. The Apollo missions to the moon have provided us with a good estimate of the density of solar-wind carbon atoms striking the moon: approximately 100,000 atoms per square centimeter per second. The number of carbon atoms reaching the earth's atmosphere is likely to be much lower because of the shielding effect of the earth's magnetic field. If we assume, however, that the earth's atmosphere receives only a tenth as many carbon atoms as the moon does, our planet would pick up more than 30 tons of solar-wind carbon per year. Multiply that by the age of the earth (4.5 billion years) and the total weight of the solar-wind carbon that may have accumulated on our planet is 135 billion tons—roughly as much as the total amount of carbon in all living matter on the earth today.

Whether or not high-energy carbon reactions were involved in the emergence of life, there are good reasons for the fact that our biology and very likely any other biology that may exist in the universe is based on carbon. The element is unique in its ability to form compounds. Its uniqueness is apparent from its special position in the periodic table of the chemical elements [see illustration on page 61]. Carbon is the smallest atom in the group of elements halfway between the electropositive elements (those that give up electrons to form chemical bonds) and the electronegative elements (those that take up electrons to form chemical bonds). Carbon has four electrons in the outermost shell of its electron cloud, and it has an equal tendency to gain or lose those four bond-forming electrons. It thus forms stable compounds with both sodium and chlorine.

As the smallest atom in its group in the periodic table carbon is also capable of forming particularly strong bonds. Because the carbon atom is small its electron cloud can approach the positively charged nucleus of another atom rather closely. The closer the approach is, the

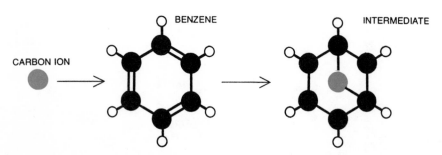

HYPOTHETICAL BRIDGED INTERMEDIATE incorporating seven atoms of carbon is believed to be formed when a high-energy carbon ion strikes a benzene molecule. The possibility of such an intermediate was not envisioned prior to studies with ion accelerators.

stronger are the bonds that are formed. The combination of these two properties—small size and electrical "halfwayness"—confers unique bond-forming abilities on carbon. One result is that carbon forms strong bonds not only with many other elements but also with itself. Carbon atoms can link up in chains, and when the chains are joined at their ends, they form rings. The chains and rings can have carbon-atom branches on them and carbon-atom links between them, with almost no restriction on the total number of carbon atoms. Add to this the ease with which carbon atoms can link up with atoms such as hydrogen, oxygen, nitrogen, phosphorus and sulfur and one can visualize the virtually limitless possibilities for carbon molecules. Living systems gain their marvelous complexity from the enormous number of

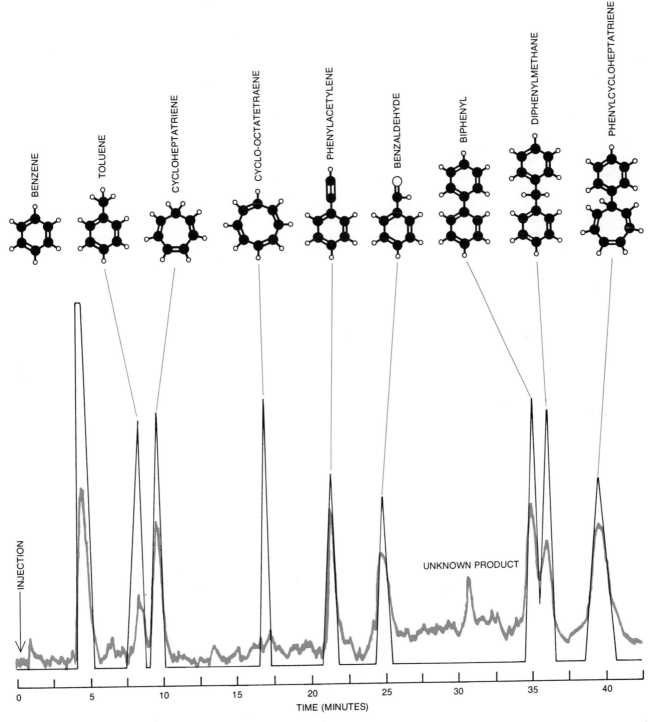

PRODUCTS FORMED BY IRRADIATION OF BENZENE with high-energy carbon-14 ions are identified by means of a gas chromatograph and a radioactivity counter. Since carbon 14 is radioactive, any new compound it becomes part of also will be radioactive. The colored line shows the radioactive peaks of the products formed by the reactions of carbon 14 and benzene. The peaks on the black line are the mass peaks of known compounds added to the sample before it entered the chromatograph. The identity of a radioactive product is established by the coincidence of a colored peak and a black peak. Additional chemical tests are carried out to confirm the identity of the products. Note that cyclo-octatetraene does not have a corresponding radioactive peak, which means little or none of this compound was formed. One major radioactive peak and several minor ones are as yet unidentified.

carbon compounds available for their construction, and it is difficult for the modern chemist to envision life based on any element but carbon. As life has evolved, many new carbon compounds have been formed. More than a million carbon compounds, both natural and synthetic, are known, but billions are possible and the universe probably harbors billions.

Although a single carbon atom is highly reactive, a mass of carbon atoms such as one finds in the graphite of a pencil lead is not particularly reactive because the carbon atoms are already combined with one another. The primary means by which chemists induce chemical reactivity is the application of heat. The higher the temperature is, the faster a molecule will move and the faster its atoms will vibrate. If the temperature is high enough, the molecule will move fast enough so that when it strikes another molecule the kinetic energy will promote a reaction between the two molecules. The high temperature may also cause a molecule to vibrate to the extent that its atoms fly apart, giving the released individual atoms a chance to enter into new combinations. In general the application of heat causes the weakest bonds to break first. At room temperature a pencil lead is inert, but if one heats it enough, its carbon atoms are jostled apart and readily combine with the oxygen in air to form carbon dioxide.

Light is another form of energy the chemist uses to promote reactions. It has long been known that sunlight promotes reactions such as bleaching or tanning, even when little or no heating is involved. We now know that the absorption of light quanta by a molecule can be even more effective than heat in putting the molecule into an energy-rich, reactive state. In contrast to what happens when heat is applied, the absorption of light can cause the breakage of a molecule's stronger bonds because the light is often selectively absorbed by those bonds.

In the past two decades chemists have been promoting reactions of carbon and of other elements at energies much higher than those that can be achieved with either heat or light, energies that are more characteristic of those found in the particles of cosmic rays and the solar wind. One method is to take advantage of the very high energy, or high velocity, of atoms created in nuclear reactions. A typical example in nature is the process that takes place when an atom of the radioactive isotope carbon 14 is born in the earth's upper atmosphere. This particular nuclear reaction is responsible for the small amount of carbon 14 found in all living things and is the basis for the widely used carbon-14 dating method. Carbon 14 is formed when neutrons, which are produced by cosmic rays that steadily bombard the earth, react with atmospheric nitrogen.

When a neutron enters the nucleus of a nitrogen atom, the resulting transient complex is unstable. It disintegrates into a proton, which is a charged hydrogen atom, and a carbon-14 atom. When the unstable complex ejects the proton in one direction, the carbon recoils in the opposite direction with an energy of 45,000 electron volts (the energy that any singly charged ion would acquire if it moved from an electrode at ground potential to an electrode that had a charge of 45,000 volts). The energy of an ordinary carbon-carbon bond is about 3.4 electron volts; therefore a 45,000-volt atom is capable of breaking 45,000 divided by 3.4, or about 13,000, carbon bonds before all its velocity is lost. Such an atom can break the strongest bond in any molecule it strikes. Hence a recoiling carbon atom may produce compounds that could not be prepared by the application of heat or light.

The general technique of recoil studies is simple. If one wants to know how recoiling carbon atoms react with the molecules of a given compound, one mixes the compound with any nitrogen-containing substance and puts the mixture in a nuclear reactor. The neutrons from the reactor collide with the nuclei of the nitrogen atoms, and the resulting carbon-14 atoms recoil through the mass of the compound until they are slowed to a velocity, or energy, at which they can form new chemical bonds. That energy, which is at most only a few tens of electron volts, is low compared with the energy of a newly created carbon-14 atom. It is, however, very high in comparison with chemical-bond energies and with the maximum energies that chemists can give to an organic compound by the application of heat, which are less than .1 electron volt per atom.

Since the recoiling carbon atoms are radioactive, the products into which they are incorporated are also radioactive. This is helpful to the chemist both for finding the new products and for finding the exact position of the carbon-14 atom in the molecule of a product. Other nuclear reactions have also been used as a source of high-energy carbon atoms. The study of the reactions of reactor-produced high-energy atoms is familiarly known as hot-atom chemistry.

Another way of imparting high energies to atoms is to use an accelerating machine such as a cyclotron. Physicists have used accelerators for several decades to examine nuclear reactions; the energies involved are in the millions or billions of electron volts. More recently chemists have been building accelerating devices that enable them to study

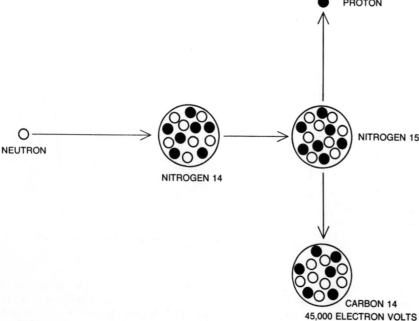

CARBON 14 IS FORMED in the earth's upper atmosphere when a neutron from cosmic rays strikes the nucleus of a nitrogen atom, which consists of seven protons (*black circles*) and seven neutrons (*open circles*). The result is an unstable nucleus consisting of seven protons and eight neutrons, which disintegrates into a proton and a carbon-14 nucleus.

	ELECTROPOSITIVE			ELECTRONEGATIVE			
GROUP I	II	III	IV	V	VI	VII

1 H HYDROGEN						
3 Li LITHIUM	4 Be BERYLLIUM	5 B BORON	6 C CARBON	7 N NITROGEN	8 O OXYGEN	9 F FLUORINE
11 Na SODIUM	12 Mg MAGNESIUM	13 Al ALUMINUM	14 Si SILICON	15 P PHOSPHORUS	16 S SULFUR	17 Cl CHLORINE

FIRST THREE ROWS OF PERIODIC TABLE are shown. (Noble-gas elements helium and neon have been omitted.) The Roman numerals designate groups of elements with similar chemical properties. Carbon is the smallest atom in the central group. Electropositive elements tend to give up electrons in their outer shell to form chemical bonds, whereas electronegative elements tend to take up electrons to fill their outer shell. Carbon, with four electrons in its outer shell, has an equal tendency to gain or lose these electrons. This confers on carbon the unique ability to form strong bonds with many other elements and also with other carbon atoms.

the reactions of atoms at the more modest energies where chemical interactions can take place [see "Chemical Accelerators," by Richard Wolfgang; SCIENTIFIC AMERICAN, October, 1968]. Such devices enable the experimenter to choose the energy, down to a few electron volts, with which an energetic atom will strike target molecules. The random destruction of the target molecules during the slowing down of a recoiling atom may thus be avoided. One such device is in use in our laboratory, and we shall describe our research with it as an illustration of the kind of information that can be obtained from energetic carbon reactions. For the progress of this work we are indebted to many former collaborators and to our current co-workers Glenn A. Fisher and Benjamin Gordon.

To form a beam of energetic carbon atoms one admits either carbon dioxide (CO_2) or methane (CH_4) containing carbon-14 atoms into a chamber where the molecules can be ionized [see illustrations on next page]. Here, between the anode and the cathode, a cloud of ions (charged molecules) is formed. The ions are expelled out of the region of their formation by an electric field of 5,000 volts. To control the ion beam one uses electrostatic lenses: metal plates of various forms on which electric charges are imposed. These lenses shape, magnify and demagnify the ion beam in ways quite analogous to the effects of optical lenses on a beam of light.

The beam that emerges from the ion source is made up of carbon in many different forms. Some of it is carbon-14 ions, some is ions incorporating oxygen (if carbon dioxide was the starting material) and some is ions incorporating hydrogen (if methane was the starting material). As the ions travel through the magnetic fields the heavier ions are bent less than the lighter ones. By adjusting the strength of the magnetic field one can select the kind of ion one wants to have emerge. Most of our work has been with carbon-14 ions. The intensity of the beam is usually about one microampere; although this does not sound like much of a current, it represents some 6×10^{12} ions per second. The beam then passes through a series of decelerating lenses, doughnut-shaped plates on which positive voltages are imposed. These lenses enable us to set the energy with which the ions strike the target at energies anywhere from their initial 5,000 electron volts down to two electron volts.

The path the ions travel must be kept at a high vacuum. Without such a vacuum the ions would strike gas molecules in the path and be scattered. In order to keep the target material from evaporating and spoiling the vacuum, the material must be held at a very low temperature. A convenient way to do so is to steadily bleed in a vapor of the target material and to freeze it on the surface of the target area, which is held at the temperature of liquid nitrogen (−196 degrees C.). This procedure has the added advantage of presenting a constantly fresh supply of target molecules to the bombarding carbon ions. Once a new product is formed it is not likely to react with additional carbon ions; it is quickly buried under the incoming target material.

The steady stream of positively charged ions on the target surface could build up positive charge that would repel subsequent incoming ions. That undesirable state of affairs is circumvented by placing near the target a small, hot tungsten filament, which provides electrons (that is, negative charges) at a rate sufficient to balance the positive charges of the incoming ions.

We learn what has happened to our energetic carbon-14 ions by determining what kinds of products incorporating radioactive carbon are formed. To do so we separate and identify the products by the simple and powerful technique of gas chromatography combined with a radioactivity detector. The radioactive products are introduced, usually as a benzene solution, into the column of the gas chromatograph: a metal tube about a quarter of an inch in diameter and several feet in length. The tube is filled with an inert solid, the particles of which are coated with an oily liquid.

The gas-chromatograph column is heated to a temperature that experience

CARBON-ION ACCELERATOR at the Lawrence Berkeley Laboratory of the University of California appears in this photograph. The ions are generated in the apparatus at left. The beam of ions is bent by a magnet and strikes the target in the box at the right.

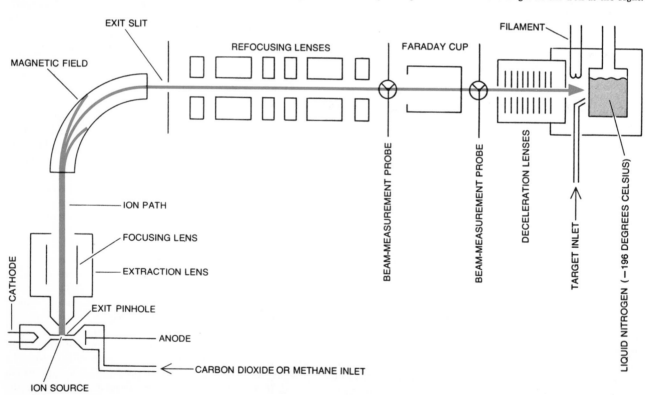

SCHEMATIC REPRESENTATION of the Berkeley carbon-ion accelerator shows how energetic carbon ions are produced. Methane or carbon dioxide containing carbon-14 atoms is admitted into the ion-source chamber, where carbon-14 and other ions are formed by the electrons emitted from the cathode. The ions are accelerated toward the magnetic field by a high positive voltage on the ion source. The ion beam passes through a magnetic field that has been adjusted to allow only the carbon-14 ions to pass through. Heavier and lighter ions strike the sides of the magnetic-field chamber. The carbon-14 ions pass through refocusing lenses and measuring devices. The Faraday cup measures the intensity of the carbon-14 beam, and two Y-shaped probes measure the beam's profile, or cross section. The interior of the accelerator is kept at a high vacuum. In order to maintain the vacuum the target material is bled in slowly as a vapor that immediately condenses on the surface of a target plate kept at −196 degrees Celsius by liquid nitrogen.

has shown gives the best separations. A sample is carried through the column by a stream of helium gas, and depending on the affinity of the constituent compounds for the oily liquid, or on their solubility in it, they move through the column at different rates. As the compounds emerge from the column their presence is recorded by a thermal-conductivity detector. This device records the difference in heat loss between a filament bathed in pure helium and another filament that is exposed to the helium stream emerging from the column. Helium is a better conductor of heat than any gaseous organic compound, so that the presence of any quantity of such a compound in the helium stream reduces the rate of heat loss from the filament. The hotter filament is a poorer conductor of electricity, and this greater resistance is converted to an electrical signal that is displayed as a line on the advancing paper strip of a recorder. The helium stream is then led to the radioactivity detector. Any radioactivity that is present is recorded by a separate pen.

From chemical knowledge or intuition we can expect that a certain compound will be formed when carbon ions strike a particular target. To determine if the compound is present we need only add a known sample of the suspected compound to the target at the conclusion of the bombardment. Because the added sample is not radioactive it appears only as a peak in the recording from the thermal-conductivity detector. If the same compound had been formed in the target during the irradiation, it would incorporate carbon 14 and would give us a peak in the radioactivity recording. Not enough of the product would have been formed, however, for it to affect the peak in the thermal-conductivity recording. Thus if we get a coincidence of a radioactive peak with the thermal-conductivity peak of a known added compound, we have provisionally identified the radioactive product, that is, we have shown it to be identical with the added compound. The procedure becomes very reliable if the coincidence of the two peaks is repeated on other columns that are packed with other kinds of oily liquids.

Finally we direct the gas stream from the gas-chromatograph column into a cold trap: a glass tube immersed in liquid nitrogen or dry ice. The low temperature freezes and traps the organic compounds but allows the helium to flow out as a gas. The frozen compounds can then be recovered and subjected to chemical tests or further gas chroma-

tography. The chemical tests can be directed not only to the question of the identity of the product but also to the location within the product molecules of the carbon-14 atoms that caused the products to be formed.

Most of our studies on the high-energy reactions of carbon have been done with the organic chemist's favorite molecule: benzene. This relatively simple hydrocarbon, C_6H_6, is the parent molecule from which a vast number of natural and synthetic compounds are formed. It has great theoretical and practical interest to organic chemists. What happens when an energetic carbon atom strikes it?

The first thing we discovered was that an accelerated carbon-14 ion reacted with the benzene to yield benzene incorporating carbon 14. The result was unexpected, because somehow the carbon 14 had to remove a carbon atom from the benzene and then remain behind as part of the benzene ring. One might imagine that carbon atoms are hard, impenetrable spheres, something

like tiny billiard balls. The incoming carbon-14 atom might strike one of the carbons in the benzene ring head on, giving up all its forward momentum and energy to the atom it had hit. That atom would be knocked away, leaving the carbon-14 atom behind as part of a new ring of six carbons.

In our experiments about one in 30 of the incoming atoms is incorporated into a new benzene molecule. That is a much higher fraction than would be expected from the billiard-ball model, in which the probability of such perfect center-to-center, complete-momentum-transfer hits is far below one in 30. From what we know about the collisions of atoms they do not in general behave like little hard spheres. The complex forces of repulsion and attraction that come into play as atoms approach one another make them somewhat soft and sticky; they can be likened to balls of putty, some harder and some softer. If one throws a putty ball at a ring of six putty balls stuck to one another, it is extremely unlikely that the thrown putty ball

BRANCHED-WIRE PROBE is moved up and down to measure the shape of the cross section of the carbon-ion beam in the ion accelerator. The Faraday cup behind the probe measures the beam's overall intensity. Cup is removed to allow the beam to reach the target.

64

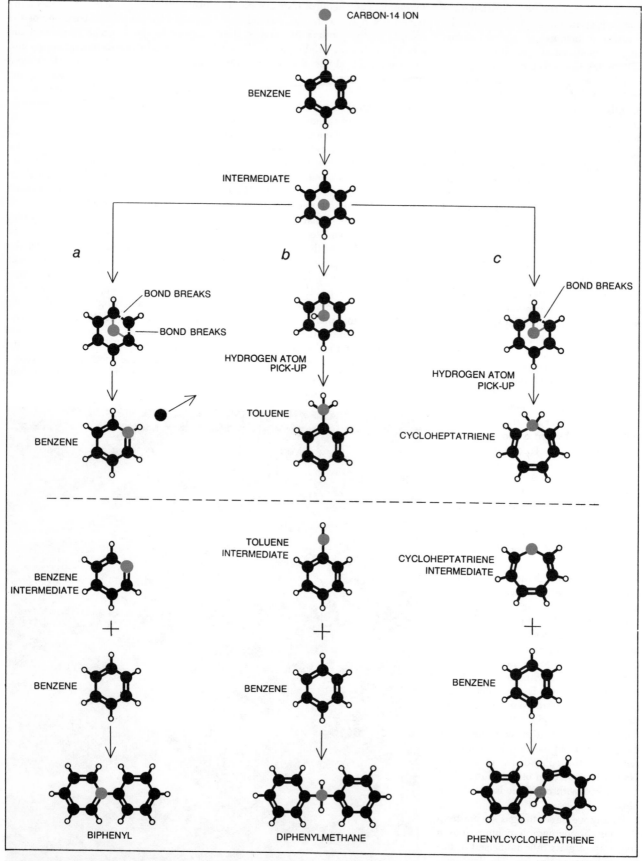

POSSIBLE MECHANISMS involved in the formation of new products when benzene is irradiated with high-energy carbon-14 ions are shown. The first step involves the formation of an energy-rich, seven-member intermediate, which then can undergo internal rearrangements to form benzene containing radioactive carbon 14 (a), toluene (b) or cycloheptatriene (c). Below the broken line are shown three possible reactive intermediates that can combine with benzene to form more complex aromatic compounds. The radioactive carbon-14 atom can appear in other positions in these products as a result of additional processes and carbon-atom rearrangements.

will replace one of the six balls to form a new six-membered ring.

Both statistics and atomic theory therefore tell us that the incoming carbon atom could not simply replace one of the benzene molecule's carbons by a colliding-billiard-ball process. So do our data, which show that the yield of benzene incorporating carbon 14 is independent of the energy with which the carbon struck the benzene. We found that one in 30 incoming carbon-14 ions is incorporated into new benzene molecules regardless of whether the carbon atoms hit at a high energy (5,000 electron volts) or a low one (two electron volts).

The carbon-carbon bond in benzene has a strength, or energy value, of more than four electron volts. Two carbon-carbon bonds must be broken in order to free one of the benzene molecule's carbon atoms. The two-electron-volt bombardment would therefore not be expected to produce a benzene with carbon 14 by the billiard-ball mechanism. Nonetheless, the reaction occurs at ion-beam energies as low as two electron volts.

In addition to benzene, one of the carbon-14-containing products of bombarding benzene with carbon-14 atoms is phenylacetylene, a molecule that has eight carbons. This brought up the possibility that the benzene containing carbon 14 was formed from an eight-carbon intermediate. For such an intermediate to be formed the incoming carbon ion would have to knock a carbon atom out of a benzene molecule, form a two-carbon fragment and go on to form a two-carbon bridge across another benzene molecule. We were able to rule out this notion by bombarding a compound, dimethyl benzene, that has two methyl (CH_3) groups attached to adjacent carbons in the benzene ring. If an intermediate with a two-carbon bridge was formed, the two carbon atoms carrying the methyl groups could be split out, leaving a new six-carbon benzene ring. On the other hand, if only a one-carbon bridge was formed, there would be no benzene with carbon 14. When we ran the experiment, we found no carbon-14 benzene among the products.

Both theoretical considerations and experimental results point to the presence of a seven-carbon intermediate, of whose existence we had no idea when we first undertook the work. It appears likely that this and other seven-carbon intermediates play a role in the reactions of energetic carbon in the solar wind; indeed, it may be that such intermediates

are formed whenever a solar-wind carbon atom strikes, with the appropriate energy, any molecule with a benzene ring.

In our experiments two compounds with seven carbon atoms, toluene and cycloheptatriene, were formed. Toluene appears to result mainly from the establishment of a new carbon-carbon bond between the slowed-down carbon-14 ion and a benzene molecule, followed by the picking up of hydrogen atoms from an adjacent benzene. Cycloheptatriene appears to result from the insertion of a carbon ion between two carbons of the benzene ring. Another product is biphenyl, which consists of two benzene rings joined together. It probably results from a reaction between a benzene that contains a carbon-14 atom and another benzene molecule. The products diphenylmethane and phenylcycloheptatriene are also formed when seven-carbon intermediates react with benzene.

The eight-carbon phenylacetylene does appear to be the result of a process in which an incoming carbon ion strips one carbon atom from a benzene and the two-carbon fragment reacts with another benzene molecule. When the energy of the carbon-14 beam is decreased below 100 electron volts, the yield of phenylacetylene drops sharply. This result suggests that at lower energies the carbon-14 ion does not have enough velocity to strip a carbon atom from the benzene molecule.

We analyzed samples of all our major products by chemical degradation procedures in order to determine the positions of the carbon 14 in the product molecules. We found that in the toluene formed from benzene bombarded by high-energy carbon-14 ions 85 percent of the carbon 14 was in the methyl group and only 15 percent was in the benzene ring. This indicates that toluene is not entirely formed from benzene by the simple establishment of a new carbon-carbon bond. There also appears to be an energy-rich, seven-carbon ring intermediate related to cycloheptatriene. This intermediate uses its excess energy to "switch" hydrogen atoms and electrons around so that some of the carbon 14 appears in the ring of the toluene product. We also measured the distribution of carbon 14 in other products. In general the percentage distribution of carbon 14 does not change over a wide range of carbon-14-ion energies; it is virtually the same for all energies from 5,000 electron volts to about six electron volts. The reason is that the carbon-14

ion must be first slowed down to five electron volts or less before it will form a bond with benzene. The higher-energy carbon-14 ions lose energy by fragmenting benzene molecules.

When the carbon-14 ion is given an energy of five electron volts, there is a sudden increase in the amount of carbon 14 appearing in the benzene ring of some of the products. In the toluene the fraction of carbon 14 in the ring jumps from 15 percent to 40 percent. It appears that at energies of five electron volts or less only first-collision reactions occur, that is, the carbon 14 does not collide with other molecules before it forms an intermediate with a benzene molecule. The five-electron-volt energy level seems to be the maximum the seven-carbon intermediate can tolerate without breaking into fragments.

At an energy level of two electron volts we also have first-collision reactions, but there is much less energy available for rearranging the position of the carbon atoms in the benzene ring. Only 6 percent of the carbon 14 is found in the ring when toluene is formed. In fact, all the reaction products we have analyzed show maximum rearrangement when they are bombarded with carbon-14 ions at five electron volts. The minimum rearrangement comes at two electron volts. These observations reveal chemical mechanisms that could only have been uncovered by studies of reactions at high energies.

Although we have identified the major products of bombarding benzene with energetic carbon ions, there are many minor products we have not yet identified. We believe some of the products are novel and are not known in contemporary carbon chemistry. It may even be possible that some of these molecules may have been important for the emergence of life. For example, a high-energy process may have been necessary for the first appearance on the prebiological earth of a particular molecule that was indispensable for the emergence of living matter. Once made, that molecule may have been able to replicate itself by autocatalysis, the process by which the presence of a particular molecule greatly speeds the production of like molecules. Our continued efforts will be toward establishing the identity of the novel high-energy products and the mechanisms by which they are formed. We are confident that such research will tell us much about the carbon chemistry that goes on in our energy-rich universe.

7

The Force between Molecules

by Boris V. Derjaguin
July 1960

Its nature has long been a mystery. Physicists in the U.S.S.R. have now succeeded both in measuring the intermolecular force and in demonstrating that it is purely electromagnetic in origin

One of the great traditions of science has been the study of the forces that hold the parts of the material world together. The tradition runs from Isaac Newton's work on gravity and celestial mechanics to the modern examination of the nucleus of the atom. In this long line of inquiry there has been, until very recently, an odd gap. The smallest units of matter, in its everyday forms, are molecules; and the force between molecules determines many characteristics of the substances with which we are familiar. Yet physicists have been unable either to measure intermolecular forces experimentally or to deduce them from theory.

These forces are not the chemical attractions, or valence bonds, that link atoms in compounds. They are forces of longer range that draw together the molecules of solids and liquids, and, to a much smaller extent, of gases. Valence forces can be "saturated"; when the available bonds of an atom are fully satisfied by attachment to its partners in a compound molecule, it exerts no force on other neighboring particles. Thus the oxygen atoms in two molecules of water do not attract each other. On the other hand, every molecule in a sample of water exerts some attraction on every other molecule.

Without these long-range forces there would be no sharp boundary between liquids and their vapors. Intermolecular forces are responsible for surface tension, capillary action, adsorption and other surface phenomena. They determine most of the properties of liquids: viscosity, heat of evaporation, solubility in other liquids. They cause colloids to coagulate. Each of these phenomena has been intensively studied for years. But without an understanding of their underlying principle there could be no fundamental and unifying theory.

In the U.S.S.R. during the past few years there have been two significant break-throughs. Irene Abrikossova and the author have succeeded in measuring the molecular force directly. Following this development the theoretical physicist E. M. Lifshitz derived a mathematical formula for the force of molecular attraction from very general principles.

To put these developments in proper perspective let us begin with some history. The earliest speculations about molecular forces were those of the 18th-century French mathematicians Alexis Claude Clairaut and Pierre Simon de Laplace. Theirs was the triumphant era when Newton's law of universal gravitation seemed capable of explaining the motions of heavenly bodies down to the minutest detail. By analogy with gravity Clairaut and Laplace assumed the existence of an attractive force acting along the line between the centers of molecules. It was apparent, however, that such a force must fall off not as the square of the distance of separation, as in the case of gravity, but more sharply. Furthermore, the constant of proportionality could be different for different molecules.

In arithmetical terms Newton's law contains the fraction G/r^2, where G is the constant of proportionality. What makes gravity comparatively easy to deal with is that G is a universal constant whose value is determined for all masses simply by measuring the force between any two known masses at any known distance (r). On the other hand, the assumed law of molecular force has the form C/r^n, where both C and n are unknown, and where C presumably differs for different kinds of molecules. It is of course impossible to pick out two individual molecules, hold them at a known distance and measure the force between them. Hence the unknowns cannot be directly determined. Nor, it turns out, can they be determined indirectly through bulk properties that depend on molecular forces. Measurements of surface tension and similar quantities cannot even settle the power law, let alone establish the constant C. No real progress could be made without a better theoretical platform from which to attack the problem.

At the end of the last century it seemed that the classical theory of electromagnetism could provide such a platform. The laws governing electric and magnetic fields and their interactions with material bodies had been worked out. Molecules were known to be composed of electrically charged particles. The visible radiation given out by molecules had just been proved to be electromagnetic waves. Perhaps electromagnetism could account completely for the intermolecular force. (It should be realized that this was just an assumption. It was also possible that some new force, previously unknown, might come into play at the molecular level, as has since proved to be the case at the nuclear level.)

One of the first workers to see clearly the implications of electromagnetic theory was the Russian physicist P. N. Lebedev, best known as the first experimenter to measure the pressure exerted by light. In 1894 he wrote, with prophetic insight: "Hertz's research into the interpretation of light waves as electromagnetic processes conceals another problem, as yet untouched. This is the problem of a radiation source, of those processes which a molecular vibrator undergoes when it emits light energy to its surroundings. On the one hand this problem leads into the field of spectro-

scopic analysis. On the other hand, almost entirely unexpectedly, it leads to one of the most complicated questions of modern physics, the science of molecular forces. This situation results from the following concepts. From the point of view of the electromagnetic theory of light one may state that two radiating molecules are two vibrators in which electromagnetic vibrations are excited. They must therefore experience forces resulting from the electromagnetic interactions of the varying electrical currents (according to Ampère's law) and varying charges (according to Coulomb's law) in them. We may therefore also assert that there must then exist molecular forces whose origin is directly related to radiative processes. . . . The most interesting as well as the most complicated case is that of a physical body in which many molecules act simultaneously on one another and are so closely packed that their vibrations are not independent. If it ever becomes possible to solve this problem completely, we will be able

to use spectroscopic data to predict the magnitude of the molecular forces due to the mutual radiation of the molecules, to calculate the temperature dependence of these forces, and by comparing these calculated quantities with experiment to solve the problem at the root of molecular physics. This problem is whether all so-called molecular forces reduce to the already known and above-mentioned action of radiation—to electromagnetic forces—or whether they involve other forces whose source is still unknown."

Lebedev came as close to the latest views on the nature of molecular forces as was possible before the development of quantum mechanics. But his statement, while ahead of its time, was very far from a quantitative law. As so often happens in science, the next step was backward. The first quantitative theories of molecular forces treated them as entirely electrostatic rather than electromagnetic. These static theories began to be developed shortly after

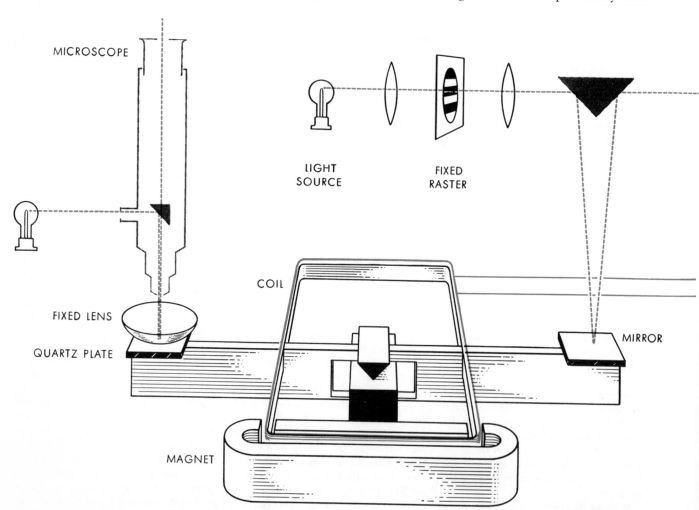

CONTROLLED BALANCE for measuring the molecular attraction between macroscopic bodies is illustrated schematically. The broken colored line at left is the path of the light beam used to de- termine the width of the gap between the bodies. The broken colored line at top right is the path of light beam by which the position of the balance arm is set and automatically stabilized. Solid

Ernest Rutherford, by his discovery of the nucleus, showed how electric charge is distributed in the atom. However, they did not reach their final form until 1930, when the German physicist Fritz London applied to them the newly discovered principles of quantum mechanics.

According to London's theory, the force between two molecules is C/r^7. The force varies inversely as the seventh power of the distance between their centers. The constant C depends on certain electrical properties of the molecules, including their "polarizability." This measures the degree to which an electric field distorts a molecule, shifting its electrons with respect to the positive nuclei of its atoms.

Unfortunately the polarizabilities of individual molecules are themselves unknown quantities, so we cannot use London's formula to calculate the absolute value of the force. If it were possible to compare molecular forces at various distances, the seventh-power law could at

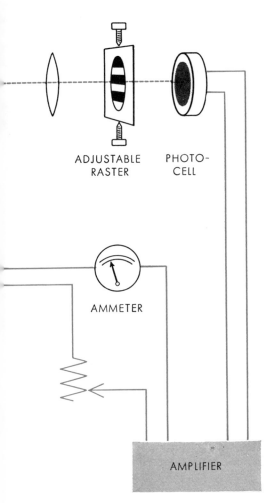

ADJUSTABLE RASTER   PHOTO-CELL

AMMETER

AMPLIFIER

colored lines trace the electrical portion of the control circuit. The operation of the system is described in detail in the text.

least be tested, but for a long time there was no way of doing this either. Attempts were made to verify the formula indirectly, for example by measuring the amount of heat energy required to vaporize a liquid (*i.e.*, pull its molecules apart). However, it can be proved that most of the energy is used up in the interval when the molecules are still within a diameter or two of each other. At such close distances London's theory is not strictly applicable, and even the concept of distance between molecules (which are not really spheres) is not clearly defined.

Despite the lack of experimental confirmation, or perhaps because of it, the London theory was generally accepted for almost 20 years. Yet it was clearly unrealistic for large distances of separation as well as for very small ones. Being an electrostatic theory, it regarded force as communicated instantaneously between molecules. In fact electric forces travel at the finite speed of light (186,-000 miles per second).

Now the "large" distances just mentioned are large only compared with molecular diameters. On the everyday scale of distance the attractive force between molecules falls away to essentially nothing at a very small separation indeed—a few ten thousandths of a millimeter. The time required by an electromagnetic impulse to travel such a distance is of the order of a ten-million billionth of a second. How can so tiny an interval make a difference in the force?

To understand this let us recall Lebedev's picture of "molecular vibrators" sending out radiation. The molecules are like radio antennas in which oscillating electric charges emit a train of electric and magnetic vibrations. When the vibrations from one antenna reach a second, they set its charges into oscillation. These oscillations become in turn a source of waves that reach back to the first antenna, exerting a force on its moving charges. Thus the two antennas interact by exchanging radiation. The strength of the interaction varies with the relative phase of the arriving waves and the oscillating charges; the interaction is stronger if the waves and particles swing back and forth together rather than being totally or partially out of step. But the relative phase depends on the distance between antennas and on the frequency of the vibrations; in other words, on the number of wavelengths in the distance of separation. If, for example, the antennas are half a wavelength apart, the emitted and absorbed waves will be exactly out of phase. At smaller

(or larger) fractions of a wavelength the shift will be correspondingly less. At very small fractions of a wavelength the phase shift is negligible.

The wavelength of most radio waves is measured in meters or centimeters. But the radiation sent out and absorbed by molecules are light waves, only a ten thousandth of a millimeter long. Thus if two molecules are separated by a ten thousandth or even a hundred thousandth of a millimeter, there will be considerable phase shift in the exchanged radiation. The shift must be considered in computing the force.

The first theoreticians to take the phase shift into account in their calculations were the Dutch physicists H. B. G. Casimir and D. Polder, who worked out an electromagnetic theory of molecular force in 1948. In their calculations they did not, of course, use the classical picture of radiation that we have just outlined but rather the ideas of quantum electrodynamics. These differ from the older view in two important respects. First, molecules do not radiate (or absorb) continuously, but in discrete amounts. Moreover, the waves they send out travel in packets or photons. Second, molecules can interact electrically without actually emitting or absorbing energy. In that case they are said to exchange "virtual" photons.

At relatively large distances the Casimir-Polder electromagnetic theory gives the formula K/r^8 for the attractive force between two molecules. Thus when the effects of phase shift are allowed for, the force decreases with the eighth power of the distance rather than with the seventh, as in London's formula. The constant K is different from London's C, but it also contains the polarizability of the molecules. Therefore it cannot be evaluated directly either.

We shall see shortly that each of the two formulas appears to be correct for an appropriate range of the distance r. London's formula applies when the distance is small enough to make the phase shift of the exchanged radiation negligible. The Casimir-Polder result describes the force at larger distances, where the phase shift cannot be disregarded.

Both theories, however, concern only pairs of isolated molecules. They do not themselves give the force of molecular attraction between two condensed bodies, each containing many molecules packed close together. Yet this is the only force one can hope to observe directly, for example by measuring the attraction between two solids separated by an extremely narrow gap. (This is precisely the measurement that my colleagues and I have finally succeeded in making.) Furthermore, it is this force that presumably causes the tiny particles of a colloid (which are nevertheless large enough to contain many molecules) to stick together and coagulate. Thus it is fundamental to the theory of the colloidal state.

If we assume that molecular forces are not influenced by interactions with neighbors, that each molecule in one dense body exerts the same force on each molecule of a nearby body as if they were so many separate pairs, then it should be possible to obtain the total force simply by addition. Some workers have made this assumption and have proceeded to compute the total force between condensed bodies both by the London and the Casimir-Polder formulas. However, there is good reason to

believe that the forces are not strictly additive, so these computations cannot give the right answer.

Clearly the individual-particle approach to molecular forces had proved unprofitable. Its formulas contained unknown quantities, and so could not be evaluated. Even if they could, there was no way of proceeding from isolated pairs of molecules to the gross samples of matter with which experimenters are obliged to deal.

This was the unsatisfactory state of theoretical knowledge when Mrs. Abrikossova and I set out in 1951 to measure the molecular attraction between a pair of solid objects. With the help of Fanny Leib, a colleague at the Academy of Sciences in Moscow, we were able to design an apparatus that met the stringent, and in some ways contradictory, requirements of the job.

In the first place we were trying to

measure a small force—as little as a ten thousandth of a gram. This in itself is not too difficult; microchemists routinely deal with far smaller forces. But in our problem the attraction to be measured shows up only when the two bodies are extremely close together—within a few ten thousandths of a millimeter, or little more than a thousandth the thickness of a human hair. Furthermore, the force varies enormously as the distance of separation is changed. It is precisely this variation that we wanted to determine. Thus we needed a way to set the width of the gap accurately at any desired distance, and to maintain the separation against the attraction force being measured. Not only that; the equilibrium had to be extremely stable so that chance variations in the gap were corrected before they could build up.

The requirements of sensitivity, stability and rapid response are not easy to reconcile. A sensitive balance tends to

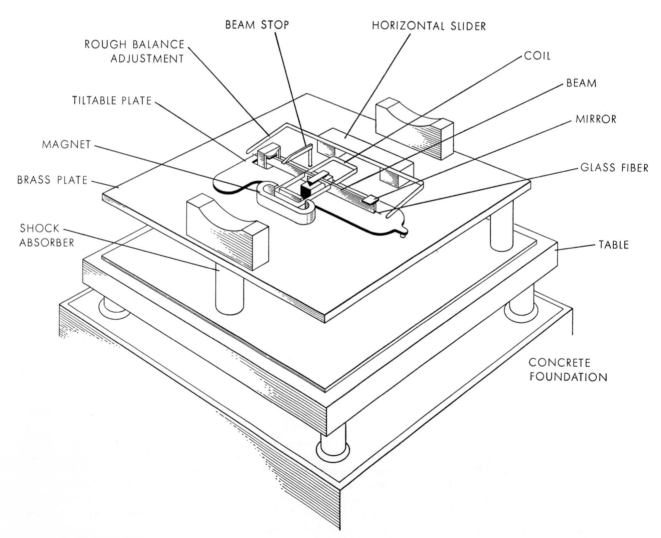

BALANCE IS MOUNTED on a brass plate connected through shock absorbers to a table, which in turn rests on a concrete foundation. Tiltable plate allows the entire balance to be rotated, changing the position of the gap with respect to the surfaces of the objects whose attraction is being measured. Glass fiber extending through the balance arm can be shifted for rough balancing.

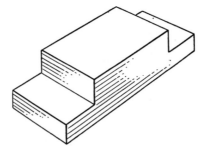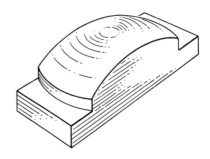

TEST BODIES for the measurement of molecular force were a highly polished flat plate (*left*) and an equally smooth spherical lens (*right*). In most trials the bodies were made of quartz, but some other materials were tested. The lens diameter was also varied.

swing widely in response to a small change in force. Our apparatus achieved these diverse aims by means of automatic control.

The bodies whose mutual attraction we measured were a flat plate and a plate with a spherical surface. (It is easier to set the distance between a sphere and a plane than between two planes, which must be held parallel.) The flat plate was fixed to one end of a balance arm three centimeters long, weighing about a tenth of a gram and balanced on a fulcrum made of a wedge-shaped agate bearing. Above the flat plate was mounted the rounded plate in an adjustable bracket [*see illustration on pages 68 and 69*].

To determine the width of the gap between the two surfaces we made use of the optical phenomenon known as Newton's rings. When light of a single color shines through a lens-shaped disk held near a flat surface, interference between the rays reflected from different surfaces produces a pattern of concentric light and dark circles [*see top illustration on next page*]. The diameter of the circles depends on the distance between the surfaces. We observed the Newton's rings in our sample through a microscope, accurately determining their diameter and thus measuring the gap width.

The automatic-control circuit, which is the heart of the arrangement, accomplishes a quadruple purpose: (1) it allows us to set the gap at any desired distance, (2) it maintains the separation by producing a force equal and opposite to the attraction, (3) through negative feedback it corrects any drift in the balance arm away from the preset position and (4) it provides a means of measuring the opposing force, and thus the attraction.

The circuit works as follows. A beam of light shines through a grid, or raster, of alternate opaque and transparent stripes. A mirror on the balance arm reflects the beam through a lens, forming an image of the grid. The image falls on a second identical grid, beyond which is a photocell. Light passing through the second grid and reaching the cell sets up an electric current that is amplified and fed into a coil of wire rigidly attached to the arm. This coil is pivoted between the poles of a magnet and thus tends to turn, like the coil in an ammeter, carrying the arm with it. The turning force depends on the amount of current passing through the coil, and is arranged to be opposite in direction to the turning force

on the arm due to molecular attraction.

The current in the coil is determined by the amount of light that reaches the photocell. Let us see what controls this. If the image of the first grid falls on the second grid in such a way that the transparent stripes of the image exactly cover the black stripes of the grid, no light gets through. The current through the coil is zero. Rotating the arm slightly shifts the angle of the mirror, thus moving the image. Now part of the transparent stripes in the image overlap the transparent stripes on the second grid, and some light passes through. The farther the arm turns, the more light passes, and the stronger the current.

In setting the gap initially at the desired width, we shift one grid by means of a micrometer screw forcing the arm to turn as the equilibrium position varies. We then read the current on an ammeter, thereby determining the force at this gap width. If a stray vibration alters the distance of separation, the current changes in such a way as to bring the arm back to its initial position. Without the control circuit, the balance arm takes several seconds to swing back and forth when it is displaced from equili-

brium. When the control is switched on, this period is reduced to a thousandth of a second.

Fluctuations of the current in response to outside vibrations set the eventual limit on the accuracy of our measurement. To minimize the disturbances we mounted the balance on a heavy pedestal, connected by means of a hydraulic shock-absorber to a cement pier sunk in the ground. Air currents were eliminated by enclosing the apparatus in an evacuated chamber. This also reduced the viscous drag of the air in the narrow gap on the motions of the balance arm.

The most annoying and persistent difficulty we encountered in getting the balance to work properly was in trying to keep the surfaces of the plates free from both electrostatic charge and dust particles. Electrostatic forces are thousands of times stronger than the molecular attraction, and would completely obscure it. Dust particles on the surfaces would change the effective gap width.

To eliminate dust we cleaned the plates with cotton dipped in ether. Charge was removed by ionizing the air in the gap with a radioactive material. Repeated treatments were often re-

OVERLAPPING GRIDS, or rasters, determine the amount of light reaching the photocell in the balance-control circuit. Image of opaque bands (*colored stripes*) on one grid falls on second grid (*circle*). Light transmitted varies from a maximum when the two sets of bands coincide (*left*) to zero when image bands wholly fill transparent spaces (*right*).

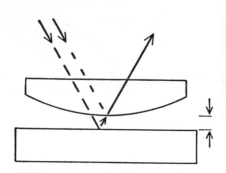

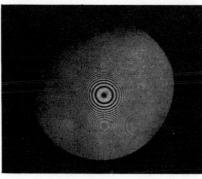

NEWTON'S RINGS (*photograph at right*), used to determine gap width in the author's experiment, are formed when monochromatic light passes through a lens and is reflected from a flat plate. Some light is also reflected from lens surface, and the rays at any given point reinforce or interfere depending on the width of the gap between lens and plate.

quired, because the surfaces tend to become charged as they are cleaned, and to get dusty when the charge is removed.

In our first experiments in 1951 we measured forces that were about 5,000 times as large as predicted by any theory. Experimenters in the Netherlands got the same sort of result at this time. When we finally were able to get rid of all electric charge, however, the spurious forces disappeared. Other workers have duplicated our results.

We measured the force between two quartz plates, between a quartz and a chromium plate and between crystals made of a mixture of thallium bromide and thallium iodide. For gaps ranging from about two to four ten-thousandths of a millimeter the force varied from approximately 20 to two ten-thousandths of a gram [*see graphs on opposite page*]. The force obviously depended on the types of molecule of which the plates were composed.

When we obtained our first successful measurements, we could compare them only with the London and Casimir-Polder theories. It was at once apparent

that the Casimir-Polder formula fit our results better than did London's. (A gap of a ten thousandth of a millimeter is, of course, large on the scale of molecular diameters.) However, we could not check the formula absolutely, for the reasons already mentioned: it contained undetermined constants, and there was no rule for applying the microscopic equation to macroscopic bodies.

When Lifshitz heard of our results, he was encouraged to attack the theoretical problem anew. In 1955 he developed a remarkable method for finding the molecular interaction of two macroscopic bodies. The method ignores the individual particles completely, and depends only on macroscopic properties, which can be measured experimentally. We can do no more here than give a hint as to the basis of his highly abstract, mathematical approach.

As every amateur radio-operator knows, the effort to receive extremely weak signals from distant stations is in the end frustrated by noise originating in the receiver itself. This noise arises out of the thermal vibrations of the mole-

cules in the wiring and other parts of the set. The resulting electromagnetic fluctuations are noticeable when a radio receiver emits a hiss. But they exist silently in any material body. The only way to suppress them completely is by cooling the bodies to absolute zero and thus removing all their thermal energy.

According to quantum theory, however, there must be fluctuations in the electromagnetic field even at absolute zero. These fluctuations are the result of the so-called zero-point energy of the electrons. By definition this energy cannot be detected or extracted from matter; it is represented entirely by the emission and absorption of virtual photons. Nevertheless it is there, and it is a means of interaction of the particles.

Lifshitz realized that such universal zero-point vibrations should in fact account for the molecular force. In his theory, however, he does not deal with discrete particles and light quanta, or photons, but with continuous matter and fields, along the lines of classical electromagnetic theory. He considers two closely spaced bodies and calculates the electro magnetic fields produced in the narrow gap between them and in the space around them by the fluctuations in the various regions of the material. From the difference between the field strength in the gap and in the surrounding space the force can be calculated.

The formulas are very complicated, but they do not contain any quantities that cannot be measured. To find the force it is necessary to know only the wavelengths absorbed by the materials (*i.e.*, their absorption spectra in the infrared, visible and ultraviolet regions) and their polarizabilities or "dielectric permeabilities." Unlike the polarizability of individual molecules the dielectric permeability of a macroscopic sample of a substance is easily determined.

Not only does the theory account for a force at absolute zero; it also shows that thermal oscillations contribute very little to the force at higher temperatures. In the range near absolute zero the force is almost independent of temperature. As the temperature is increased, the force does vary, but only through an indirect effect. Changing the temperature changes the quantum state of the electrons in the bodies, altering their absorption spectra and hence the force.

It is surely remarkable that a purely molecular effect can be calculated by ignoring the very existence of particles. (They are ignored except for one point:

NONADDITIVITY of molecular force on the macroscopic scale is demonstrated by the fact that the total force of attraction between the large block and the small ones is changed by shifting the small blocks from the arrangement shown at left to the one shown at right.

The theory is valid only for gap widths substantially greater than the molecular diameter.) But there is an even greater surprise. Whereas it is impossible, as we have seen, to pass mathematically from a microscopic, two-molecule theory to the macroscopic situation, the reverse transition is straightforward. The attraction between a pair of isolated molecules comes out of the Lifshitz theory as a special limiting case. For small distances it reduces to the London seventh-power law, and for larger distances, to the Casimir-Polder eighth-power law. And these equations are reached more easily by the roundabout method of Lifshitz than by the earlier direct calculations!

Furthermore, the calculations settle definitely the question of whether the force is additive. It is not. The attraction between two bodies is not the simple sum of the attraction between all pairs of their molecules. In fact, even on the macroscopic scale the force does not add [*bottom illustration on opposite page*].

In the case of macroscopic samples of matter the force depends on many factors, including the shapes of the bodies. When two flat, parallel plates are separated by a gap much smaller than the wavelengths of their main absorption lines, the complex formula reduces to a comparatively simple expression: The force varies inversely as the cube of the gap width. When the separation is much greater than the absorption wavelength, the force varies inversely as the fourth power of the distance. (These formulas correspond respectively to the inverse seventh- and eighth-power laws for pairs of individual molecules.)

As soon as Lifshitz announced his results, we applied them to a calculation of the force between the bodies we had used in our experiments. The curves at right represent the theoretical predictions. As can be seen, our experimental points cluster quite closely around the curves. This agreement between theory and experiment is particularly convincing because the theory contains no undetermined constants; the comparison is absolute and applies successfully to several different substances.

Although our experiments verify the Lifshitz formulas only for the special case of "large" gaps, they tend to support the whole theory. And so the electromagnetic nature of molecular forces and their relation to spectra, so long ago postulated by Lebedev, have now been demonstrated. He would be glad to know that the attraction between molecules does not "involve other forces whose source is still unknown."

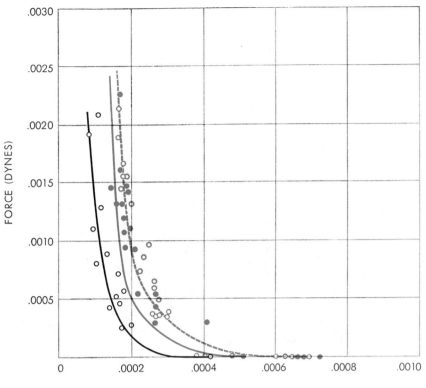

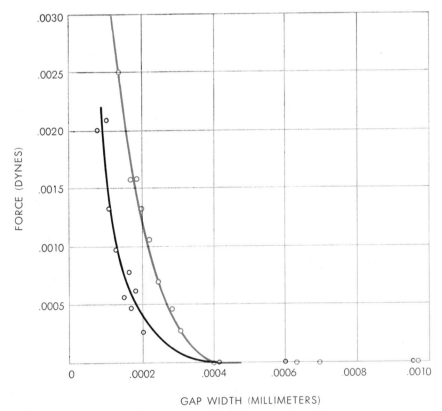

EXPERIMENT AND THEORY are compared in these graphs. The curves show the variation of the molecular force of attraction with distance for the bodies used in the author's experiments, according to the Lifshitz theory. The points represent experimental values. The upper graph refers to experiments on a flat plate and a spherical lens 10 centimeters in radius. Data for quartz bodies are in black; for thallium halide, in solid color; for a quartz and chromium body, in broken colored line and open dots. Lower graph shows data for a 10-centimeter quartz sphere (*black*) and for a 26-centimeter sphere (*color*).

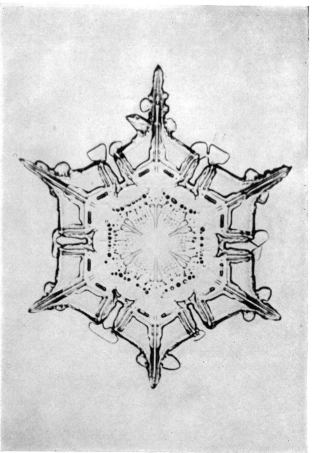

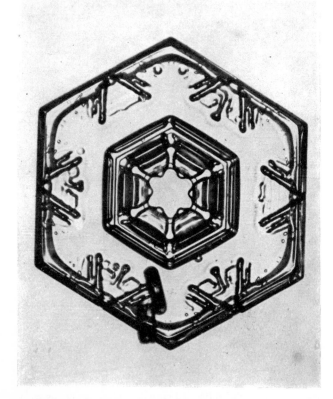

SNOW CRYSTALS are enlarged about 50 diameters in these photomicrographs by Vincent J. Schaefer of the Munitalp Foundation in Schenectady, N. Y. The hexagonal symmetry of the crystals is due to their molecular structure *(see diagram on page 78)*.

Water

by Arthur M. Buswell and Worth H. Rodebush
April 1956

*Although we take its properties for granted, they are most
unusual. As an example, it has the rare property of being
lighter as a solid than a liquid. If it were not, lakes would
freeze from the bottom up!*

Water is the only common liquid on our planet. Next to air it is the substance with which we are most intimately acquainted. Because it is so familiar, we are apt to overlook the fact that water is an altogether peculiar substance. Its properties and behavior are quite unlike those of any other liquid. To take just one example, water has the rare property of being denser as a liquid than as a solid, and it is probably the only substance that attains its greatest density at a few degrees above the freezing point (four degrees centigrade). The consequences of this behavior are of great importance to life on our planet. When ice forms on a lake, for example, its lower density (only nine tenths that of liquid water) keeps it on top and it acts as an insulating blanket to retard cooling of the underlying water. As a result lakes in the temperate zones do not freeze solid to the bottom but leave a zone for the winter survival of aquatic life. On the other hand, the same peculiar property of water has fatal consequences for living cells. When water in the cells freezes and becomes less dense, its expansion damages or breaks up the cells.

Even the elements of which water is composed—oxygen and hydrogen—are chemically exceptional. Both are unusually reactive. Oxygen is our chief source of energy, being responsible for the respiration of living organisms and the combustion of fuels. Hydrogen, unique in the fact that it has no enclosing shell but only a single electron, is able to attach itself to other atoms not only by means of its electron (a valence bond) but also by virtue of the attraction of its unoccupied, positively charged side for an electron in a second atom. This attachment is known as the hydrogen bond. In water the two hydrogen atoms attached to each oxygen atom can become linked to other atoms as well by means of these so-called hydrogen bonds. As a consequence the H_2O molecules are joined together, so that water should be considered not a collection of separate molecules but a united association. In effect the whole mass of water in a vessel is a single molecule.

The best method of detecting hydrogen bonds is to study water with the infrared spectrograph. We have found that the hydrogen bond absorbs radiation most strongly at a wavelength of about three microns, which is in the near infrared region of heat radiation—*i.e.*, close to the visible light spectrum. Liquid water absorbs this radiation so powerfully that if our eyes were sensitive to the infrared, water would appear jet black. There is some absorption even in the visible spectrum at the red end. The fact that water absorbs red light accounts for its characteristic blue-green color.

Water's 33 Components

One of the shocks to our familiar notions about water is that its formula is not simply H_2O. Nor is it a single substance. The beginning of this disillusionment came in 1934 when Harold Urey discovered "heavy water." Urey found that the purest water contained besides hydrogen and oxygen another substance like hydrogen but with an atomic weight of two, or twice that of hydrogen. This substance, which is now called deuterium, combines with oxygen to form the compound D_2O. By now, we know, of course, that there is a third isotope of hydrogen, called tritium, and three isotopes of oxygen: 0-16, 0-17 and 0-18. Thus the purest water that can be prepared in the laboratory is made up

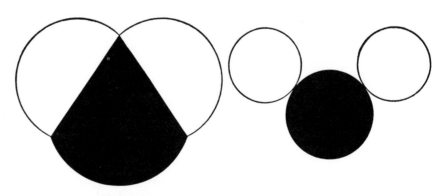

MOLECULE of water consists of one oxygen atom (*black*) and two hydrogen atoms (*white*). The distance between the center of the oxygen atoms and the center of each of the hydrogen atoms is .9 Angstrom unit (one Angstrom unit: .00000001 centimeter). The angle formed by the two hydrogen atoms is 105 degrees. These dimensions are fitted together in the drawing at left. In the more schematic drawing at right the size of the atoms has been reduced. This representation of the molecule is used in the following drawings.

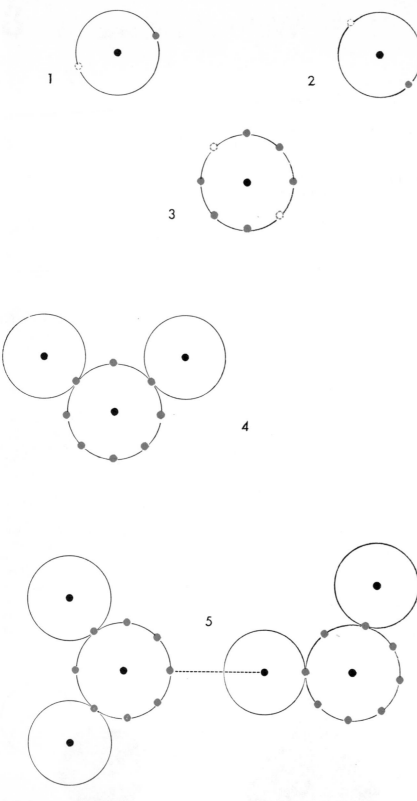

ELECTRONS AND PROTONS of the water molecule account for most of its physical and chemical properties. The hydrogen atom (1 *and* 2 *in this highly schematic picture*) consists of a positively charged proton (*black dot*) and a negatively charged electron (*colored dot*). The oxygen atom (3) has eight electrons, six of which are arranged in an outer shell. Because hydrogen shell has room for one more electron (*broken colored circle*), and the outer shell of oxygen has room for two more electrons, the atoms have an affinity for each other. In the water molecule (4) the electrons of the hydrogen atoms are shared by the oxygen atom. Because the positively charged proton of the hydrogen atom now sticks out from the water molecule, it has an attraction for the negatively charged electrons of a neighboring water molecule (5). This relatively weak force (*broken line*) is called a hydrogen bond.

of six different isotopes, which may be combined in 18 different ways. If we add the various kinds of ions into which the addition or removal of an electron may transform water's atoms, we find that pure water contains no fewer than 33 substances [*see top of page 79*].

Of course the amounts of the isotopes other than common hydrogen and common oxygen (0-16) are tiny. Tritium and oxygen 17 appear only in the minutest traces, and deuterium is present to the extent of about 200 parts per million and oxygen 18 about 1,000 parts per million. However, the properties of heavy water, particularly the D_2O variety, have attracted wide interest and have been extensively studied.

D_2O has a slightly higher boiling point than H_2O (101.4 degrees C.), freezes at a substantially higher temperature (3.8 degrees C.), and is somewhat more viscous than ordinary water. Its physiological properties are surprising. In animals and plants it appears to be entirely inert and useless. Seeds will not sprout in D_2O, and rats given only D_2O to drink will die of thirst.

The largest use of heavy water is as a moderator in nuclear reactors, but it is also widely employed in theoretical research, especially in organic and biological chemistry. If compounds containing active hydrogen are treated with D_2O, deuterium will replace the hydrogen and the compound will show changes in chemical properties resulting from the lesser reactivity of deuterium.

It is interesting to find that the amount of D_2O in natural water appears to be the same whether the water comes from an alpine glacier or the bottom of the ocean, from willow wood or mahogany.

Tritium is more ephemeral and more variably distributed. It is formed in the highest layers of the atmosphere by the bombardment of cosmic rays, and falls in rain and snow [see "Tritium in Nature," by Willard F. Libby; SCIENTIFIC AMERICAN, April, 1954]. Since tritium is radioactive, with a half-life of 12.5 years, it disappears after a time from water which has been out of contact with the atmosphere. Wines, and water in wells, can be dated by their tritium content. An interesting well in the Urbana-Champaign area was found to be devoid of detectable tritium, which means that at least 50 years have elapsed since the water fell as rain.

The functions of water in nature are innumerable. It is the solvent par excellence. It is the medium in which life originated and in which all organisms still exist. The living cell consists largely

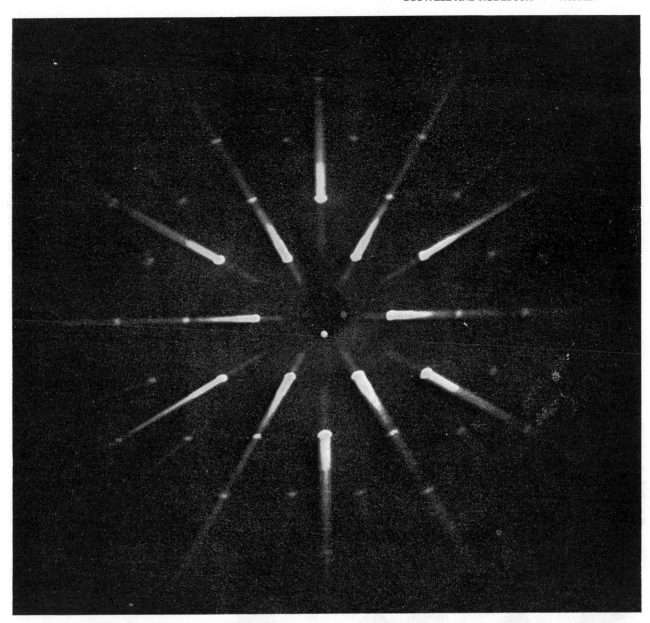

X-RAY DIFFRACTION photograph of ice was made with a precession camera by I. Fankuchen and his colleagues at the Polytechnic Institute of Brooklyn. The position of the spots in the photograph is related to the symmetry of the crystal (*see diagrams on next page*).

of water and literally floats in water. Considering how predominantly living matter is made up of this fluid, the extent to which it takes on a solid shape is surprising indeed.

Water plays a fundamental role in the protein molecule, the basic material of living matter. Proteins have a structure which places them in the class of substances known to chemists as plastics. In order that a plastic may possess flexibility and other desirable physical properties, it must contain a fluid called a plasticizer. Water is a "plasticizer" for proteins. In the chainlike protein molecule the hydrogen bonds of water provide secondary links which fix the pattern of the molecule. Removal of water alters the pattern and "denatures" the protein. Fortunately the process is reversible, so that water in the cells can restore the pattern. On the other hand, the hydrogen bonds of hydrides other than water, such as ammonia or hydrogen cyanide, form a stable denatured configuration which freezes the protein in a dead pattern. This is why all the hydrides except water are extremely toxic. Their action is somewhat like that of a virus, which corrupts the true protein structure into a strange distorted pattern.

Water's Structure

To understand the behavior of water we must understand its structure. It is far from simple. The best approach is a study of the structure of ice. The arrangement of the oxygen and hydrogen atoms in the ice crystal has been determined by X-rays and other means. The two hydrogens are bonded to the oxygen approximately at right angles to each other, more exactly, at an angle of 105 degrees [*see illustrations on page 75*]. If the angle were 109 degrees, the frozen water molecules would form a cubic lattice, as in the diamond crystal. But in this case such a structure would be unstable because of the strain on the distorted bonds. The exact arrangement of the molecules in the ice crystal is not known with certainty; we know that they form a hexagonal structure, which

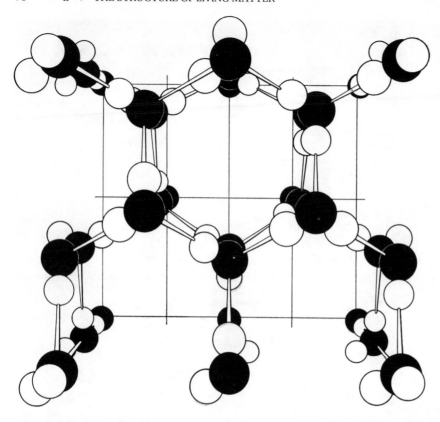

ICE consists of water molecules in this arrangement. The top drawing shows a model of ice seen from one direction. The bottom drawing shows the same model seen as if the reader had turned the top drawing forward on a horizontal axis in the plane of the page. Some hydrogens have been omitted from the molecules which touch the grid. Each hydrogen in each molecule is joined to an oxygen in a neighboring molecule by a hydrogen bond (*rods*). In actuality the molecules of ice are packed more closely together; here they have been pulled apart to show the structure. In a similar model of liquid water the molecules would be much more loosely organized, farther apart and joined by more hydrogen bonds.

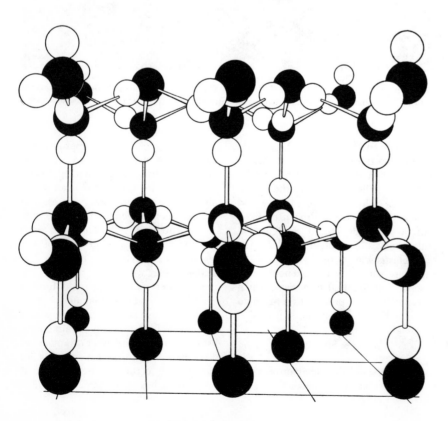

is exhibited on the macroscopic scale in the form of snowflakes. Each molecule is surrounded by four nearest neighbors, so that the group has one molecule at the center and the other four at the corners of a tetrahedron. The molecules and groups of molecules are joined together by hydrogen bonds.

The forces of attraction between the molecules in ice or water produce a strong inward pressure. As we shall see, this accounts for some of water's peculiar properties. In the form of ice, its open structure resembles a bridge arch under heavy downward stress. When the temperature of the ice rises to zero centigrade, the thermal agitation of the molecules is sufficient to cause the ice structure to collapse, and the water becomes fluid. It is well known that the application of pressure from outside will make ice melt at a lower temperature; evidently this reinforces the internal pressure within the ice and assists its collapse. Contrariwise, we can assume that if the internal pressure is reduced in some way, the melting point of ice will rise. Calculations indicate that if this pressure were entirely eliminated, ice would not melt until its temperature reached 15 degrees or more centigrade (59 degrees Fahrenheit).

According to X-ray determinations, the average distance between the center of one oxygen atom and the center of the next in the ice crystal is 2.72 Angstrom units (an Angstrom being one hundred-millionth of a centimeter). When ice melts to liquid water, the hydrogen bonds are stretched and the molecules move farther apart: the distance between oxygens is increased to about 2.9 Angstroms on the average. This stretching would open the structure further and make water less dense were it not for the fact that in the fluid the molecules crowd together in more compact groups. Each molecule is now surrounded by five or more neighbors instead of only four.

The chaotic disorder in which water molecules exist in the liquid state is difficult to picture. Their arrangement shifts continually. The angle between the two hydrogen atoms in the water molecule no longer remains fixed near a right angle but becomes variable, so that the molecule is flexible. Each oxygen atom now attracts by electrical forces not two extra hydrogen atoms as in ice, but three or more. Thus we may find an oxygen atom surrounded by five or six hydrogens and a hydrogen atom surrounded by as many as three oxygens. In the closely knit, flexible structure the hydrogens constantly shift their posi-

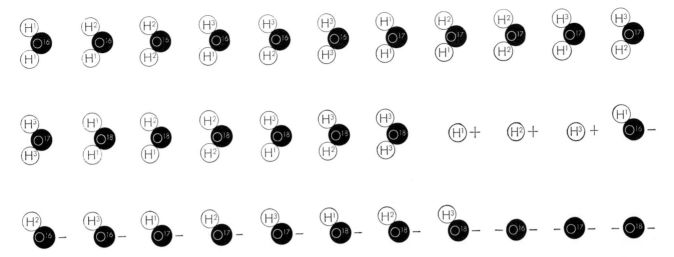

WATER IS NOT H₂O but a mixture of 33 different substances. Eighteen of these are combinations of three isotopes of hydrogen and three of oxygen. The three hydrogen isotopes are ordinary hydrogen (H^1), deuterium (H^2) and tritium (H^3). The three oxygen isotopes are ordinary oxygen (O^{16}), oxygen 17 and oxygen 18. The remaining substances are various ions (*plus or minus signs*).

tions and displace one another. Each such displacement is propagated in a chain or zipper fashion throughout the liquid. This has consequences which affect the viscosity, dielectric constant and electrical conductivity of water.

Water's Properties

In an ordinary unassociated liquid such as benzene the molecules flow by sliding around one another. In water the motion is rolling rather than sliding. Since the molecules are connected by hydrogen bonds, at least one bond must be broken before any flow can occur. From the fact that the molecules are bonded together, it might be expected theoretically that the viscosity of water should be comparatively high. However, each hydrogen bond in water is shared on the average between two other molecules, and one of these weakened bonds is easily broken. The greater viscosity of ice is due to the fact that each hydrogen is bonded to only a single oxygen atom from another molecule, and this firmer bond must be broken before any movement can occur.

The dielectric constant of a liquid is a measure of its capacity to neutralize the attraction between electrical charges. For example, when sodium chloride dissolves in water, the positively charged sodium and negatively charged chlorine ions are separated. They are kept apart because water has a high dielectric constant—the highest of any common liquid.

It reduces the force of attraction between the oppositely charged ions in solution to not much more than 1 per cent of the original value. The reason for water's strong neutralizing action lies in the arrangement of its molecules. In an aggregation of water molecules a hydrogen atom does not share its electron equally with the oxygen atom to which it is attached: the electron is closer to the oxygen atom than to the hydrogen. As a result the hydrogen atoms are positively charged and the oxygen negatively charged. Now when a substance is dissolved, separating into ions, the oxygen atoms are attracted to the positive ions and the hydrogens to the negative ones. Consequently water molecules surrounding a positive

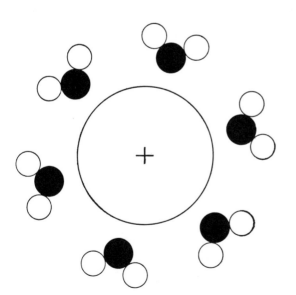

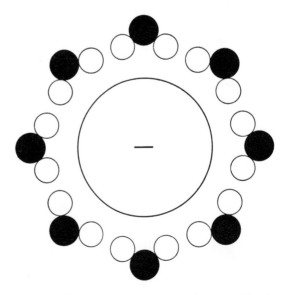

IONS (*circles labeled with plus and minus signs*) in water are kept apart because they polarize the water molecules around them. Because the oxygen atom of the water molecule has more negative charge than the hydrogen atoms, it is attracted to a positive ion (*left*). Because the hydrogen atoms have more positive charge than the oxygen atom, they are attracted to a negative ion (*right*).

ion are oriented with their oxygens next to the ion, and molecules around a negative ion turn their hydrogens toward the ion. Thus the water molecules act as cages which separate and neutralize the ions. This explains why water is so effective a solvent for electrolytes (substances which dissociate into ions) such as sodium chloride.

Water is generally supposed to be a good conductor of electricity. Every lineman knows the danger of handling high-voltage electrical lines when standing on a moist surface. Actually the conductivity is due to impurities dissolved in the water. Water is such a good solvent for electrolytes, including carbon dioxide from the atmosphere, that any moist surface may be assumed to be a good conductor. But pure water (which is difficult to keep pure—it must be kept out of contact with air and in a vessel of an inert material such as quartz) is a very good insulator indeed. The reason is that while the hydrogen and oxygen atoms in a water molecule are in a sense charged, or ionized, they cannot move about separately because they are attached to each other, and hence cannot carry an electric current.

One of the anomalous properties of liquid water is its high specific heat, or heat-holding capacity. The specific heat of a substance is the quantity of heat required to raise the temperature of one gram one degree centigrade. The specific heat of liquid water is more than twice as great as that of ice. The explanation is that the liquid's ionized oxygen and hydrogen atoms, though held together, behave like free ions in their capacity to vibrate in response to heat. Thus they can absorb as much energy as if the ions were really free.

The strong bonding of water molecules accounts for the fact that water has unusually high melting and boiling points. It also explains why it is so difficult to vaporize ice. To do this we must break all the hydrogen bonds holding the molecules together. Calculations indicate that the total energy of the hydrogen bonds in one mole of water (18 grams) is equivalent to 6,000 calories.

Hydrates

For more than 60 years physical chemists have studied water largely in terms of solutions of electrolytes. This study has produced considerable information about electrolytes and ions, but not a great deal about the properties of water itself. Strangely enough, in recent years we have learned much more about water by examining its behavior with substances which for all practical purposes are insoluble in water!

This behavior was called to the attention of chemists in a dramatic fashion by certain surprising natural phenomena. One was the fact that corn sometimes showed frost effects when the temperature was 40 degrees F., well above freezing. Another was the discovery that pipelines carrying natural gas often became clogged with a slushy "snow," containing water, at temperatures as high as 68 degrees F. The plain indication was that these freeze-ups were due to the water. But this raised some startling and interesting questions. What made water freeze at these high temperatures? How could water combine, or become "bound," with substances which were all but insoluble in it? The mystery was not lessened when it was discovered that even the noble gases such as argon and krypton, which refuse all chemical reactions, could join with water to form a quasi compound.

Let us look at these questions in the light of what we have learned about water's structure and properties. Ten years ago in Illinois we began a study of the water-solubility of certain hydrocarbons. Methane gas will serve as an example. The methane molecule does not form ions in water, nor does it accept the hydrogen bonds. There is very little attraction between it and the water molecule. It is, however, slightly soluble in water, and the dissolving methane molecules form compounds with water—"hydrates"—in which several water molecules are joined to one of methane.

The reaction liberates 10 times as much heat as when methane dissolves in hexane, although it is much more soluble in hexane than in water. This fact becomes even more surprising on close examination. The methane molecule occupies more than twice the volume of a water molecule. To form this relatively large cavity for itself on dissolving, a great deal of energy would be required: it should be somewhat greater than the heat of vaporization of water—say 10,000 calories per mole. How could so much energy be provided? The forces of attraction between methane and water are apparently too slight to supply any appreciable part of such an amount.

There is an alternative possibility. The presence of the methane may drastically change the water structure itself.

HYDRATE is formed when a foreign molecule in water is electrically neutral and just the right size for the water molecules to collect around it in crystalline cage. This cage can then grow to a much larger crystal. It is part of a repeating unit of 136 molecules.

Let us suppose that the dissolved methane molecule is surrounded by an envelope of 10 or 20 water molecules. The formation of such a structure would account for the heat liberated. In the space occupied by the methane molecule the attractive force on the water molecules, and hence the inward pressure, would disappear. Under these conditions, as we have seen, water will freeze at a higher temperature. Thus the molecules at the interface between the methane and water molecules may crystallize into "ice." The frozen hydrates may accumulate and separate out of the solution.

This hypothesis is known as the "iceberg" theory. It is supported by the fact that practically all the nonelectrolytic substances tested have been found to form solid crystalline hydrates. In contrast, electrolytes show little tendency to form them.

All this leads to an entirely new concept of solubility. Chemists have long supposed that solubility always involves attractive forces. But it now appears that the dissolution of a nonelectrolyte is due not to an attraction between the substance and water but to a lack of attraction. The nonionic substances combine with water because they remove internal pressure and thereby permit formation of a crystalline compound.

In order to understand the formation of these hydrates, it is necessary to consider their molecular structure in detail. They tend to fall into groups according to the number of water molecules they contain.

The ground work for the study of the hydrates structure was laid by M. von Stackelberg in Germany 10 years ago. He showed by X-ray studies that this structure was cubic, in contrast to the hexagonal structure of ice. W. F. Claussen of our laboratory recently attacked the problem of building such cubic structures, each containing a gas molecule, into a repeating lattice. It turns out that there are two possible cubic lattices, one of which was proposed and worked out by Linus Pauling of the California Institute of Technology. This has a spacing of 12 Angstroms between molecules while the other has 17 Angstroms. The smaller lattice contains 46 water molecules and the larger one 136. The holes for gas molecules in the smaller lattice have 12 or 14 walls, while those in the larger one have 12 or 16 sides. These holes are of different sizes and make possible a bewildering array of hydrates. The different sized holes can be filled only with different sized molecules, and not all the holes in a lattice need be filled. The model explains the actual

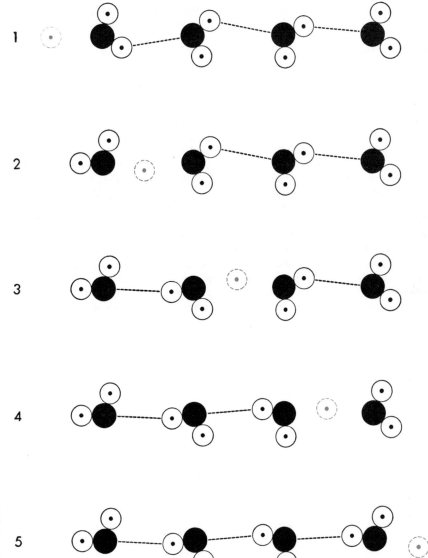

RAPID MIGRATION OF HYDROGEN IONS through water is explained by the assumption that the ions do not actually travel through the water but are passed from one molecule to the next by a process of exchange. Here hydrogen ion is represented by colored dot surrounded by a broken circle. In the first horizontal row the hydrogen ion approaches a water molecule. In the second row the ion has taken the place of one hydrogen atom of the molecule, expelling the atom as a new ion. In the third row the new ion repeats this process.

composition of hydrates with remarkable precision.

The importance of this type of hydrate to the processes of life cannot be overemphasized. These processes occur mainly at the interfaces between water and protein molecules. Water has a very strong tendency to crystallize there, for the protein molecule contains large nonionic, or nonpolar groups. Any hydrate so formed has a lower density than ice; consequently its formation can cause a large, destructive expansion.

The freezing of corn at a temperature of 40 degrees F. becomes understandable in terms of the formation of a hydrate. Winter wheat, on the other hand,

forms hydrates slowly as temperatures drop through the fall, and under these conditions the hydrate acts as an effective antifreeze protecting the cells from damage.

The frozen food industry uses rapid freezing to avoid the formation of large crystals which would damage the plant cells. But it might be well to explore the possibility of the opposite approach. Very slow cooling of living plant foods might form hydrates which would prevent damage from ice crystals when the plant was frozen.

Let us return now to see how the structure of water may be modified when an electrolyte, say a salt, goes into solution. The only direct physical clue we have lies in the behavior of the salt ions in conducting an electric current. The rate of motion of the ions will depend in part on the resistance they encounter in the liquid, and this in turn will depend on the size of the moving particles. If water molecules are attached to an ion firmly enough to move along with it, they will of course increase the apparent size of the ion. Studies of the mobilities of various ions show that positive ions smaller than potassium carry such a cage of water molecules with them. The positively charged ion attracts the oxygens of two water molecules quite strongly, and if the volume of the ion plus its two water molecules is not greater than that of a methane molecule, a cage of hydrogen-bonded water molecules will form around this group as a nucleus. Positive ions larger than potassium fail to pick up such a cage. The same is true of most, but not all, negative ions.

The positively charged hydrogen ion and the negatively charged hydroxyl ion (OH) are surrounded by cages, and yet they show the highest mobility in carrying a current. We must conclude that they manage in some way to escape from the cage. Actually the mechanism is not hard to picture: they continually form new cages as they travel by a process called proton transfer. Under the influence of an electric field a hydrogen ion may jump from one water molecule to the next. When this has occurred, the hydrogen on the farther side of the water molecule takes up its part in the race like a relay runner and jumps to the next water molecule. Thus a succession of protons, each doing its bit, carries the current. The motion is rapid, because each proton moves only a short step. Transfer of the proton also explains the conduction of electricity by hydroxyl ions. When a proton jumps toward the right, say, and joins a hydroxyl ion, it leaves a hydroxyl ion on its left. The effect is the same as if a hydroxyl itself moved to the left.

Water, then, is not simple H_2O but a unique and complicated material with distinct and varied chemical properties. It has a definite though changing physical structure which depends on the orientation of its molecules with respect to one another, and to the molecules of dissolved substances. Since the behavior of all living nature and much of the inanimate world is inseparably linked to the peculiar characteristics of this liquid, the study of water substance can tell us a great deal about fundamental aspects of the world in which we live.

9

The Structure of Cell Membranes

by C. Fred Fox
February 1972

The thin, sturdy envelope of the living cell consists of lipid, phosphate and protein. The proteins act as both gatekeepers and active carriers, determining what passes through the membrane

Every living cell is enclosed by a membrane that serves not only as a sturdy envelope inside which the cell can function but also as a discriminating portal, enabling nutrients and other essential agents to enter and waste products to leave. Called the cytoplasmic membrane, it can also "pump" substances from one side to the other against a "head," that is, it can extract a substance that is in dilute solution on one side and transport it to the opposite side, where the concentration of the substance is many times higher. Thus the cytoplasmic membrane selectively regulates the flux of nutrients and ions between the cell and its external milieu.

The cells of higher organisms have in addition to a cytoplasmic membrane a number of internal membranes that isolate the structures termed organelles, which play various specialized roles. For example, the mitochondria oxidize foodstuffs and provide fuel for the other activities of the cell, and the chloroplasts conduct photosynthesis. Single-cell organisms such as bacteria have only a cytoplasmic membrane, but its structural diversity is sufficient for it to serve some or all of the functions carried out by the membranes of organelles in higher cells. It is clear that any model formulated to describe the structure of membranes must be able to account for an extraordinary range of functions.

Membranes are composed almost entirely of two classes of molecules: proteins and lipids. The proteins serve as enzymes, or biological catalysts, and provide the membrane with its distinctive functional properties. The lipids provide the gross structural properties of the membrane. The simplest lipids found in nature, such as fats and waxes, are insoluble in water. The lipids found in membranes are amphipathic, meaning that one end of the molecule is hydrophobic, or insoluble in water, and the other end hydrophilic, or water-soluble. The hydrophilic region is described as being polar because it is capable of carrying an ionic (electric) charge; the hydrophobic region is nonpolar.

In most membrane lipids the nonpolar region consists of the hydrocarbon chains of fatty acids: hydrocarbon molecules with a carboxyl group (COOH) at one end. In a typical membrane lipid two fatty-acid molecules are chemically bonded through their carboxyl ends to a backbone of glycerol. The glycerol backbone, in turn, is attached to a polar-head group consisting of phosphate and other groups, which often carry an ionic charge [see *illustration on next page*]. Phosphate-containing lipids of this type are called phospholipids.

When a suspension of phospholipids in water is subjected to high-energy sound under suitable conditions, the phospholipid molecules cluster together to form closed vesicles: small saclike structures called liposomes. The arrangement of phospholipids in the walls of both liposomes and biological membranes has recently been deduced with the help of X-ray diffraction, which can reveal the distance between repeating groups of atoms. An X-ray diffraction analysis by M. F. Wilkins and his associates at King's College in London indicates that two parallel arrays of the polar-head groups of lipids are separated by a distance of approximately 40 angstroms and that the fatty-acid tails are stacked parallel to one another in arrays of 50 or more phospholipid molecules.

The X-ray data suggest a structure for liposomes and membranes in which the phospholipids are arranged in two parallel layers [see *illustrations on page 85*]. The polar heads are arrayed externally on the bilayer surfaces, and the fatty-acid tails are pointed inward, perpendicular to the plane of the membrane surface. This model of phospholipid structure in membranes is identical with one proposed by James F. Danielli and Hugh Davson in the mid-1930's, when no precise structural data were available. It is also the minimum-energy configuration for a thin film composed of amphipathic molecules, because it maximizes the interaction of the polar groups with water.

Unlike lipids, proteins do not form orderly arrays in membranes, and thus their arrangement cannot be assessed by X-ray diffraction. The absence of order is not surprising. Each particular kind of membrane incorporates a variety of protein molecules that differ widely in molecular weight and in relative numbers; a membrane can incorporate from 10 to 100 times more molecules of one type of protein than of another.

Since little can be learned about the disposition of membrane proteins from a general structural analysis, investigators have chosen instead to study the orientation of one or a few species of the proteins in membranes. In the Danielli-Davson model the proteins are assumed to be entirely external to the lipid bilayer, being attached either to one side of the membrane or to the other. Although information obtained from X-ray diffraction and high-resolution electron microscopy indicates that this is probably true for the bulk of the membrane protein, biochemical studies show that the Danielli-Davson concept is an oversimplification. The evidence for alternative locations has been provided chiefly by Marc Bretscher of the Medical Research Council laboratories in Cambridge and by Theodore L. Steck, G. Franklin and Donald F. H. Wallach of the Harvard Medical School. Their results suggest that certain proteins penetrate the lipid bilayer and that others extend all the way through it.

Bretscher has labeled a major protein of the cytoplasmic membrane of human red blood cells with a radioactive substance that forms chemical bonds with the protein but is unable to penetrate the membrane surface. The protein was labeled in two ways [*see illustration on pages 86 and 87*]. First, intact red blood cells were exposed to the label so that it became attached only to the portion of the protein that is exposed on the outer surface of the membrane. Second, red blood cells were broken up before the radioactive label was added. Under these conditions the label could attach itself to parts of the protein exposed on the internal surface of the membrane as well as to parts on the external surface.

The two batches of membrane, labeled under the two different conditions, were treated separately to isolate the protein. The purified protein from the two separate samples was degraded into definable fragments by treatment with a proteolytic enzyme: an enzyme that cleaves links in the chain of amino acid units that constitutes a protein. A sample from each batch of fragments was now placed on the corner of a square of filter paper for "fingerprinting" analysis. In this technique the fragments are separated by chromatography in one direction on the paper and by electrophoresis in a direction at right angles to the first. In the chromatographic step each type of fragment is separated from the others because it has a characteristic rate of travel across the paper with respect to the rate at which a solvent travels. In the electrophoretic step the fragments are further separated because they have char-

acteristic rates of travel in an imposed electric field.

Once a separation had been achieved the filter paper was laid on a piece of X-ray film so that the radioactively labeled fragments could reveal themselves by exposing the film. When films from the two batches of fragments were developed, they clearly showed that more labeled fragments were present when both the internal and the external surface of the cell membrane had been exposed to the radioactive label than when the outer surface alone had been exposed. This provides strong evidence that the portion of the protein that gives rise to the additional labeled fragments is on the inner surface of the membrane.

Steck and his colleagues obtained similar results with a procedure in which they prepared two types of closed-membrane vesicle, using as a starting material the membranes from red blood cells. In one type of vesicle preparation (right-side-out vesicles) the outer membrane surface is exposed to the external aqueous environment. In the other type of preparation (inside-out vesicles) the inner surface of the membrane is exposed to the external aqueous environment. When the two types of vesicle are treated with a proteolytic enzyme, only those proteins exposed to the external aqueous environment should be degraded. Steck found that some proteins are susceptible to digestion in both the right-side-out and inside-out vesicles, indicating that these proteins are exposed on both membrane surfaces. Other proteins are susceptible to proteolytic digestion in right-side-out vesicles but not in inside-out vesicles. Such proteins are evidently located exclusively on only one side of the membrane. This information lends credence to the concept of sidedness in

membranes. Such sidedness had been suspected for many years because the inner and outer surfaces of cellular membranes are thought to have different biological functions. The development of a technique for preparing vesicles with right-side-out and inside-out configurations should be extremely useful in determining on which side of the membrane a given species of protein resides and thus functions.

Daniel Branton and his associates at the University of California at Berkeley have developed and exploited the technique of freeze-etch electron microscopy to study the internal anatomy of membranes. In freeze-etch microscopy a suspension of membranes in water is frozen rapidly and fractured with a sharp blade. Wherever the membrane surface runs parallel to the plane of fracture much of the membrane will be split along the middle of the lipid bilayer. A thin film of platinum and carbon is then evaporated onto the surface of the fracture. This makes it possible to examine the anatomy of structures in the fracture plane by electron microscopy.

The electron micrographs of the fractured membrane reveal many particles, approximately 50 to 85 angstroms in diameter, on the inner surface of the lipid bilayer. These particles are not observed if the membrane samples are first treated with proteolytic enzymes, indicating that the particles probably consist of protein [*see illustration on page 89*]. From quantitative estimates of the number of particles revealed by freeze-etching, Branton and his colleagues have suggested that between 10 and 20 percent of the internal volume of many biological membranes is protein.

Somewhere between a fifth and a quarter of all the protein in a cell is physically associated with membranes. Most of the other proteins are dissolved in the aqueous internal environment of the cell. In order to dissolve membrane proteins in aqueous solvents detergents must be added to promote their dispersion. One might therefore expect membrane proteins to differ considerably from soluble proteins in chemical composition. This, however, is not the case.

The amino acids of which proteins are composed can be classified into two groups: polar and nonpolar. S. A. Rosenberg and Guido Guidotti of Harvard University analyzed the amino acid composition of proteins from a number of membranes and found that they contain about the same percentage of polar and nonpolar amino acids as one finds in the soluble proteins of the common colon

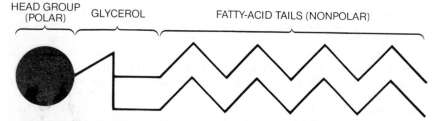

HEAD GROUP (POLAR)   GLYCEROL   FATTY-ACID TAILS (NONPOLAR)

TYPICAL MEMBRANE LIPID is a complex molecular structure, one end of which is hydrophilic, or water-soluble, and the other end hydrophobic. Such a substance is termed amphipathic. The hydrophilic, or polar, region consists of phosphate and other constituents attached to a unit of glycerol. The polar-head group, when in contact with water, often carries an electric charge. The glycerol component forms a bridge to the hydrocarbon tails of two fatty acids that constitute the nonpolar region of the lipid. In this highly schematic diagram the zigzag lines represent hydrocarbon chains; each angle is occupied by a carbon atom and two associated hydrogen atoms. The terminal carbon of each chain is bound to three hydrogen atoms. Phosphate-containing amphipathic lipids are called phospholipids.

bacterium *Escherichia coli*. Thus differences in amino acid composition cannot account for the water-insolubility of membrane proteins.

Studies conducted by L. Spatz and Philipp Strittmatter of the University of Connecticut indicate that the most likely explanation for the water-insolubility of membrane proteins is the arrangement of their amino acids. Spatz and Strittmatter subjected membranes of rabbit liver cells to a mild treatment with a proteolytic enzyme. The treatment released the biologically active portion of the membrane protein: cytochrome b_5. In a separate procedure they solubilized and purified the intact cytochrome b_5 and treated it with the proteolytic enzyme. This treatment also released the water-soluble, biologically active portion of the molecule, together with a number of small degradation products that were insoluble in aqueous solution. The biologically active portion of the molecule, whether obtained from the membrane or from the purified protein, was found to be rich in polar amino acids. The protein fragments that were insoluble in water, on the other hand, were rich in nonpolar amino acids. These observations suggest that many membrane proteins may be amphipathic, having a nonpolar region that is embedded in the part of the membrane containing the nonpolar fatty-acid tails of the phospholipids and a polar region that is exposed on the membrane surface.

We are now ready to ask: How do substances pass through membranes? The nonpolar fatty-acid-tail region of a phospholipid bilayer is physically incompatible with small water-soluble substances, such as metal ions, sugars and amino acids, and thus acts as a barrier through which they cannot flow freely. If one measures the rate at which blood sugar (glucose) passes through the phospholipid-bilayer walls of liposomes, one finds that it is far too low to account for the rate at which glucose penetrates biological membranes. Information of this kind has given rise to the concept that entities termed carriers must be present in biological membranes to facilitate the passage of metal ions and small polar molecules through the barrier presented by the phospholipid bilayer.

Experiments with biological membranes indicate that the hypothetical carriers are highly selective. For example, a carrier that facilitates the transport of glucose through a membrane plays no role in the transport of amino acids or other sugars. An interesting experimental

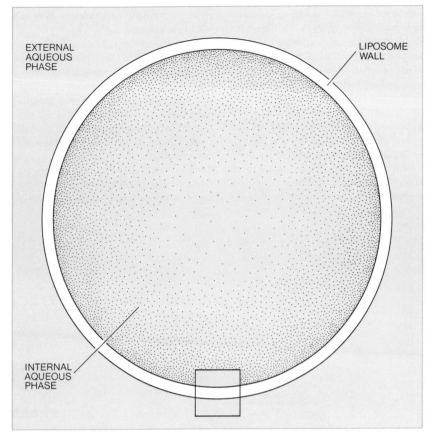

ARTIFICIAL MEMBRANE-ENCLOSED SAC, known as a liposome, is created by subjecting an aqueous suspension of phospholipids to high-energy sound waves. X-ray diffraction shows that the phospholipids in the liposome assume an orderly arrangement resembling what is found in the membranes of actual cells. Area inside the square is enlarged below.

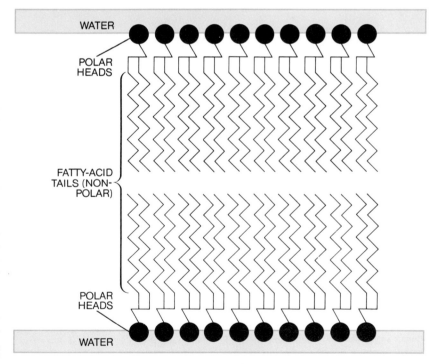

CROSS SECTION OF LIPOSOME WALL shows how the membrane is formed from two layers of lipid molecules. The polar heads of amphipathic lipids face toward the aqueous solution on each side while the nonpolar fatty-acid tails face inward toward one another.

system for measuring selective ion transport was developed by A. D. Bangham, M. M. Standish and J. C. Watkins of the Agricultural Research Council in Cambridge, England, and by J. B. Chappell and A. R. Crofts of the University of Cambridge. As a model carrier they used valinomycin, a nonpolar, fat-soluble antibiotic consisting of a short chain of amino acids (actually 12); such short chains are termed polypeptides to distinguish them from true proteins, which are much larger. Valinomycin combines with phospholipid-bilayer membranes and makes

them permeable to potassium ions but not to sodium ions.

The change in permeability is conveniently studied by measuring the change in electrical resistance across a phospholipid bilayer between two chambers containing a potassium salt in aqueous solution. The experiment is performed by introducing a sample of phospholipid into a small hole between the two chambers. The lipid spontaneously thins out until the chambers are separated by only a thin membrane consisting of a phospholipid bilayer. Electrodes

are then placed in the two chambers to measure the resistance across the membrane.

The resistance across a phospholipid bilayer in the absence of valinomycin is several orders of magnitude higher than the resistance across a typical biological membrane: 10 million ohms centimeter squared compared with between 10 and 10,000. This indicates that phospholipid-bilayer membranes are essentially impermeable to small hydrophilic ions. If a small amount of valinomycin (10^{-7} gram per milliliter of salt solution) is intro-

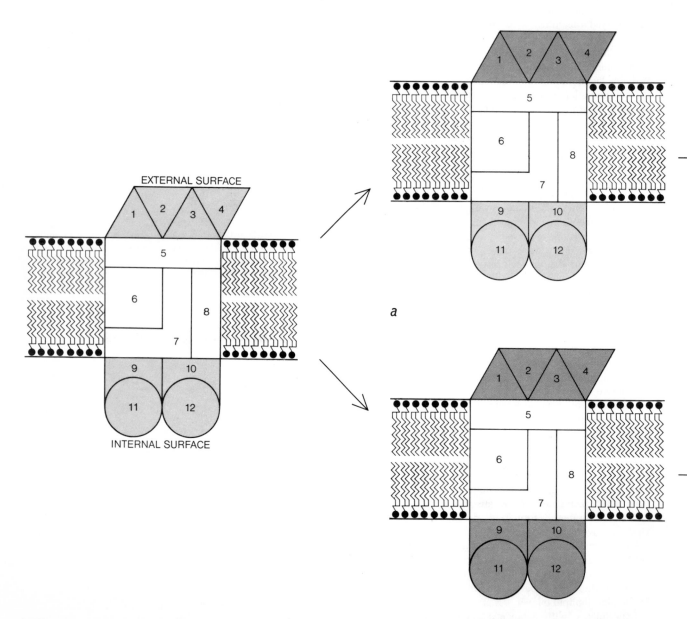

LOCATION OF PROTEINS IN MEMBRANES can be inferred by attaching radioactive labels to the proteins. These diagrams depict an experiment in which a major protein in the membrane of red blood cells was labeled (*a*). When intact cells (*top sequence*) are exposed to the radioactive substance, only the portion of the protein on the outside wall picks up the label (*color*). When the cells are broken before labeling (*bottom sequence*), the radioactive la-

bel is able to reach portions of the protein that are exposed to the internal as well as to the external surfaces of the membrane. This can be demonstrated by isolating and purifying the protein labeled under the two conditions. The protein is then broken up into defined fragments (*numbered shapes*) by treating it with a proteolytic enzyme (*b*). Portions of the two batches of fragments are spotted on the corners of filter paper for "fingerprinting" (*c*). This is a

duced into the chambers containing the potassium solution, the resistance falls by five orders of magnitude and the permeability of the phospholipid bilayer to potassium ions rises by a like amount. The permeability of the experimental membrane now essentially duplicates the permeability of biological membranes.

If the experiment is repeated with a sodium chloride solution in the chambers, one finds that the addition of valinomycin causes only a slight change in resistance. Hence valinomycin meets two of the most important criteria for a bio-

logical carrier: it enhances permeability and it is highly selective for the transported substance. The question that now arises is: How does valinomycin work?

First of all, valinomycin is nonpolar. Thus it is physically compatible with and can dissolve in the portion of the bilayer that contains the nonpolar fatty-acid tails. Second, valinomycin can evidently diffuse between the two surfaces of the bilayer. S. Krasne, George Eisenman and G. Szabo of the University of California at Los Angeles have shown that the enhancement of potassium-ion transport by

valinomycin is interrupted when the bilayer is "frozen" by lowering the temperature. Third, valinomycin must bind potassium ions in such a way that the ionic charge is shielded from the nonpolar region of the membrane. Finally, valinomycin itself must have a selective binding capacity for potassium ions in preference to sodium or other ions.

With valinomycin as a model for carrier-mediated transport, one can postulate three essential steps: recognition of the ion, diffusion of the ion through the membrane, and its release on the other

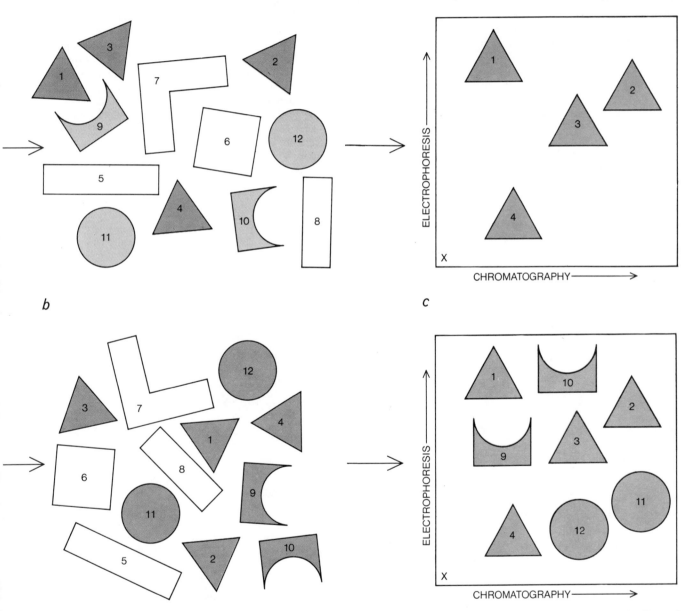

b

c

technique that combines chromatography with electrophoresis. By chromatography alone protein fragments would migrate at different rates depending primarily on their solubility in the solvent system. Electrophoresis involves establishing an electric-potential gradient along one axis of the filter paper. Since various fragments have different densities of electric charge they are further separated. A piece of X-ray film is then placed over each sheet of filter

paper. Radiation from the labeled fragments exposes the film and reveals where the various fragments have come to rest. A comparison of the X-ray films produced in the parallel experiments shows that more protein fragments are labeled when the red blood cells are broken before labeling and that the additional fragments (*9, 10, 11, 12*) must represent portions of the original protein that extend through the membrane and penetrate the inner surface.

side. In the first step some part of the valinomycin molecule, embedded in the membrane, "recognizes" the potassium ion as it approaches the surface of the membrane and captures it. In the second step the complex consisting of valinomycin and the potassium ion diffuses through the membrane. Finally, on reaching the opposite surface of the membrane the potassium ion is dissociated from the complex and is released.

The argument to this point can be summarized in a few words. The fundamental structure of biological membranes is a phospholipid bilayer, the phospholipid bilayer is a permeability barrier and carriers are needed to breach it. In addition, the membrane barrier must often be breached in a directional way. In a normally functioning cell hundreds of kinds of small molecule must be present at a higher concentration inside the cell than outside, and many other small molecules must be present at a lower concentration inside the cell than outside. For example, the concentration of potassium ions in human cells is more than 100 times greater than it is in the blood that bathes them. For sodium ions the concentrations are almost exactly reversed. The maintenance of these differences in concentration is absolutely essential; even slight changes can result in death.

Although the model system based on valinomycin provides considerable insight into the function and selectivity of carriers, it sheds no light on the transport mechanism that can pump a substance from a low concentration on one side of the membrane to a higher concentration on the other. Our understanding of concentrative transport (or, as it is usually termed, active transport) owes much to the pioneering effort of Georges Cohen, Howard Rickenberg, Jacques Monod and their associates at the Pasteur Institute in Paris. The Pasteur group studied the transport of milk sugar (lactose) through the cell membrane of the bacterium *Escherichia coli*. Genetic experiments suggested that the carrier for lactose transport was a protein. Studies of the rate of transport revealed that the transport process behaves like a reaction catalyzed by an enzyme, giving further support to the idea that the carrier is a protein. The Pasteur group also found that the lactose-transport system is capable of active transport, producing a lactose concentration 500 times greater inside the cell than outside. The active-transport process depends on the expenditure of metabolic energy; poisons that block energy metabolism destroy the ability of the cell to concentrate lactose.

A model that accounts for many (but not all) of the properties of the active-transport system that are typified by the lactose system postulates the existence of a carrier protein that can change its shape. The protein is visualized as resembling a revolving door in the membrane wall [*see the illustration on page 90*]. The "door" contains a slot that fits the target substance to be transported. The slot normally faces the cell's external environment. When the target substance enters the slot, the protein changes shape and is thereby enabled to rotate so that the slot faces into the cell. When the target substance has been discharged into the cell, the protein remains with its slot facing inward until the cell expends energy to rotate the protein so that the slot again faces outward.

Working with Eugene P. Kennedy at the Harvard Medical School in 1965, I succeeded in identifying the lactose-transport carrier. We found, as we had expected, that it is a protein with an enzyme-like ability to bind lactose. Since then a number of other transport carriers have been identified, and all turn out to be proteins. The lactose carrier resides in the membrane and is hydrophobic; thus it is physically compatible

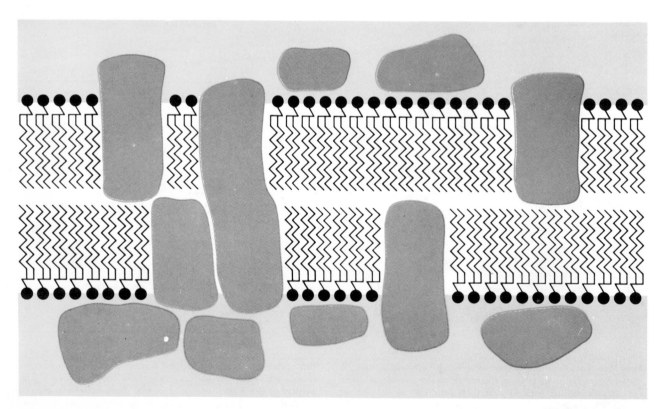

ANATOMY OF BIOLOGICAL MEMBRANE is suggested in this schematic diagram. Phospholipid molecules stacked side by side and back to back provide the basic structure. The gray shapes represent protein molecules. In some cases several proteins (for example the five at the left) are bound into a single functional complex. Proteins can occupy all possible positions with respect to the phospholipid bilayer: they can be entirely outside or inside, they can penetrate either surface or they can extend through the membrane.

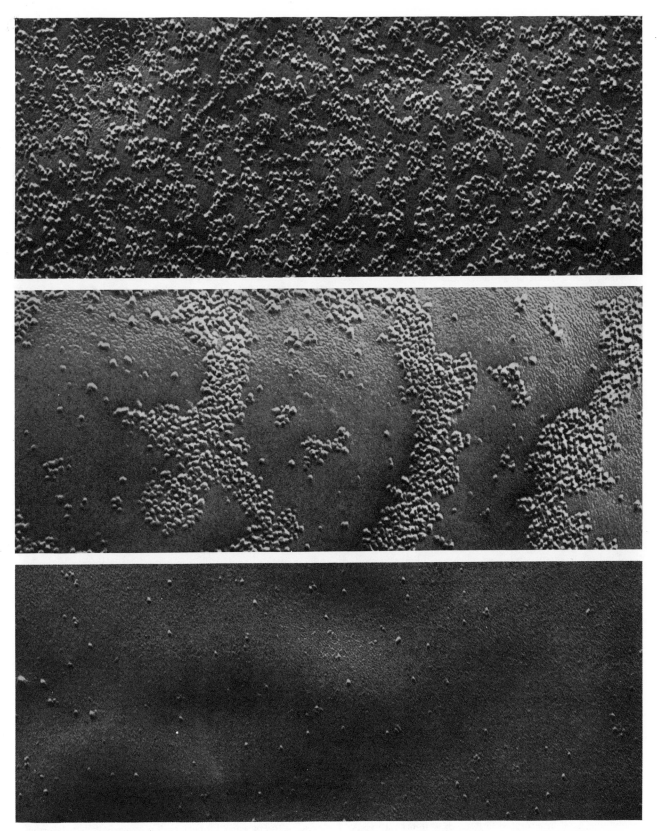

EVIDENCE FOR PROTEINS within the bilayer structure of cell membranes is provided by freeze-etch electron microscopy. A suspension of membranes in water is frozen and then fractured with a sharp blade. The fracture will often split a membrane in the middle along a plane parallel to the surface. After platinum and carbon vapors are deposited along the fracture surface the specimen can be studied in the electron microscope. The micrograph at the top shows many particles 50 to 85 angstroms in diameter embedded in a fractured membrane from rabbit red blood cells. The other two views show how the number of particles is greatly reduced if the membrane is first treated with a proteolytic enzyme that digests 45 percent (*middle*) or 70 percent (*bottom*) of the original membrane protein. The missing particles have presumably been digested by the enzyme. The membrane preparations are enlarged some 95,000 diameters in these micrographs made by L. H. Engstrom in Daniel Branton's laboratory at the University of California at Berkeley.

with the nonpolar-lipid phase of the membrane.

In 1970 Ron Kaback and his associates at the Roche Institute of Molecular Biology observed that the energy that drives the active transport of lactose and dozens of other low-molecular-weight substances in *E. coli* is directly coupled to the biological oxidation of metabolic intermediates such as D-lactic acid and succinic acid. How energy derived from the oxidation of D-lactic acid can be used to drive active transport is one of the more interesting unsolved problems in membrane biology.

Since transport carriers must be mobile within the membrane in order to move substances from one surface to the other, one might guess that the region of the membrane containing the fatty-acid tails should not have a rigid crystalline structure. X-ray diffraction studies indicate that the fatty acids of membranes in fact do have a "liquid crystalline" structure at physiological temperature, that is, around 37 degrees Celsius. In other words, the fatty acids are not aligned in a rigid crystalline lattice. The techniques of electron paramagnetic resonance and nuclear magnetic resonance can be used to study the flexibility of the fatty-acid side chains in membranes. Several investigators, notably Harden M. McConnell and his associates at Stanford University, have concluded that the fatty acids of membranes are quasi-fluid in character.

Membranes incorporate two classes of fatty acids: saturated molecules, in which all the available carbon bonds carry hydrogen atoms, and unsaturated molecules, in which two or more pairs of hydrogen atoms are absent (with the result that two or more pairs of carbon atoms have double bonds). The fluid character of membranes is largely determined by the structure and relative proportion of the unsaturated fatty acids. In phospholipids consisting only of saturated fatty acids the fatty-acid tails are aligned in a rigidly stacked crystalline array at physiological temperatures. In phospholipids consisting of both saturated and unsaturated fatty acids the fatty acids are packed in a less orderly fashion and thus are more fluid. The double bonds of unsaturated fatty acids give rise to a structural deformation that interrupts the ordered stacking necessary for the formation of a rigid crystalline structure [*see illustration on next page*].

My colleagues and I at the University of Chicago (and later at the University of California at Los Angeles) and Peter Overath and his associates at the Uni-

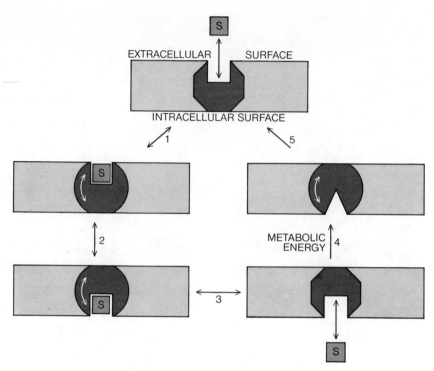

MECHANISM OF "ACTIVE" TRANSPORT may involve a carrier protein (*dark gray*) with the properties of a revolving door. A carrier protein can capture a substance, *S*, that exists outside the membrane in dilute solution and transport it to the inside of the cell, where the concentration of *S* is greater than it is outside. When *S* is bound to the protein, the protein changes shape (*1*), thus enabling it to rotate (*2*). When *S* becomes detached and enters the cell (*3*), the protein returns to its immobile form. Metabolic energy must be expended (*4*) to alter the protein's shape so that it can rotate and again present its binding site to the cell exterior (*5*). Other protein carriers have the capacity to transport substances from low concentration inside the cell to solutions of higher concentration outside the cell.

versity of Cologne have varied the fatty-acid composition of biological membranes to study the effects of fatty-acid structure on transport. When the membrane lipids are rich in unsaturated fatty acids, transport can proceed at rates up to 20 times faster than it does when the membrane lipids are low in unsaturated fatty acids. These experiments show that normal membrane function depends on the fluidity of the fatty acids.

The temperature at which cells live and grow can have a pronounced effect on the amount of unsaturated fatty acid in their membranes. Bacteria grown at a low temperature have membranes with a greater proportion of unsaturated fatty acid than those grown at a higher temperature. This adjustment in fatty-acid composition is necessary if the membranes are to function normally at low temperature. A similar adjustment can take place in higher organisms. For example, there is a temperature gradient in the legs of the reindeer; the highest temperature is near the body, the lowest is near the hooves. To compensate for this temperature gradient the cells near the hooves have membranes whose lipids are enriched in unsaturated fatty acids.

Although, as we have seen, phospholipids can spontaneously form bilayer films in water, this process only provides a physical rationale as to why the predominant structure in membranes is a phospholipid bilayer. The events leading to the assembly of a biological membrane are far more complex. The cells of higher organisms contain a number of unique membrane structures. They differ widely in lipid composition, and each type of membrane has its own unique complement of proteins. The diversity in protein composition and in the location of proteins within membranes explains the functional diversity of different types of membrane. Rarely does a single species of protein exist in more than one type of membrane.

Since all membrane proteins are synthesized at approximately the same cellular location, what is it that determines that one type of protein will be incorporated only into the cytoplasmic membrane and that another type will turn up only in a mitochondrial membrane? At present this question can be answered only by conjecture tinctured with a few facts. Two general hypotheses for membrane assembly can be offered. One possibility is that new pieces of membrane are made from scratch by a self-assembly mechanism in which all the components of a new piece of membrane come together spontaneously. This new piece could then be inserted into an existing membrane. A second possibility is that newly made proteins are simply inserted at random into a preexisting membrane.

Recent studies in my laboratory at the University of California at Los Angeles and in the laboratories of Philip Siekevitz and George E. Palade at Rockefeller University support the second hypothesis. That is all well and good, but what determines why a given protein is incorporated only into a given kind of membrane? Although this must be answered by conjecture, it is known that many proteins are specifically bound to other proteins in the same membrane. Such protein-protein interactions are not uncommon; many of the functional entities in membranes are complexes of several proteins. Thus the proteins in a membrane may provide a template that is recognized by a newly synthesized protein and that helps to insert the newly synthesized protein into the membrane. In this way old membrane could act as a template for the assembly of new membrane. This might explain why different membranes incorporate different proteins.

Why, then, do different membranes have different lipid compositions? The answers to this question are even more obscure. In general lipids are synthesized within the membrane; the enzymes that catalyze the synthesis are part of the membrane. Some lipids, however, are made in one membrane and then shuttled to another membrane that has no inherent capacity to synthesize them. Since there is an interchange of lipids between various membranes, it seems unlikely that the variations in lipid composition in different membranes can be explained by dissimilarities in the synthetic capacity of a given membrane for a given type of lipid. There are at least two possible ways of accounting for differences in lipid composition. One possibility is that different membranes may destroy different lipids at different rates; another is that the proteins of one species of membrane may selectively bind one type of lipid, whereas the proteins of another species of membrane may bind a different type of lipid. It is obvious from this discussion that concrete evidence on the subject of membrane assembly is scant but that the problems are well defined.

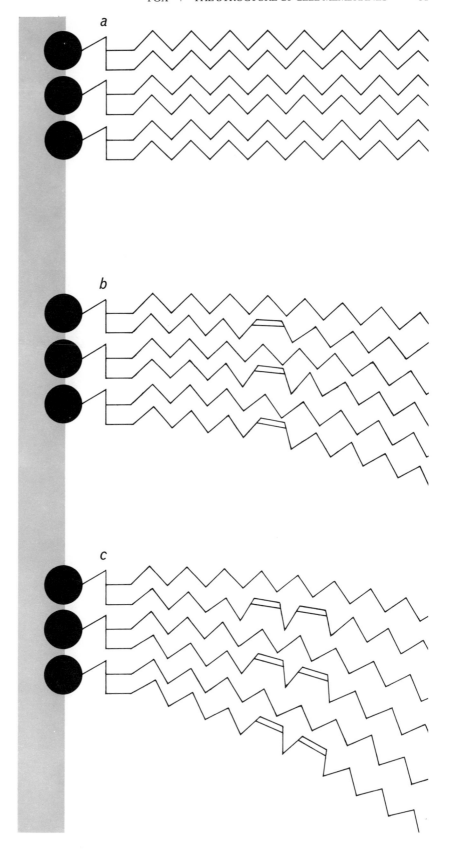

VARIATION IN FATTY-ACID COMPOSITION can disrupt the orderly stacking of phospholipids in a biological membrane. In a lipid layer composed entirely of saturated fatty acids (a) the fatty-acid chains contain only single bonds between carbon atoms and thus nest together to form rigid structures. In a lipid layer containing unsaturated fatty acids with one double bond (b) the double bonds introduce a deformation that interferes with orderly stacking and makes the fatty-acid region somewhat fluid. When fatty acids with two double bonds are present (c), the deformation and the consequent fluidity are greater still.

III

RADIANT ENERGY
AND LIFE

III

RADIANT ENERGY
AND LIFE

INTRODUCTION

In the first two sections of this volume we considered the chemical nature of the biosphere and the forces that hold molecules and biopolymers together. We will now examine how different environmental forces interact with and affect biological matter. In this section we will discuss the effects of radiant energy on the biosphere.

Absorption of radiant energy by a molecule puts the molecule in a more energetic state, and it becomes more reactive. The extra energy can be dissipated in a variety of ways, notably by chemical reaction with another molecule. The energy can also be transferred to a neighboring molecule in a nonreactive collision; the original excited molecule then returns to its unexcited "ground" state. Or, the energy can be reradiated as light in a process called fluorescence. Some molecules, called pigments, absorb energy in the visible-light portion of the electromagnetic spectrum. To exert any effect, the radiant energy must be absorbed, so biological sensitivity to visible light is confined to those parts of the living organism that contain pigment molecules.

Details of the nature of electromagnetic radiation and how it interacts with molecules are the subject of the article, "The Chemical Effects of Light" (page 98), by Gerald Oster. As you read this article, note especially that only a narrow band of the available spectrum of radiation is utilized in biological reactions such as photosynthesis and vision. The region of the spectrum that is absorbed by plants corresponds roughly to the region used in animal vision, namely, the visible region. The reasons that the range of wavelengths between 4000 and 7000 angstroms is biologically so active are discussed in several of the articles in this section.

The overall effects of radiant energy on the biosphere, however, are due to a much wider range of the solar radiation spectrum. Some radiation is absorbed by the inanimate portions of the biosphere. Oxygen, ozone, carbon dioxide, and water absorb ultraviolet and/or infrared radiation. The energy absorbed by the atmosphere and ocean surface generates winds, rain, and ocean currents. As we have seen, these phenomena are of great ecological importance in providing for the cycling of the elements of the biosphere. The radiation absorbed by the atmosphere also functions to maintain the ambient temperatures of the earth's surface. Note especially the discussions of the role of ozone in the upper atmosphere. This subject has become very important in the last few years, since it was discovered that certain industrial gases introduce halogens into the upper atmosphere.

Chlorine, in particular, when irradiated by short-wavelength ultra-violet radiation, can catalyze the decomposition of ozone molecules.

The total amount of ozone in the upper atmosphere is in a steady state. A finite rate of synthesis from oxygen, which absorbs ultra-violet radiation, is counterbalanced by a finite rate of decay, again due to ultraviolet radiation absorbed by ozone itself. Chlorine atoms increase the rate of decay while leaving the rate of synthesis unaffected. This could eventually cause a significant reduction in the total amount of ozone present. As a result, more short-wavelength ultra-violet radiation would reach sea level. The biological effects of this radiation are discussed in the article, "Ultraviolet Radiation and Nucleic Acid" (page 119), by R. A. Deering.

The article, "Life and Light" (page 106), by George Wald deals more directly with the absorption of light by biomolecules. Some of the energized biological pigment molecules undergo substantial chemical modifications and can initiate chemical transformations in adjacent molecules. The absorbed energy is finally stabilized in the form of a covalent chemical bond. This is the essence of a photosynthetic reaction. In other cases the pigment molecule changes its shape owing to a change in a bond angle rather than a change in the type of bond. If such a molecule is in intimate association with a delicate biological structure, such as a membrane, the change in shape can cause a substantial strain in the membrane, with a consequent modification of its properties. This is probably what occurs in the membranes of the light-receptor cells of the eye. The strain is believed to cause a nerve impulse to be generated.

Light is utilized biologically for two rather different purposes. The most obvious and important of these is to supply energy for photosynthesis. The other purpose is to enable organisms to gather information about the environment, or, in biological terms, to sense the environment. Wald emphasizes the basic difference between an energy-gathering process such as photosynthesis and a light-triggered reaction such as vision. In the latter, light energy serves only to trigger a chain of events within the living system rather than provide the total energy for the chemical activity in question.

In recent years Walther Stoeckenius and his coworkers at the University of California, San Francisco, found a purple pigment very similar to rhodopsin (the "visual purple" in animal eyes) in a simple bacterium. Called *bacteriorhodopsin*, it is a component of the outer membranes of these bacteria, where it might function in an energy-gathering capacity rather than to trigger a neurological signal, as it does in the retina.

Just as this book was being prepared for printing, Stoeckenius published an exciting new article, "The Purple Membrane of Salt-loving Bacteria" (Offprint 1340), which describes his work in detail. It is now clear that bacteriorhodopsin acts as the light-sensitive molecule in a new (to man) photosynthetic process—the first to be discovered that does not depend on chlorophyll. It appears that changes in the topology of biological membranes in which light-absorbing molecules are embedded play an important role in the biological response to light. The consequences of such structural changes are the redistribution of ions, the formation of concentration and electrical gradients, and, eventually, either the synthesis of ATP of the generation of an action potential.

By a long evolutionary process, living matter has created a variety of structures that can process incoming light signals. These enable living organisms not only to perceive objects in their vicinity (what we call seeing) but also to sense the changes in the seasons of the year. It will become apparent from several of the articles in this section that vision is only one aspect of the information gathered by living organisms from incoming radiation. Wald's discussion of the range of radiant energy that is practically available at sea level and that is theoretically utilizable by living matter is most instructive. The role of light in biological processes is further emphasized in the intriguing speculation on the possibility of life on a dark planet.

Previously I mentioned the possible increase in ultraviolet radiation reaching sea level due to the effect of pollutants on the ozone layer of the atmosphere. Many skin cancers are believed to be mutations that are caused by ultraviolet irradiation. It is therefore expected that the number of skin cancers will increase greatly if the ozone layer is diminished. How does ultraviolet radiation cause biological damage? R. A. Deering, in the article already mentioned, addresses himself to at least one aspect of this question. Nucleic acids, the most vital macromolecules of the cell, absorb ultraviolet radiation. The chemical changes that occur in these molecules upon irradiation illustrate a photochemical effect and the nature of biological mutations induced by radiation.

Note also the existence of mechanisms in living cells for repairing damaged DNA. Some of the repair enzymes were found to be specific for UV-irradiated DNA, with which they form a molecular complex. When the complex is irradiated with blue light, the pyrimidine dimers of the irradiated DNA dissociate. The repaired DNA is then released from the complex. Although short-wavelength UV does not penetrate to sea level at the present, and has probably not done so since oxygen accumulated in the atmosphere, a specific enzyme that binds only UV-damaged DNA nevertheless exists in many living cells.

It is interesting to speculate whether this enzyme is a relic of an ancient, cyclic photochemical reaction similar to the cyclic process that occurs in the retina. Short-wavelength UV is absorbed by a DNA molecule, and the damaged molecule now binds to the enzyme, where a second photochemical step restores the DNA to its original state. Such a photobiological reaction could have functioned as a sensing and triggering device. What its role is at present is not known.

I also mentioned the different roles that light plays in the biosphere. Among these was the use of light by organisms as a clue to the state of their environment. It is thus interesting to read in Richard J. Wurtman's article, "The Effects of Light on the Human Body" (page 125), that man too is subliminally affected by light. These effects are again of various kinds. One of them is apparently a direct biochemical effect on calcium metabolism, brought about at least in part by the synthesis of vitamin D_3 in irradiated skin. A more subtle effect is that exerted on the functions of the pineal body, which in turn controls many other endocrine glands. The dim light that penetrates the skull is sufficient to trigger these important reactions.

Several interesting studies concerning the effects of light on the development and sexual maturity of birds and mammals have been described in the article, "Nonvisual Light Reception" (Offprint 1243), by Michael Menaker.

The conclusions drawn by Wurtman are pertinent to our well-being. It appears that not only the quantity but also the spectral quality of light are significant for normal physiological functioning of our bodies. Wurtman recommends that we be as alert to the quality of light in our indoor environment as we are to the chemical quality of our outdoor environment.

The use of light in the plant world is also much more intriguing than it appears at first glance. We are all aware of the role of light in providing the energy required for photosynthesis. We now know that plants also utilize light signals for a variety of other physiological adaptations to their environment. Some of these interesting roles of light in the life of plants are discussed by W. L. Butler and Robert J. Downs in their article, "Light and Plant Development" (page 135).

Phytochrome, the plant pigment that is responsible for the action of red and far-red light on the development of plants, has been isolated and purified since the article was written. It turned out to be an open-chain tetrapyrrole compound associated with a protein molecule. It is related to the red and blue pigments that are found in red and blue-green algae. Phytochrome, however, does not account for all the morphogenetic effects of light on plants. Other, blue-light-absorbing pigments are apparently involved in the initiation of many physiological reactions, such as phototropism and flowering.

At present, research is being conducted in an attempt to explain the molecular mechanism by which light absorption by such pigments initiates physiological processes in plants. In at least some cases it has been found that phytochrome is associated with the cytoplasmic membrane. The orientation of the molecules within the membrane is found to change upon illumination. Changes in membrane properties could initiate a variety of other processes—for example, by facilitating the permeability of the membrane to some metabolites or ions. We see in these plant systems the sensitivity of biological systems to the spectral quality of the light in their environment. This is important in the ecology of organisms that live in caves, deep waters, or under a forest canopy, where the light spectrum is different from direct sunlight.

In his article (already mentioned), George Wald discussed the biological significance of the entire electromagnetic spectrum. He defined the range between 320 and 1100 nanometers (formerly called millimicrons) as the photobiological range. Radiation of longer wavelengths was dismissed for two reasons. One is the low energy of the quanta and the other is the fact that most infrared radiation from the sun is absorbed by water vapor and CO_2 in the atmosphere.

It is therefore interesting to read the article, "The Infrared Receptors of Snakes" (page 143), by R. Igor Gamow and John F. Harris. The approximately 10-micrometer (10,000-nanometer) wavelengths that certain snakes can "see" are energetically quite weak, corresponding roughly to the energies of hydrogen bonds. The nature of the specialized chemical receptor for this radiation is not yet known. Gamow and Harris suggest that the mitochondria of the nerve ending of the infrared receptor organ may be the primary receptors. These cellular organelles can be isolated and studied outside the living cells. It would be exciting to find that the mitochondria of this organ differ in their biochemical properties from the mitochondria of other cells in the snake.

Snakes are not the only organisms that sense and use signals from infrared sources. Many insects are guided to their warm-blooded hosts by the infrared emission from these animals. Thus the sensitivity of biological systems to radiant energy includes a much wider range of the spectrum than that which we call visible light.

10

The Chemical Effects of Light

by Gerald Oster
September 1968

Visible light triggers few chemical reactions (except in living cells), but the photons of ultraviolet radiation readily break chemical bonds and produce short-lived molecular fragments with unusual properties

Our everyday world endures because most substances, organic as well as inorganic, are stable in the presence of visible light. Only a few complex molecules produced by living organisms have the specific property of responding to light in such a way as to initiate or participate in chemical reactions ["How Light Interacts with Living Matter," by Sterling B. Hendricks, September 1968]. Outside of living systems only a few kinds of molecules are sufficiently activated by visible light to be of interest to the photochemist.

The number of reactive molecules increases sharply, however, if the wavelength of the radiant energy is shifted slightly into the ultraviolet part of the spectrum. To the photochemist that is where the action is. Thus he is primarily concerned with chemical events that are triggered by ultraviolet radiation in the range between 180 and 400 nanometers. These events usually happen so swiftly that ingenious techniques have had to be devised to follow the molecular transformations that take place. It is now routine, for example, to identify molecular species that exist for less than a millisecond. Species with lifetimes measured in microseconds are being studied, and new techniques using laser pulses are pushing into the realm where lifetimes can be measured in nanoseconds and perhaps even picoseconds.

The photochemist is interested in such short-lived species not simply for their own sake but because he suspects that many, if not most, chemical reactions proceed by way of short-lived intermediaries. Only by following chemical reactions step by step in fine detail can he develop plausible models of how chemical reactions proceed in general. From such studies it is often only a short step to the development of chemical processes and products of practical value.

When a quantum of light is absorbed by a molecule, one of the electrons of the molecule is raised to some higher excited state. The excited molecule is then in an unstable condition and will try to rid itself of this excess energy by one means or another. Usually the electronic excitation is converted into vibrational energy (vibration of the atoms of the molecule), which is then passed on to the surroundings as heat. Such is the case, for example, with a tar roof on a sunny day. An alternative pathway is for the excited molecule to fluoresce, that is, to emit radiation whose wavelength is slightly longer than that of the exciting radiation. The bluish appearance of quinine water in the sunlight is an example of fluorescence; the excitation is produced by the invisible ultraviolet radiation of the sun.

The third way an electronically excited molecule can rid itself of energy is the one of principal interest to the photochemist: the excited molecule can undergo a chemical transformation. It is the task of the photochemist to determine the nature of the products made, the amount of product made per quantum absorbed (the quantum yield) and how these results depend on the concentrations of the starting materials. His next step is to combine these data with the known spectroscopic and thermodynamic properties of the molecules involved to make a coherent picture. It must be admitted, however, that only the simplest photochemical reactions are understood in detail.

There is also a fourth way an excited molecule can dissipate its energy: the molecule may be torn apart. This is called photolysis. As might be expected, photolysis occurs only if the energy of the absorbed quantum exceeds the energy of the chemical bonds that hold the molecule together. The energy required to photolyse most simple molecules corresponds to light that lies in the ultraviolet region [*see illustration on page 100*]. For example, the chlorine molecule is colored and thus absorbs light in the visible range (at 425 nanometers), but it has a low quantum yield of photolysis when exposed to visible light. When it is exposed to ultraviolet radiation at 330 nanometers, on the other hand, the quantum yield is close to unity: each quantum of radiation absorbed ruptures one molecule.

Albert Einstein proposed in 1905 that one quantum of absorbed light leads to the photolysis of one molecule, but it required the development of quantum mechanics in the late 1920's to explain why the quantum yield should depend on the wavelength of the exciting light. James Franck and Edward U. Condon, who carefully analyzed molecular excitation, pointed out that when a molecule makes a transition from a ground state to an electronically excited state, the transition takes place so rapidly that the interatomic distances in the molecule do not have time to change. The reason is that the time required for transition is much shorter than the period of vibration of the atoms in the molecule.

To understand what happens when a molecule is excited by light it will be helpful to refer to the illustration on the opposite page. The lower curve represents the potential energy of a vibrating diatomic molecule in the ground state. The upper curve represents the potential energy of the excited molecule, which is also vibrating. The horizontal lines in the lower portion of each curve indicate the energy of discrete vibrational levels. If the interatomic distance

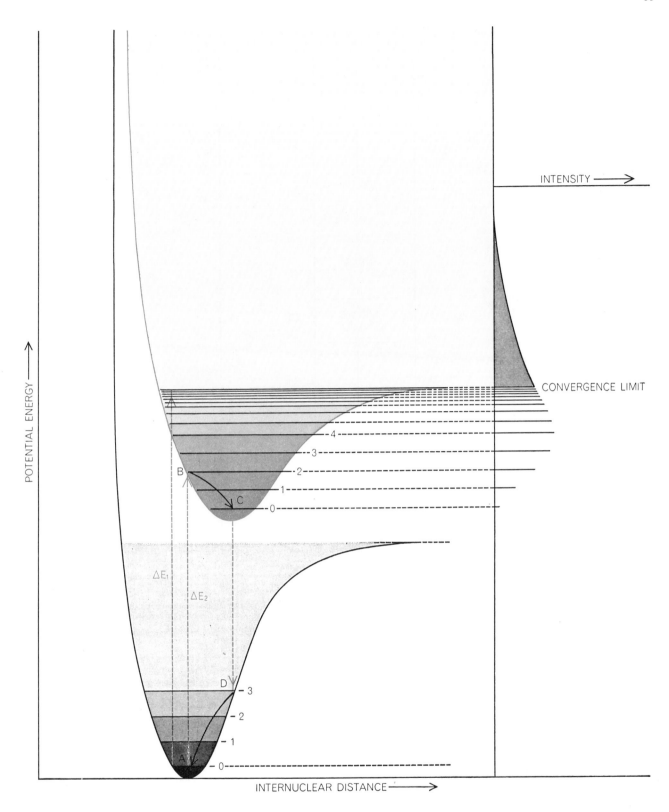

RESPONSE OF SIMPLE MOLECULES TO PHOTONS can be followed with the help of potential-energy curves. The lower curve
represents the potential energy of a typical diatomic molecule in
the ground state; the upper curve represents its potential energy
in the first excited electronic state. Because the two atoms of the
molecule are constantly vibrating, thus changing the distance between atomic nuclei, the molecule can occupy different but discrete energy levels (*horizontal lines*) within each electronic state.
The molecule in the lowest ground state can be dissociated, or
photolysed, if it absorbs a photon with an energy equal to or greater than ΔE_1. This is the energy required to carry the molecule to
or beyond the "convergence limit." The length of the horizontal
lines at the right below that limit represents the probability of
transition from the ground electronic state to a particular vibrational level in the excited electronic state. Thus a photon with an
energy of ΔE_2 will raise the molecule to the second level (B) of
that state. There it will vibrate, ultimately lose energy to surrounding molecules and fall to C. It can now emit a photon with somewhat less energy than ΔE_2 and fall to D. This is called fluorescence. After losing vibrational energy molecule will return to A.

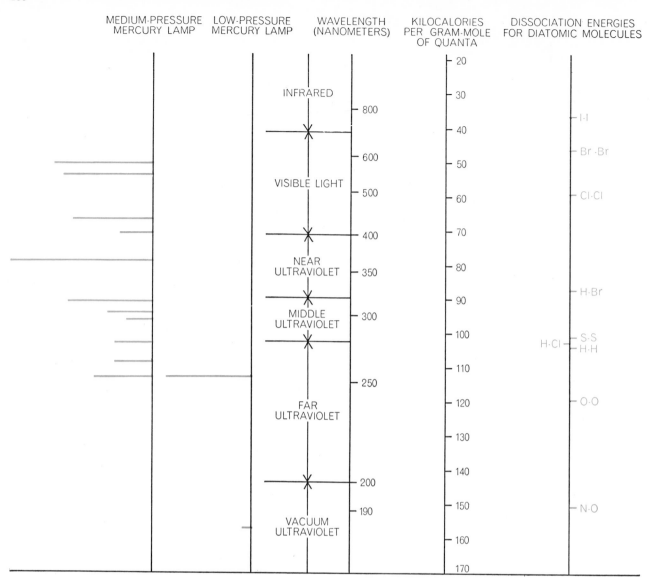

| MEDIUM-PRESSURE MERCURY LAMP | LOW-PRESSURE MERCURY LAMP | WAVELENGTH (NANOMETERS) | KILOCALORIES PER GRAM-MOLE OF QUANTA | DISSOCIATION ENERGIES FOR DIATOMIC MOLECULES |

DISSOCIATION ENERGIES of most common diatomic molecules are so high that the energy can be supplied only by radiation of ultraviolet wavelengths. The principal exceptions are molecules of chlorine, bromine and iodine, all of which are strongly colored, indicating that they absorb light. The energy carried by a quantum of radiation, or photon, is directly proportional to its frequency, or inversely proportional to its wavelength. There are 6.06×10^{23} photons in a gram-mole of quanta. This is the number required to dissociate a gram-mole of diatomic molecules (6.06×10^{23} molecules) if the quantum yield is unity. A gram-mole is the weight in grams equal to the molecular weight of a molecule, thus a gram-mole of oxygen (O_2) is 32 grams. The principal emission wavelengths of two commonly used types of mercury lamp are identified at the left. Lengths of the bars are proportional to intensity.

becomes large enough in the ground state, the molecule can come apart without ever entering the excited state. The curve for the excited state is displaced to the right of the curve for the ground state, indicating that the average interatomic distance (the minimum in each curve) is somewhat greater in the excited state than it is in the ground state. That is, the excited molecule is somewhat "looser."

The molecule can pass from the ground state to one of the levels of the excited state by absorbing radiation whose photon energy is equal to the energy difference between the ground state and one of the levels of the excited state. Provided that the quantum of radiation is not too energetic the molecule will remain intact and continue to vibrate. After a brief interval it will emit a quantum of fluorescent radiation and drop back to the ground state. Because the emission occurs when the excited molecule is at the lowest vibrational level, the emitted energy is less than the absorbed energy, hence the wavelength of the fluorescent radiation is greater than that of the absorbed radiation.

When the absorbed radiation exceeds a certain threshold value, the molecule comes apart; it is photolysed. At this point the absorption spectrum, shown at the right side of the illustration, becomes continuous, because the molecule is no longer vibrating at discrete energy levels. As long as the molecule is intact only discrete wavelengths of light can be absorbed.

It is possible for the excited state to pass to the ground state without releasing a quantum of radiation, in which case the electronic energy is dissipated as heat. Franck and Condon explained that this was accomplished by an overlapping, or crossing, of the two potential-energy curves, so that the excited molecule slides over, so to speak, to the ground state, leaving the molecule in an abnormally high state of vibration. This vibrational energy is then readily transferred to surrounding molecules.

As far as life on the earth is concerned, the most important photolytic reaction in nature is the one that creates a canopy of ozone in the upper atmosphere. Ozone is a faintly bluish gas whose molecules consist of three atoms of oxygen; ordinary oxygen molecules contain two atoms. Ozone absorbs broadly in the middle- and far-ultraviolet regions with a maximum at 255 nanometers. Fortunately ozone filters out just those wavelengths that are fatal to living organisms.

Ozone production begins with the photolysis of oxygen molecules (O_2), which occurs when oxygen strongly absorbs ultraviolet radiation with a wavelength of 190 nanometers. The oxygen atoms released by photolysis may simply recombine or they may react with other oxygen molecules to produce ozone (O_3). When ozone, in turn, absorbs ultraviolet radiation from the sun, it is either photolysed (yielding O_2 and O) or it contributes to the heating of the atmosphere. A dynamic equilibrium is reached in which ozone photolysis balances ozone synthesis.

Early in this century physical chemists were presented with a photolytic puzzle. It was observed that when pure chlorine and hydrogen are exposed to ultraviolet radiation, the quantum yield approaches one million, that is, nearly a million molecules of hydrogen chloride (HCl) are produced for each quantum of radiation absorbed. This seemed to contradict Einstein's postulate that the quantum yield should be unity. In 1912 Max Bodenstein explained the puzzle by proposing that a chain reaction is involved [*see upper illustration at right*].

The chain reaction proceeds by means of two reactions, following the initial photolysis of chlorine (Cl_2). The first reaction, which involves the breaking of the fairly strong H-H bond, creates a small energy deficit. The second reaction, which involves the breaking of the weaker Cl-Cl bond, makes up the deficit with energy to spare. Breaking the H-H bond requires 104 kilocalories per gram-mole (the equivalent in grams of the molecular weight of the reactants, in this case H_2). Breaking the Cl-Cl bond requires only 58 kilocalories per gram-mole. In both of the reactions that break these bonds HCl is produced, yielding 103 kilocalories per gram-mole. Consequently the first reaction has a deficit of one kilocalorie per gram-mole and the

second a surplus of 45 (103 − 58) kilocalories per gram-mole. The two reactions together provide a net of 44 kilocalories per gram-mole. Thus the chain reaction is fueled, once ultraviolet radiation provides the initial breaking of Cl-Cl bonds.

The chain continues until two chlorine atoms happen to encounter each other to form chlorine molecules. This takes place mainly at the walls of the reaction vessel, which can dissipate some of the excess electronic excitation energy of the chlorine atoms and allow chlorine mole-

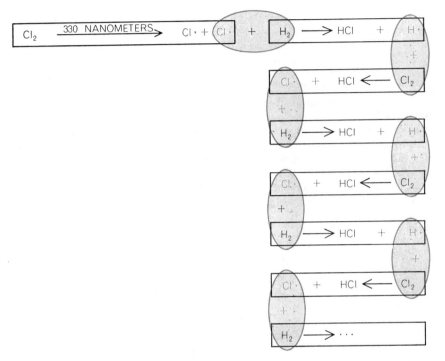

CHAIN REACTION is produced when pure chlorine and hydrogen are exposed to ultraviolet radiation. A wavelength of 330 nanometers is particularly effective. Such radiation is energetic enough to dissociate chlorine molecules, which requires only 58 kilocalories per gram-mole, but it is too weak to dissociate hydrogen molecules, which requires 104 kilocalories per gram-mole. The formation of HCl in the subsequent reactions provides 103 kilocalories per gram-mole. Since 104 kilocalories are needed for breaking the H-H bond, the reaction of atomic chlorine (Cl·) and H_2 involves a net deficit of one kilocalorie per gram-mole. However, the next reaction in the chain, involving H· and Cl_2, provides a surplus of 45 kilocalories (103 − 58). This energy surplus keeps the chain reaction going.

PHOTOLYSIS OF ACETONE, which yields primarily ethane and carbon monoxide, is a much studied photochemical reaction. It was finally understood by postulating the existence of short-lived free radicals, fragments that contain unsatisfied valence electrons.

cules to form. The free atoms may also be removed by impurities in the system.

Bromine molecules will likewise undergo a photochemical reaction with hydrogen to yield hydrogen bromide. The quantum yield is lower than in the chlorine-hydrogen reaction because atomic bromine reacts less vigorously with hydrogen than atomic chlorine does. Bromine atoms react readily, however, with olefins (linear or branched hydrocarbon molecules that contain one double bond). Each double bond is replaced by two bromine atoms. This is the basis of the industrial photobromination of hydrocarbons. Bromination can also be carried out by heating the reactants in the presence of a catalyst, but the product itself may be decomposed by such treatment. The advantage of the photochemical process is that the products formed are not affected by ultraviolet radiation.

An important industrial photochlorin-

ation process has been developed by the B. F. Goodrich Company. There it was discovered that when polyvinyl chloride is exposed to chlorine in the presence of ultraviolet radiation, the resulting plastic withstands a heat-distortion temperature 50 degrees Celsius higher than the untreated plastic does. As a result this inexpensive plastic can now be used as piping for hot-water plumbing systems.

A much studied photolytic reaction is one involving acetone (C_2H_6CO). When it is exposed to ultraviolet radiation, acetone gives rise to ethane (C_2H_6) with a quantum yield near unity, together with carbon monoxide and a variety of minor products, depending on the wavelength of excitation. The results can be explained by schemes that involve free radicals—fragments of molecules that have unsatisfied valence elec-

trons. Photolysis of acetone produces the methyl radical (CH_3) and the acetyl radical (CH_3CO). Two methyl radicals combine to form ethane [see lower illustration on page 101].

W. A. Noyes, Jr., of the University of Rochester and others assumed the existence of these free radicals in order to explain the end products of the photolysis. Because the lifetime of free radicals may be only a ten-thousandth of a second, they cannot be isolated for study. Since the end of World War II, however, the technique of flash spectroscopy has been developed for recording their existence during their brief lifetime.

Flash spectroscopy was devised at the University of Cambridge by R. G. W. Norrish and his student George Porter, who is now director of the Royal Institution. They designed an apparatus [see illustration below] in which a sample is illuminated with an intense burst of ultraviolet to create the photolytic products. A small fraction of a second later weaker light is beamed into the reaction chamber; at the far end of the chamber the light enters a spectrograph, which records whatever wavelengths have not been absorbed. The absorbed wavelengths provide clues to the nature of the short-lived species produced by photolysis. In 1967 Norrish and Porter shared the Nobel prize in chemistry with Manfred Eigen of the University of Göttingen, who had also developed techniques for studying fast reactions.

Flash spectroscopy has greatly increased chemists' knowledge about the "triplet state," an excited state that involves the pairs of electrons that form chemical bonds in organic molecules. Normally the spins of the paired electrons are antiparallel, or opposite to each other. When exposed to ultraviolet radiation, the molecules are raised to the first excited state and then undergo a nonradiative transition to an intermediate state in which the spins of two electrons in the same state are parallel to each other. This is the triplet state. If it is again exposed to ultraviolet or visible radiation, the triplet state exhibits its own absorption spectrum, which lies at a longer wavelength than the absorption spectrum of the normal ground state, or state of lowest energy [see illustration at left].

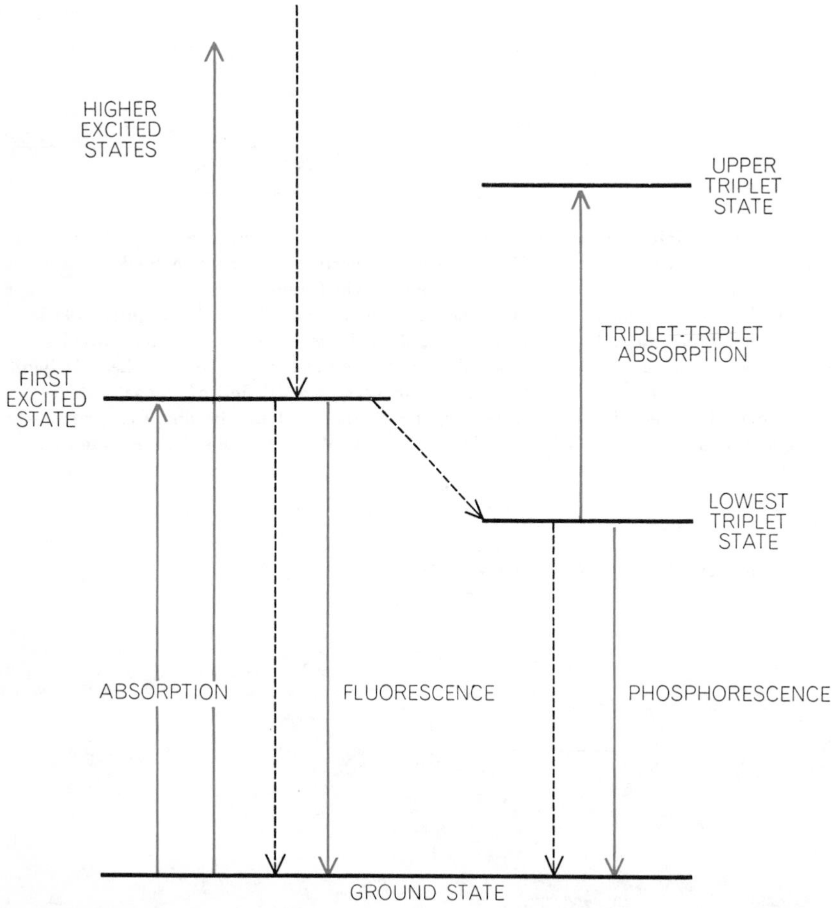

TRIPLET STATE has become an important concept in understanding the photochemical reactions of many organic molecules. Like all molecules, they can be raised to an excited state by absorption of radiation. They can also return to the ground state by normal fluorescence: reemission of a photon. Alternatively, they can drop to the triplet state without emission of radiation. (Broken lines indicate nonradiative transitions.) The existence of this state can be inferred from the wavelength of the radiation it is then able to absorb in passing to a higher triplet state. The triplet state arises when the spins of paired electrons point in the same direction rather than in the opposite direction, as they ordinarily do.

The concept of the triplet state in organic molecules is due mainly to the work of G. N. Lewis and his collaborators at the University of California at Berkeley in the late 1930's and early 1940's. These workers found that when dyes (notably fluorescein) are dissolved in a rigid medium such as glass and are exposed to a strong light, the dyes change color. When the light is removed, the dyes revert to their normal color after a second or so. This general phenomenon is called photochromism. Lewis deduced the existence of the triplet state and ascribed its fairly long duration to the time required for the parallel-spin electrons to become uncoupled and to revert to their normal antiparallel arrangement.

In 1952 Porter and M. W. Windsor used flash spectroscopy to search for the triplet state in the spectra of organic molecules in ordinary fluid solvents. They were almost immediately successful. They found that under such conditions the triplet state has a lifetime of about a millisecond.

In his Nobel prize lecture Porter said: "Any discussion of mechanism in organic photochemistry immediately involves the triplet state, and questions about this state are most directly answered by means of flash photolysis. It is now known that many of the most important photochemical reactions in solution, such as those of ketones and quinones, proceed almost exclusively via the triplet state, and the properties of this state therefore become of prime importance."

While studying the photochemistry of dyes in solution, my student Albert H. Adelman and I, working at the Polytechnic Institute of Brooklyn, demon-

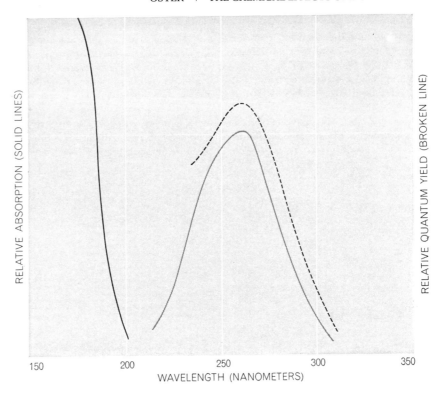

ABSORPTION SPECTRUM OF OZONE (*solid curve in color*) peaks at about 250 nanometers in the ultraviolet. As a happy consequence, the canopy of ozone in the upper atmosphere removes the portion of the sun's radiation that would be most harmful to life. The biocidal effectiveness of ultraviolet radiation is shown by the broken black line. The solid black curve is the absorption spectrum of molecular oxygen. For reasons not well understood, ultraviolet radiation of 200 nanometers does not penetrate the atmosphere.

strated that the chemically reactive species is the triplet state of the dye.

Specifically, when certain dyes are excited by light in the presence of electron-

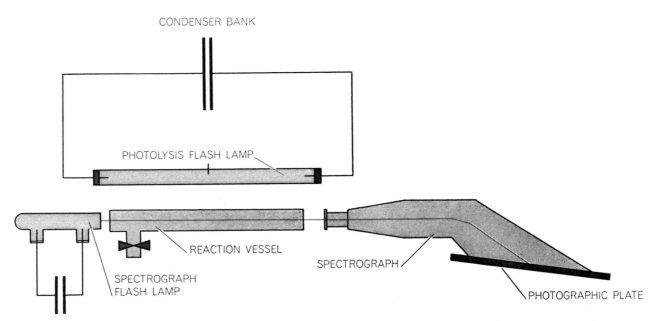

APPARATUS FOR FLASH PHOTOLYSIS was devised by R. G. W. Norrish and George Porter at the University of Cambridge. With it they discovered the short-lived triplet state that follows the photolysis of various kinds of molecules, organic as well as inorganic. The initial dissociation is triggered by the photolysis flash lamp, which produces an intense burst of ultraviolet radiation. A millisecond or less later another flash lamp sends a beam of ultraviolet radiation through the reaction vessel. Free radicals in the triplet state absorb various wavelengths ("triplet-triplet" absorption) and the resulting spectrum is recorded by the spectrograph.

donating substances, the dyes are rapidly changed into the colorless ("reduced") form. Our studies showed that the reactive state of the dye—the triplet state—has a lifetime of about a tenth of a millisecond. The dye is now a powerful reducing agent and will donate electrons to other substances, with the dye being returned to its oxidized state [*see illustration on page 105*]. In other words, the dye is a photosensitizer for chemical reductions; visible light provides the energy for getting the reaction started.

In the course of these studies I discovered that free radicals are created when dyes are photoreduced. The free radicals make their presence known by causing vinyl monomers to link up into polymers. The use of free radicals for bringing about polymerization of monomers is well known in industry. It occurred to me that adding suitable dyes to monomer solutions would provide the basis for a new kind of photography. In such a solution the concentration of free radicals would be proportional to the intensity of the visible light and thus the degree of polymerization would be controlled by light. It has turned out that very accurate three-dimensional topographical maps can be produced in plastic by this method.

The use of dyes as photosensitizing agents is, of course, fundamental to photography. In 1873 Hermann Wilhelm Vogel found that by adding dyes to silver halide emulsions he could make photographic plates that were sensitive to visible light. At first such plates responded only to light at the blue end of the spectrum. Later new dyes were found that extended the sensitivity farther and farther toward the red end of the spectrum, making possible panchromatic emulsions. Photographic firms continue to synthesize new dyes in a search for sensitizers that will act efficiently in the infrared part of the spectrum. The nature of the action of sensitizers in silver halide photography is still obscure, nearly 100 years after the effect was first demonstrated. The effect seems to depend on the state of aggregation of the dye absorbed to the silver halide crystals.

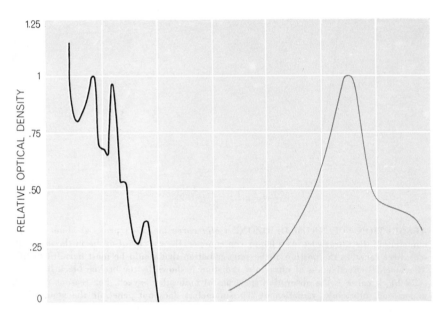

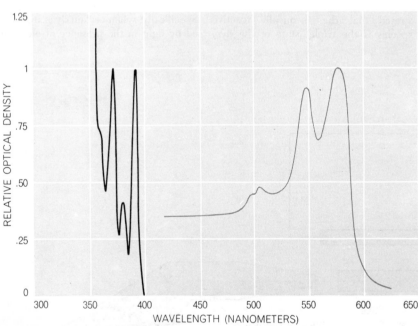

TRIPLET-TRIPLET ABSORPTION OF VISIBLE LIGHT has been observed in the author's laboratory at the Polytechnic Institute of Brooklyn. His equipment sends a beam of ultraviolet radiation into samples embedded in a plastic matrix in one direction and visible light at right angles to the ultraviolet radiation. The visible absorption spectra are then recorded in the presence of ultraviolet radiation. The black curves at the left in these two examples show the absorption of the electronic ground state. The colored curves at the right show the absorption of visible wavelengths that raises the excited molecule from the lowest triplet state to upper triplet states. The top spectra were produced by chrysene, the lower spectra by 1,2,5,6-dibenzanthracene. Both are aromatic coal-tar hydrocarbons.

The reverse of photoreduction—photooxidation—can also be mediated by dyes, as we have found in our laboratory. Here again the reactive species of the dye is the dye in the triplet state. We have found that the only dyes that will serve as sensitizers for photooxidation are those that can be reduced in the presence of light.

The oxidized dye—the dye peroxide—is a powerful oxidizing agent. In the process of oxidizing other substances the dye is regenerated [*see illustration page 105*]. My student Judith S. Bellin and I have demonstrated this phenomenon, and we have employed dye-sensitized photooxidation to inactivate some biological systems. These systems include viruses, DNA and ascites tumor cells. That dyes are visible-light sensitizers for biological inactivation was first demonstrated in 1900 by O. Raab, who observed that a dye that did not kill a culture of protozoa did so when the culture was placed near a window.

The inactivation that results from dye sensitization is different from the inactivation that results when biological systems are exposed to ultraviolet radiation. Here the inactivation often seems to re-

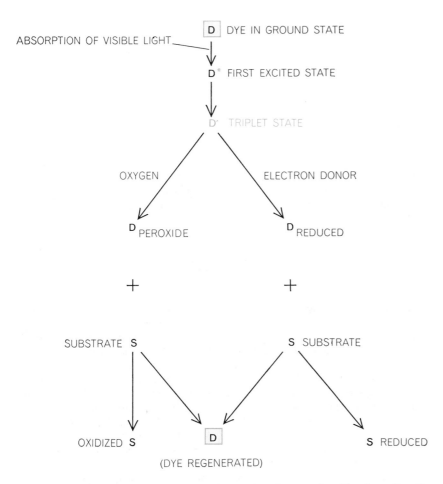

UNUSUAL PROPERTIES OF TRIPLET STATE have been explored by the author. Certain dyes in the triplet state can act either as strong oxidizing or as strong reducing agents, depending on the conditions to which the triplet state itself is exposed. In the presence of a substance that donates electrons (i.e., a reducing agent), the dye is reduced and can then donate electrons to some other substance (*substrate* S). In the presence of an oxidizing agent, the dye becomes highly oxidized and can then oxidize, or remove electrons from, a substrate. In both cases the dye is regenerated and returns to its normal state. The author's studies show that the reactive state of the dye lives only about .1 millisecond.

sult from the production of dimers: the cross-linking of two identical or similar chemical subunits. Photodimerization is implicated, for example, in the bactericidal action of ultraviolet radiation. It has long been known that the bactericidal action spectrum (the extent of killing as a function of wavelength) closely parallels the absorption spectrum of DNA, the genetic material. If dried-down films of DNA are irradiated with ultraviolet, they become cross-linked. According to one view the cross-linking occurs by means of the dimerization of thymine, one of the constituent groups of DNA.

Although this may well be the mode of action of ultraviolet radiation, my own feeling is that insufficient consideration has been given to the photolysis of the disulfide bonds of the proteins in bacteria. This bond is readily cleaved by ultraviolet radiation and has an absorption spectrum resembling that of DNA. Disulfide bonds are vital in maintaining the structure and activity of proteins; their destruction by ultraviolet radiation could also account for the death of bacteria.

In using dyes as sensitizers for initiating chemical reactions we are taking our first tentative steps into a realm where nature has learned to work with consummate finesse. Carbon dioxide and water are completely stable in the presence of visible light. Inside the leaves of plants, however, the green dye chlorophyll, when acted on by light, mediates a sequence of chemical reactions that dissociates carbon dioxide and water and reassembles their constituents into sugars and starches. A dream of photochemists is to find a dye, or sensitizer, that will bring about the same reactions in a nonliving system. There is reason to hope that such a system could be a good deal simpler than a living cell.

11

Life and Light

by George Wald
October 1959

Life depends on a narrow band in the electromagnetic spectrum. This is a consequence of the way in which molecules react to radiation, and must hold true not only on earth but elsewhere in the universe.

All life on this planet runs on sunlight, that is, on photosynthesis performed by plants. In this process light supplies the energy to make the organic molecules of which all living things are principally composed. Those plants and animals which are incapable of photosynthesis live as parasites on photosynthetic plants. But light—that form of radiant energy which is visible to the human eye—comprises only a narrow band in the spectrum of the radiant energy that pervades the universe. From gamma rays, which may be only one ten-billionth of a centimeter long, the wavelengths of electromagnetic radiation stretch through the enormous range of 10^{16}—10,000 million million times—up to radio waves, which may be miles in length. The portion of this spectrum that is visible to man is mainly contained between the wavelengths 380 to 760 millimicrons (a millimicron is ten millionths of a centimeter). By using very intense artificial sources one can stretch the limits of human vision somewhat more widely: from about 310 to 1,050 millimicrons. The remarkable fact is that, lying altogether within this slightly wider range of wavelengths, and mainly enclosed between 380 and 760 millimicrons, we also find the vision of all other animals, the bending of plants toward light, the oriented movements of animals toward or away from light and, most important, all types of photosynthesis. This is the domain of photobiology.

Why these wavelengths rather than others? I believe that this choice is dictated by intrinsic factors which involve the general role of energy in chemical reactions, the special role that light energy plays in photochemical reactions, and the nature of the molecules that mediate the utilization of light by living organisms. It is not merely a tautology

to say that photobiology requires the particular range of wavelengths we call light. This statement must be as applicable everywhere in the universe as here. Now that many of us are convinced that life exists in many places in the universe (it is hard to see how to avoid this conclusion), we have good reason to believe that everywhere we should find photobiology restricted to about the same range of wavelengths. What sets this range ultimately is not its availability, but its suitability to perform the tasks demanded of it. There cannot be a planet on which photosynthesis or vision occurs in the far infrared or far ultraviolet, because these radiations are not appropriate to perform these functions. It is not the range of available radiation that sets the photobiological domain, but rather the availability of the proper range of wavelengths that decides whether living organisms can develop and light can act upon them in useful ways.

We characterize light by its wave motion, identifying the regions of the spectrum by wavelength or frequency [see *illustration at top of pages 108 and 109*]. But in its interactions with matter—its absorption or emission by atoms and molecules—light also acts as though it were composed of small packets of energy called quanta or photons. These are in fact a class of ultimate particles, like protons and electrons, though they have no electric charge and very little mass. Each photon has the energy content: $E = hc/\lambda$, in which h is Planck's universal constant of action (1.58×10^{-34} calorie seconds), c is the velocity of light (3×10^{10} centimeters per second in empty space) and λ is the wavelength. Thus, while the intensity of light is the rate of delivery of photons, the work that a single photon can do (its energy content) is inversely proportional

to its wavelength. With the change in the energy of photons, from one end of the electromagnetic spectrum to the other, their effects upon matter vary widely. For this reason photons of different wavelengths require different instruments to detect them, and the spectrum is divided arbitrarily on this basis into regions called by different names.

In the realm of chemistry the most useful unit for measuring the work that light can do is the "einstein," the energy content of one mole of quanta (6.02×10^{23} quanta). One molecule is excited to enter into a chemical reaction by absorbing one quantum of light; so one mole of molecules can be activated by absorbing one mole of quanta. The energy content of one einstein is equal to 2.854×10^7 gram calories, divided by the wavelength of the photon expressed in millimicrons. With this formula one can easily interconvert wavelength and energy content, and so assess the chemical effectiveness of electromagnetic radiations.

Energy enters chemical reactions in two separate ways: as energy of activation, exciting molecules to react; and as heat of reaction, the change in energy of the system resulting from the reaction. In a reacting system, at any moment, only the small fraction of "hot" molecules react that possess energies equal to or greater than a threshold value called the energy of activation. In ordinary chemical reactions this energy is acquired in collisions with other molecules. In a photochemical reaction the energy of activation is supplied by light. Whether light also does work on the reaction is an entirely separate issue. Sometimes, as in photosynthesis, it does so; at other times, as probably in vision, it seems to do little or no work.

Almost all ordinary ("dark") chemical

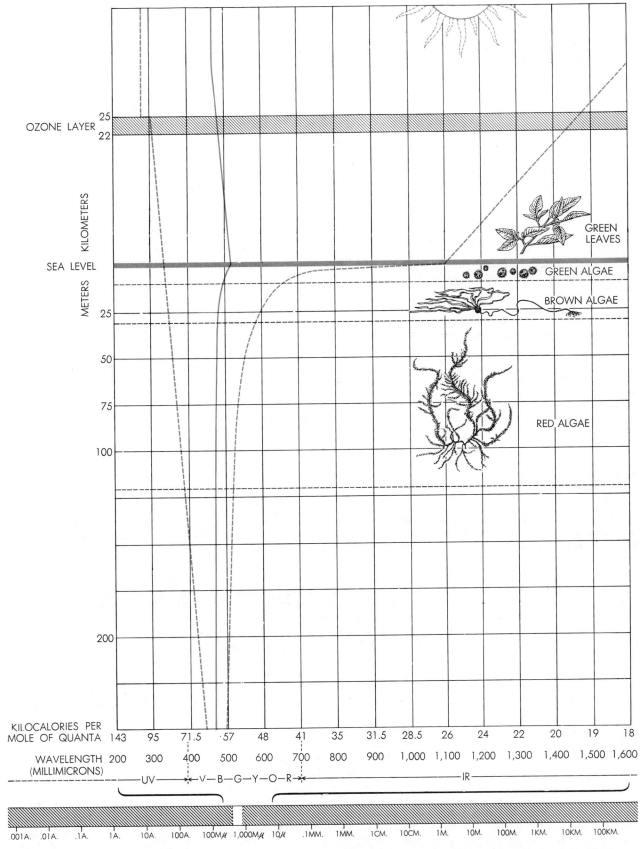

KILOCALORIES PER MOLE OF QUANTA: 143   95   71.5   57   48   41   35   31.5   28.5   26   24   22   20   19   18

WAVELENGTH (MILLIMICRONS): 200   300   400   500   600   700   800   900   1,000   1,100   1,200   1,300   1,400   1,500   1,600

UV — V – B – G – Y – O – R — IR

.001A.   .01A.   .1A.   1A.   10A.   100A.   100MM.   1,000MM.   10M.   .1MM.   1MM.   1CM.   10CM.   1M.   10M.   100M.   1KM.   10KM.   100KM.

SPECTRUM OF SUNLIGHT at the earth's surface is narrowed by atmospheric absorption to the range of wavelengths (from 320 to 1,100 millimicrons) that are effective in photobiological processes. The sunlight reaching the domain of life in the sea is further narrowed by absorption in the sea water. The solid colored line locates the wavelengths of maximum intensity; the broken colored lines, the wavelength-boundaries within which 90 per cent of the solar energy is concentrated at each level in the atmosphere and ocean. The letters above the spectrum of wavelengths at bottom represent ultraviolet (UV), violet (V), blue (B), green (G), yellow (Y), orange (O), red (R) and infrared (IR). Other usages in the chart are explained in the illustration at top of next two pages.

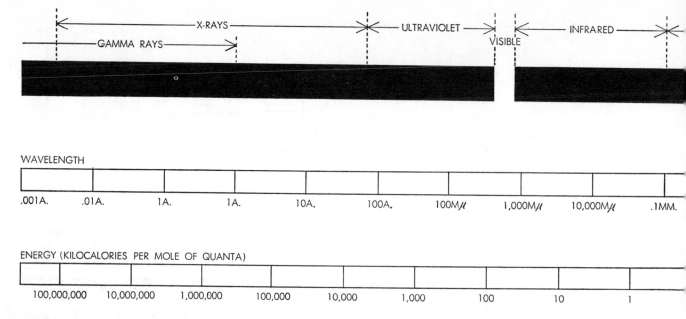

ELECTROMAGNETIC SPECTRUM is divided by man into qualitatively different regions (*top bar*), although the only difference between one kind of radiation and another is difference in wavelength (*middle bar*). From gamma rays, measured here in angstrom units, or hundred millionths of centimeter (A.), through light waves, measured here in millimicrons, or ten millionths of a centi-

reactions involve energies of activation between 15 and 65 kilogram calories (kilocalories) per mole. This is equivalent energetically to radiation of wavelengths between 1,900 and 440 millimicrons. The energies required to break single covalent bonds—a process that, through forming free radicals, can be a potent means of chemical activation—almost all fall between 40 and 90 kilocalories per mole, corresponding to radiation of wavelengths 710 to 320 millimicrons. Finally, there is the excitation of valence electrons to higher orbital levels that activates the reactions classified under the heading of photochemistry; this ordinarily involves energies of about 20 to 100 kilocalories per mole, corresponding to the absorption of light of wavelengths 1,430 to 280 millimicrons. Thus, however one approaches the activation of molecules for chemical reactions, one enters into a range of wavelengths that coincides approximately with the photobiological domain.

Actually photobiology is confined within slightly narrower limits than photochemistry. Radiations below 300 millimicrons (95 kilocalories per mole) are incompatible with the orderly existence of such large, highly organized molecules as proteins and nucleic acids. Both types of molecule consist of long chains of units bound to one another by primary valences. Both types of molecule, however, are held in the delicate and specific configurations upon which their functions in the cell depend by the relatively

weak forces of hydrogen-bonding and van der Waals attraction.

These forces, though individually weak, are cumulative. They hold a molecule together in a specific arrangement, like zippers. Radiation of wavelengths shorter than 300 millimicrons unzips them, opening up long sections of attachment, and permitting the orderly arrangement to become random and chaotic. Hence such radiations denature proteins and depolymerize nucleic acids, with disastrous consequences for the cell. For this reason about 300 millimicrons represents the lower limit of radiations capable of promoting photoreactions, yet compatible with life.

From this point of view we live upon a fortunate planet, because the radiation that is useful in promoting orderly chemical reactions comprises the great bulk of that of our sun. The commonly stated limit of human vision—400 to 700 millimicrons—already includes 41 per cent of the sun's radiant energy before it reaches our atmosphere, and 46 per cent of that arriving at the earth's surface. The entire photobiological range—300 to 1,100 millimicrons—includes about 75 per cent of the sun's radiant energy, and about 83 per cent of that reaching the earth.

From about 320 to 1,100 millimicrons—virtually the photobiological range—the sun's radiation reaches us with little modification. The atmosphere directly above us causes an attenuation, mainly by scattering rather than absorption of

light, which is negligible at 700 millimicrons and increases exponentially toward shorter wavelengths, so that at 400 millimicrons the radiation is reduced by about half. In the upper atmosphere, however, a layer of ozone, at a height of 22 to 25 kilometers, begins to absorb the sun's radiation strongly at 320 millimicrons, and at 290 millimicrons forms a virtually opaque screen. It is only the presence of this layer of ozone, removing short-wave antibiotic radiation, that makes terrestrial life possible.

At long wavelengths the absorption bands of water vapor cut strongly into the region of solar radiation from 720 to 2,300 millimicrons. Beyond 2,300 millimicrons the infrared radiation is absorbed almost completely by the water vapor, carbon dioxide and ozone of the atmosphere. The sun's radiation, therefore, which starts toward the earth in a band reaching from about 225 to 3,200 millimicrons, with its maximum at about 475 millimicrons, is narrowed by passing through the atmosphere to a range of about 310 to 2,300 millimicrons at the earth's surface.

The differential absorption of light by water confines more sharply the range of illumination that reaches living organisms in the oceans and in fresh water. The infrared is removed almost immediately in the surface layers. Cutting into the visible spectrum, water attenuates very rapidly in succession the red, orange, yellow and green. The short-wavelength limit is also gradually drawn

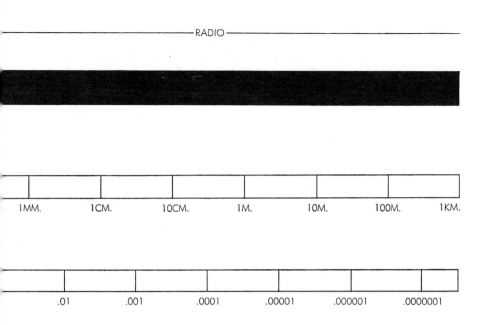

RADIO

| 1MM. | 1CM. | 10CM. | 1M. | 10M. | 100M. | 1KM. |

| .01 | .001 | .0001 | .00001 | .000001 | .0000001 |

meter (Mμ), the waves range upward in length to the longest radio waves. The difference in wavelength is associated with a decisive difference in the energy conveyed by radiation at each wavelength. This energy content (*bottom bar*) is inversely proportional to wavelength.

matter once every 300 years. All the oxygen in our atmosphere, having been bound by various oxidation processes, is renewed by photosynthesis once in about 2,000 years.

In the original accumulation of this capital of carbon dioxide and oxygen, early in the history of the earth, it is thought that the process of photosynthesis itself profoundly modified the character of the earth's atmosphere and furnished the essential conditions for the efflorescence and evolution of life. Some of the oldest rock formations have lately been discovered to contain recognizable vestiges of living organisms, including what appear to have been photosynthetic forms. So for example iron gunflint cherts found in southern Ontario contain microscopic fossils, among which appear to be colonial forms of blue-green algae. These deposits are estimated to be at least 1.5 billion years old, so that if this identification can be accepted, photosynthesis has existed at least that long on this planet.

It now seems possible that the original development of the use of light by organisms, through the agency of chlorophyll pigments, may have involved not primarily the synthesis of new organic matter, but rather the provision of stores of chemical energy for the cell. A few years ago the process called photosynthetic phosphorylation was discovered, and has since been intensively explored,

in, so that the entire transmitted radiation is narrowed to a band centered at about 475 millimicrons, in the blue.

Photosynthesis

Each year the energy of sunlight, via the process of photosynthesis, fixes nearly 200 billion tons of carbon, taken up in the form of carbon dioxide, in more complex and useful organic molecules: about 20 billion tons on land and almost 10 times this quantity in the upper layers of the ocean. All the carbon dioxide in our atmosphere and all that is dissolved in the waters of the earth passes into this process, and is completely renewed by respiration and the decay of organic

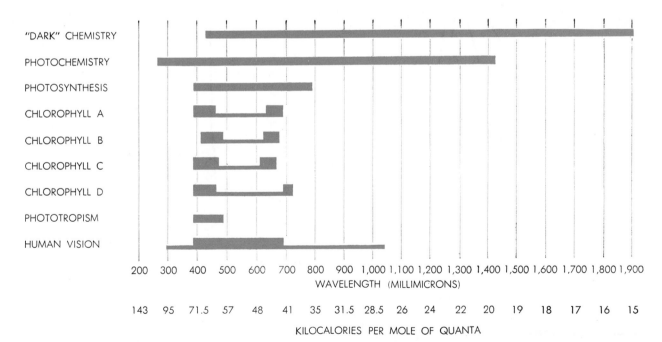

ENERGY CONTENT OF LIGHT is matched to the energy requirements of chemistry and photobiological processes and to the absorption spectra of photoreactive substances. The thicker segments of the bars opposite the chlorophylls indicate the regions of maximum absorption of light in each case, and the thicker segment in the bar opposite human vision indicates the normal boundaries.

mainly by Daniel I. Arnon of the University of California. By a still-unknown mechanism light forms the terminal high-energy phosphate bonds of adenosine triphosphate (ATP), which acts as a principal energy-carrier in the chemistry of the cell. One of the most interesting features of this process is that it is anaerobic; it neither requires nor produces oxygen. At a time when our atmosphere still lacked oxygen, this process could have become an efficient source of ATP. Among the many things ATP does in cells one of the most important is to supply the energy for organic syntheses. This direct trading of

the energy of sunlight for usable chemical energy in the form of ATP would therefore already have had as by-product the synthesis of organic structures. Mechanisms for performing such synthesis directly may have been a later development, leading to photosynthesis proper.

The essence of the photosynthetic process is the use of the energy of light to split water. The hydrogen from the water is used to reduce carbon dioxide or other organic molecules; and, in photosynthesis as performed by algae and higher plants, the oxygen is released into the atmosphere.

We owe our general view of photosynthesis in great part to the work of C. B. van Niel of the Hopkins Marine Station of Stanford University. Van Niel had examined the over-all reactions of photosynthesis in a variety of bacteria. Some of these organisms—green sulfur bacteria—require hydrogen sulfide to perform photosynthesis; van Niel discovered that in this case the net effect of photosynthesis is to split hydrogen sulfide, rendering the hydrogen available to reduce carbon dioxide to sugar, and liberating sulfur rather than oxygen. Still other bacteria—certain nonsulfur purple bacteria, for example—require organic

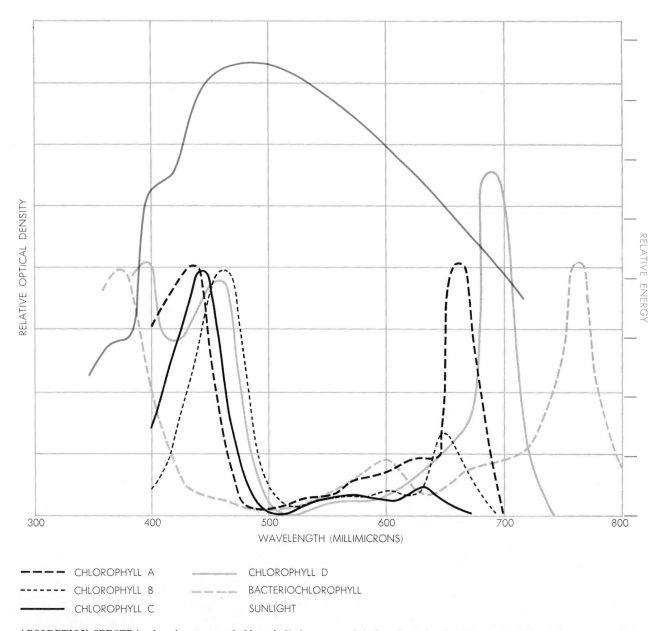

- - - -	CHLOROPHYLL A	——————	CHLOROPHYLL D
- - - - -	CHLOROPHYLL B	— — —	BACTERIOCHLOROPHYLL
——————	CHLOROPHYLL C		SUNLIGHT

ABSORPTION SPECTRA of various types of chlorophyll show the regions of the spectrum in which these substances absorb sunlight most effectively, measured on scale of relative optical density at left. Paradoxically the chlorophylls absorb best at the ends of the spectrum of sunlight, where energy, shown on scale at right, falls off steeply from the maximum around middle of the spectrum.

substances in photosynthesis. Here van Niel found that the effect of photosynthesis is to split hydrogen from these organic molecules to reduce carbon dioxide, liberating in this case neither oxygen nor sulfur but more highly oxidized states of the organic molecules themselves. Finally there are forms of purple bacteria that use molecular hydrogen directly in photosynthesis to reduce carbon dioxide, and liberate no by-product.

The efficiency of photosynthesis in algae and higher green plants is extraordinarily high—just how high is a matter of continuing controversy. The work of reducing one mole of carbon dioxide to the level of carbohydrate is in the neighborhood of 120 kilocalories. This energy requirement, though the exact figure is approximate, cannot be evaded through any choice of mechanism. Thanks to the selective absorption of the green chlorophyll pigment, light is made available for this process in quanta whose energy content is 41 or 42 kilocalories per mole, corresponding to quanta of red light of wavelength about 680 millimicrons. It is apparent, therefore, that several such quanta are required to reduce one molecule of carbon dioxide. If the energy of light were used with perfect efficiency, three quanta might perhaps suffice.

About 35 years ago the great German biochemist Otto Warburg performed experiments which appeared to show that in fact about four quanta of light of any wavelength in the visible spectrum are enough to reduce a molecule of carbon dioxide to carbohydrate. This might have meant an efficiency of about 75 per cent. Later a variety of workers in this country and elsewhere insisted that when such experiments are performed more critically, from eight to 12 quanta are required per molecule of carbon dioxide reduced. This discrepancy led to one of the bitterest controversies in modern science.

Many of us have grown tired of this controversy, which long ago bogged down in technical details and fruitless recriminations. I think it significant, however, that a number of recent, non-Warburgian, investigations have reported quantum demands of about six, and in at least one case the reported demand was as low as five. These numbers represent very high efficiencies (50 to 60 per cent), though not quite as high as Warburg prefers to set them.

Investigators have now turned from the question of efficiency to a more fruitful study of the specific uses to which quanta are put in photosynthesis. This is yielding estimates of quantum demand related to specific mechanisms rather than to controversial details of experimentation.

To reduce one molecule of carbon dioxide requires four hydrogen atoms and apparently three high-energy phosphate bonds of ATP. If we allow one quantum for each hydrogen atom (a point not universally conceded), that yields directly a quantum demand of four. If the ATP can be supplied in other ways, for example by respiration, four may be enough. If, however, light is needed also to supply ATP, by photosynthetic phosphorylation, then more quanta are needed; how many is not yet clear. Yet if one quantum were to generate one phosphate bond, the theoretical quantum-demand of photosynthesis, with all the energy supplied by light, would be four plus three, or seven. That would represent a high order of efficiency in the conversion of the energy of light to the energy of chemical bonds.

It is curious to put this almost obsessive concentration on the efficiency of photosynthesis together with what I think to be one of the most remarkable facts in all biology. Chlorophylls, the pigments universally used in photosynthesis, have absorption properties that seem just the opposite of what is wanted in a photosynthetic pigment. The energy of sunlight as it reaches the surface of the earth forms a broad maximum in the blue-green to green region of the spectrum, falling off at both shorter and longer wavelengths. Yet it is precisely in the blue-green and green, where the energy of sunlight is maximal, that the chlorophylls absorb light most poorly; this, indeed, is the reason for their green color. Where the absorption by chlorophyll is maximal—in widely separated bands in the violet and red—the energy of sunlight has fallen off considerably [see illustration on opposite page].

After perhaps two billion years of selection, involving a process whose efficiency is more important than that of any other process on earth, this seems an extraordinarily poor performance. It is a curious fact to put together with Warburg's comment (at one point in the quantum-demand controversy) that in a perfect nature, photosynthesis also is perfect. I think that the question it raises may be put more usefully as follows: What properties do the chlorophylls have that are so profoundly advantageous for photosynthesis as to override their disadvantageous absorption spectra?

We have the bare beginnings of an answer; it is emerging from a deeper understanding of the mechanism of photosynthesis, in particular as it is expressed in the structure and function of chlorophyll itself. Chlorophyll a, the type of chlorophyll principally involved in the photosynthesis of algae and higher plants, owes its color, that is, its capacity for absorbing light, to the possession of a long, regular alternation of single and double bonds, the type of arrangement called a conjugated system [see illustration on next two pages]. All pigments, natural and synthetic, possess such conjugated systems of alternate single and double bonds. The property of such systems that lends them color is the possession of particularly mobile electrons, called pi electrons, which are associated not with single atoms or bonds but with the conjugated system as a whole. It requires relatively little energy to raise a pi electron to a higher level. This small energy-requirement corresponds with the absorption of radiation of relatively long wavelengths, that is, radiation in the visible spectrum; and also with a high probability, and hence a strong intensity, of absorption.

In chlorophyll this conjugated system is turned around upon itself to form a ring of rings, a so-called porphyrin nucleus, and this I think is of extraordinary significance. On the one hand, as the illustration on these two pages shows, it makes possible a large number of rearrangements of the pattern of conjugated single and double bonds in the ring structure. Each such arrangement corresponds to a different way of arranging the external electrons, without moving any of the atoms. The molecule may thus be conceived to resonate among and be a hybrid of all these possible arrangements. In such a structure the pi electrons can not only oscillate, as in a straight-chain conjugated system; they can also circulate.

The many possibilities of resonance, together with the high degree of condensation of the molecule in rings, give the chlorophylls a peculiar rigidity and stability which I think are among the most important features of this type of structure. Indeed, porphyrins are among the most inert and stable molecules in the whole of organic chemistry. Porphyrins, apparently derived from chlorophyll, have been found in petroleum, oil shales and soft coals some 400 million years old.

This directs our attention to special features of chlorophyll, which are directly related to its functions in photosynthesis. One such property is not to utilize the energy it absorbs immediately in

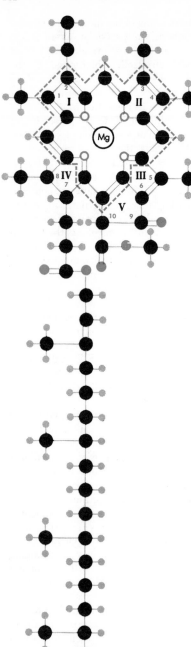

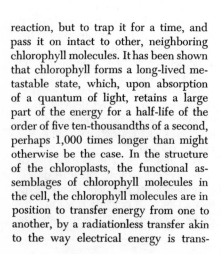

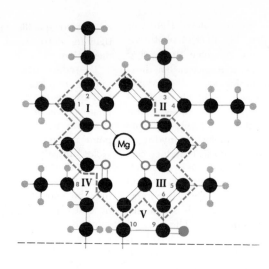

● CARBON   (Mg) MAGNESIUM

● OXYGEN   ○ NITROGEN

● HYDROGEN

CHLOROPHYLL MOLECULE, diagrammed in its entirety at left, owes its photobiological activity to the rigid and intricate porphyrin structure at top. The arrangement of the bonds in this structure may resonate among the configurations

reaction, but to trap it for a time, and pass it on intact to other, neighboring chlorophyll molecules. It has been shown that chlorophyll forms a long-lived metastable state, which, upon absorption of a quantum of light, retains a large part of the energy for a half-life of the order of five ten-thousandths of a second, perhaps 1,000 times longer than might otherwise be the case. In the structure of the chloroplasts, the functional assemblages of chlorophyll molecules in the cell, the chlorophyll molecules are in position to transfer energy from one to another, by a radiationless transfer akin to the way electrical energy is trans-

ferred in an induction motor. This capacity for transferring the energy about, so that it virtually belongs to a region of the chloroplast rather than to the specific molecule of chlorophyll that first absorbed it, makes possible the high efficiency of photosynthesis. While photosynthesis is proceeding rapidly, many chlorophyll molecules, having just reacted, are still in position to absorb light, but not to utilize it. In this way large amounts of absorbed energy that would otherwise be degraded into heat are retained and passed about intact until used photosynthetically.

One sign of the capacity to retain the energy absorbed as light and pass it on relatively intact is the strong fluorescence exhibited by chlorophyll. This green or blue-green pigment fluoresces red light; and however short the wavelengths that are absorbed—that is, however large the quanta—the same red light is fluoresced, corresponding to quanta of energy content about 40 kilocalories per mole. This is the quantity of energy that is passed from molecule to molecule in the chloroplast and eventually made available for photosynthesis.

The generally inert structure of chlorophyll must somewhere contain a chemically reactive site. Such a site seems to exist in the five-membered carbon ring, usually designated ring V in the structural diagrams. James Franck of the University of Chicago some years ago called attention to the possibility that it is here that the reactivity of chlorophyll is localized. Recent experiments by Wolf

Vishniac and I. A. Rose of Yale University, employing the radioactive isotope of hydrogen (tritium), have shown that chlorophyll, both in the cell and in solution, can take up hydrogen in the light though not in the dark, and can transfer it to the coenzyme triphosphopyridine nucleotide, which appears to be principally responsible for transferring hydrogen in photosynthesis. There is some evidence to support Franck's suggestion that the portion of chlorophyll involved in these processes is the five-membered ring.

Chlorophyll thus possesses a triple combination of capacities: a high receptivity to light, an inertness of structure permitting it to store the energy and relay it to other molecules, and a reactive site equipping it to transfer hydrogen in the critical reaction that ultimately binds hydrogen to carbon in the reduction of carbon dioxide. I would suppose that these properties singled out the chlorophylls for use by organisms in photosynthesis in spite of their disadvantageous absorption spectrum.

Photosynthetic organisms cope with the deficiencies of chlorophyll in a variety of ways. In 1883 the German physiologist T. W. Engelmann pointed out that in the various types of algae other pigments must also function in photosynthesis. Among these are the carotenoid pigments in the green and brown algae, and the phycobilins, phycocyanin and phycoerythrin (related to the animal bile-pigments) in the red and blue-green algae. Engelmann showed that each type

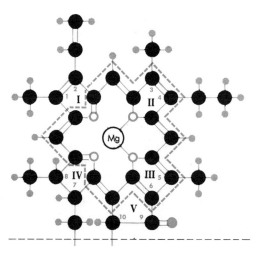

diagrammed in the middle and at the right. These and other possible configurations of the bonds help to make it possible for the chlorophyll molecule to trap and store energy which is conveyed to it by light quanta.

of alga photosynthesizes best in light of the complementary color: green algae in red light, brown algae in green light, red algae in blue light. He pointed out that this is probably the basis of the layering of these types of algae at various depths in the ocean.

All these pigments act, however, by transferring the energy they absorb to one another and eventually to chlorophyll *a;* whatever pigments have absorbed the light, the same red fluorescence of chlorophyll *a* results, with its maximum at about 670 to 690 millimicrons. The end result is therefore always the same: a quantum with an energy content of about 40 kilocalories per mole is made available to chlorophyll *a* for photosynthesis. The accessory pigments, including other varieties of chlorophyll, perform the important function of filling in the hole in the absorption spectrum of chlorophyll.

Still another device helps to compensate for the failure of chlorophyll to absorb green and blue-green light efficiently: On land and in the sea the concentration of chlorophyll and the depth of the absorbing layer are maximized by plant life. As a result chlorophyll absorbs considerable energy even in the wavelengths at which its absorption is weakest. Leaves absorb green light poorly, yet they do absorb a fraction of it. One need only look up from under a tree to see that the cover of superimposed leaves permits virtually no light to get through, green or otherwise. The lower leaves on a tree, though plentifully supplied with chloro-

plasts, may receive too little light to contribute significantly to photosynthesis. By being so profligate with the chlorophylls, plants compensate in large part for the intrinsic absorption deficiencies of this pigment.

Phototropism

The phototropism of plants—their tendency to bend toward the light—is excited by a different region of the spectrum from that involved in photosynthesis. The red wavelengths, which are most effective in photosynthesis, are wholly ineffective in phototropism, which depends upon the violet, blue and green regions of the spectrum. This relationship was first demonstrated early in the 19th century by a worker who reported that when he placed a flask of port wine between a growing plant and the light from a window, the plant grew about as well as before, but no longer bent toward the light. Recently, more precise measurements with monochromatic lights have shown that the phototropism of both molds and higher plants is stimulated only by light of wavelengths shorter than approximately 550 millimicrons, lying almost completely within the blue-green, blue and violet regions of the spectrum.

Phototropism must therefore depend on yellow pigments, because only such pigments absorb exclusively the short wavelengths of the visible spectrum. All types of plant that exhibit phototropism appear to contain such yellow pigments, in the carotenoids. In certain instances the carotenoids are localized specifically in the region of the plant that is phototropically sensitive. The most careful measurements of the effectiveness of various wavelengths of light in stimulating phototropism in molds and higher plants have yielded action spectra which resemble closely the absorption spectra of the carotenoids that are present.

A number of lower invertebrates—for example, hydroids, marine organisms that are attached to the bottom by stalks —bend toward the light by differential growth, just as do plants. The range of wavelengths which stimulate this response is also about the same as that in plants. It appears that here also carotenoids, which are usually present in considerable amount, may be the excitatory agents. Phototactic responses, involving motion of the whole animal toward or away from the light, also abound throughout all groups of invertebrates. Unfortunately no one has yet correlated accurately the action spectra for such re-

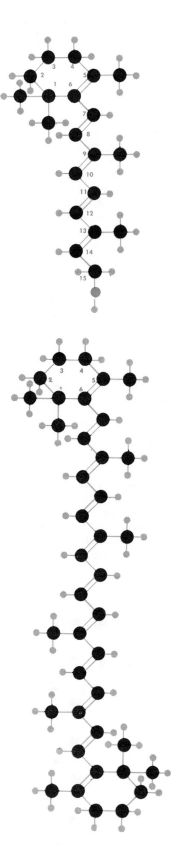

CAROTENE MOLECULE *(bottom)* **is probable light-receptor in phototropism and is synthesized by plants. In structure it is a double vitamin A molecule** *(top)*. **Vitamin A, in turn, is precursor of retinine molecule** *(illustration on page 10)*, **which mediates vision.**

sponses with the absorption spectra of the pigments that are present, so that no rigorous identification of the excitatory pigments can be made at present. This is a field awaiting investigation.

Vision

Only three of the 11 major phyla of animals have developed well-formed, image-resolving eyes: the arthropods (insects, crabs, spiders), mollusks (octopus, squid) and vertebrates. These three types of eye are entirely independent developments. There is no connection among them, anatomical, embryological or evolutionary. This is an important realization, for it means that three times, in complete independence of one another, animals on this planet have developed image-forming eyes.

It is all the more remarkable for this reason that in all three types of eye the chemistry of the visual process is very nearly the same. In all cases the pigments which absorb the light that stimulates vision are made of vitamin A, in the form of its aldehyde, retinene, joined with specific retinal proteins called opsins. Vitamin A ($C_{20}H_{29}OH$) has the structure of half a beta-carotene ($C_{40}H_{56}$), with a hydrogen and a hydroxyl radical (OH) added at the broken double bond.

Thus animal vision not only employs substances of the same nature as the carotenoids involved in phototropism of plants; there is also a genetic connection. Animals cannot make vitamin A *de novo*, but derive it from the plant carotenoids consumed in their diet. All photoreception, from phototropism in lower and higher plants to human vision, thus appears to depend for its light-sensitive pigments upon the carotenoids.

The role of light in vision is fundamentally different from its role in photosynthesis. The point of photosynthesis is to use light to perform chemical work, and the more efficiently this conversion is accomplished, the better the process serves its purpose. The point of vision is excitation; there is no evidence that the light also does work. The nervous structures upon which the light acts, so far as we know, are ready to discharge, having been charged through energy supplied by internal chemical reactions. Light is required only to trigger their responses.

Because this distinction is not always understood, attempts are frequently made to force parallels between vision and photosynthesis. In fact, these processes differ so greatly in their essential natures that no deep parallelism can be expected. The problem of quantum demand, for example, raises entirely different issues in vision as compared with photosynthesis. In photosynthesis one is interested in the minimum number of quanta needed to perform a given chemical task. In vision the problem hinges not on energetic efficiency but on differential sensitivity. The light intensities within which animals must see range from starlight to noonday sunlight; the latter is about a billion times brighter than the former. It is this enormous range of intensities that presents organisms with their fundamental visual problem: how to see at the lowest intensities without having vision obliterated by glare at the highest.

In the wholly dark-adapted state a vertebrate rod, the receptor principally involved in night vision, can respond to the absorption of a single quantum of visible light. To be sure, in the human eye, in which this relationship has been studied most completely, this minimal response of a single rod does not produce a visual sensation. In the dark-adapted state, seeing requires that at least five such events occur almost simultaneously within a small area of retina. This arrangement is probably designed to place the visual response above the "noise level" of the retina. From careful electrophysiological measurements it seems that a retina, even in total darkness, transmits a constant barrage of randomly scattered spontaneous responses to the brain. If the response of a single rod entered consciousness, we should be seeing random points of light flickering over the retina at all times.

The eye's extraordinary sensitivity to light is lost as the brightness of the illumination is increased. The threshold of human vision, which begins at the level of a few quanta in the dark-adapted state, rises as the brightness of the light increases until in bright daylight one million times more light may be needed just to stimulate the eye. But the very low quantum-efficiency in the light-adapted condition nonetheless represents a high visual efficiency.

The statement that the limits of human vision are 380 and 760 millimicrons is actually quite arbitrary. These limits are the wavelengths at which the visual sensitivity has fallen to about a thousandth of its maximum value. Specific investigations have pursued human vision to about 312 millimicrons in the near ultraviolet, and to about 1,050 in the near infrared.

In order to see at 1,050 millimicrons, however, 10,000 million times more light energy is required if cones are being stimulated, and over a million million times more energy if rods are being stimulated. This result came out of measurements made in our laboratory at Harvard University during World War II in association with Donald R. Griffin and Ruth Hubbard. As we exposed our eyes to flashes of light in the neighborhood of 1,000 millimicrons, we could not only see the flash but feel a momentary flush of heat on the cornea of the eye. At about 1,150 millimicrons, just a little farther into the infrared than our ex-

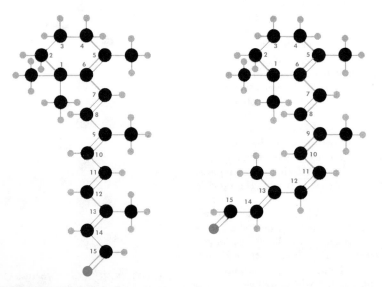

RETINENE MOLECULE is the active agent in the pigments of vision. Upon absorbing energy of light the geometry of the molecule changes from the so-called *cis* arrangement at left to the *trans* arrangement at right. This change in structure triggers process of vision.

periments had taken us, the radiation should have become a better stimulus as heat than as light.

The ultraviolet boundary of human vision, as that of many other vertebrates, raises a special problem. Ordinarily our vision is excluded from the ultraviolet, not primarily because the retina or its visual pigments are insensitive to that portion of the spectrum, but because ultraviolet light is absorbed by the lens of the eye. The human lens is yellow in color and grows more deeply yellow with age. One curious consequence of this arrangement is that persons who have had their lenses removed in the operation for cataract have excellent ultraviolet vision.

One may wonder how it comes about that man and many other vertebrates have been excluded from ultraviolet vision by the yellowness of their lenses. Actually this effect is probably of real advantage. All lens systems made of one material refract shorter wavelengths more strongly than longer wavelengths, and so bring blue light to a shorter focus than red. This phenomenon is known as chromatic aberration, or color error, and even the cheapest cameras are corrected for it. In default of color correction the lens seems to do the next best thing; it eliminates the short wavelengths of the spectrum for which the color error is greatest.

One group of animals, however, makes important use of the ultraviolet in vision. These are the insects. The insect eye is composed of a large number of independent units, the ommatidia, each of which records a point in the object, so that the image as a whole is composed as a mosaic of such points. Projection by a lens plays no part in this system, and chromatic aberration is of no account.

How does it happen that whenever vision has developed on our planet, it has come to the same group of molecules, the A vitamins, to make its light-sensitive pigments? I think that one can include plant phototropism in the same question, and ask how it comes about that all photoreception, animal and plant, employs carotenoids to mediate excitation by light. We have already asked a similar question concerning the chlorophylls and photosynthesis; and what chlorophylls are to photosynthesis, carotenoids are to photoreception.

Both the carotenoids and chlorophylls owe their color to the possession of conjugated systems. In the chlorophylls these are condensed in rings; in the carotenoids they are mainly in straight

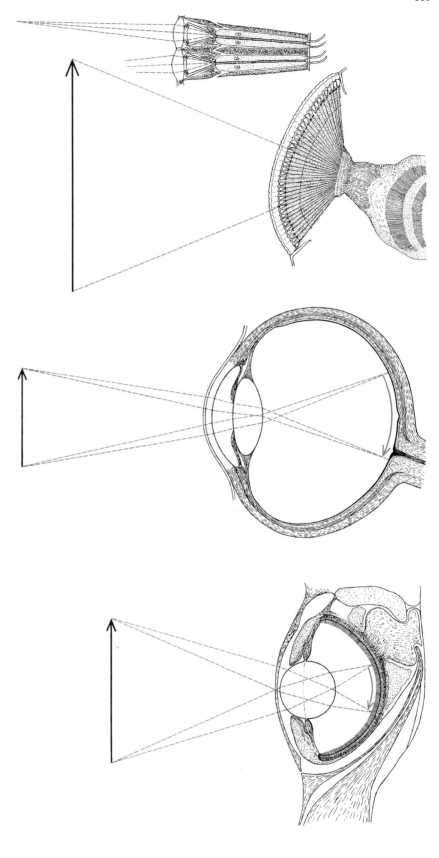

EYES of three kinds have evolved quite independently in three phyla: insects (*top*), vertebrates (*center*) and mollusks (*bottom*). In all three types of eye, however, the chemistry of vision is mediated by retinene derived from the carotenoids synthesized by plants.

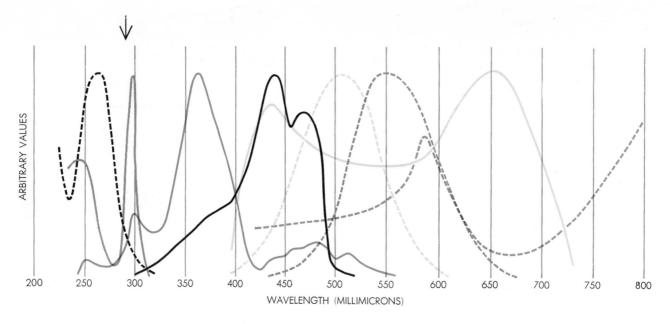

ARBITRARY VALUES

WAVELENGTH (MILLIMICRONS)

A ----- E ——
B —— F -----
C —— G -----
D —— H -----

PHOTOBIOLOGICAL PROCESSES are activated by different regions of the spectrum: killing of a bacillus (A), sunburn in human skin (B), insect vision (C), phototropism in an oat plant (D), photosynthesis in wheat (E), human "night" vision (F), human "day" vision (G), photosynthesis in a bacterium (H). Arrow at left marks limit of solar short waves.

chains. The chlorophylls fluoresce strongly; the carotenoids, weakly or not at all. Much of the effectiveness of the chlorophylls in photosynthesis is associated with a high capacity for energy transfer; there is as yet no evidence that such energy transfer has a place in vision.

I think that the key to the special position of the carotenoids in photoreception lies in their capacity to change their shapes profoundly on exposure to light. They do this by the process known as *cis-trans* isomerization. Whenever two carbon atoms in a molecule are joined by a single bond, they can rotate more or less freely about this bond, and take all positions with respect to each other. When, however, two carbon atoms are joined by a double bond, this fixes their position with respect to each other. If now another carbon atom is joined to each of this pair, both the new atoms may attach on the same side of the double bond (the *cis* position) or on opposite sides, diagonally (the *trans* position). These are two different structures, each of them stable until activated to undergo transformation—isomerization—into the other.

Carotenoids, possessing as they do long straight chains of conjugated double bonds, can exist in a great variety of such *cis-trans* or geometrical isomers. No other natural pigments approach them in this regard. Porphyrins and other natural pigments may have as

many or more double bonds, but are held in a rigid geometry by being bound in rings.

Cis-trans isomerization involves changes in shape. The all-*trans* molecule is relatively straight, whereas a *cis* linkage at any point in the chain represents a bend. In the composition of living organisms, which depends in large part on the capacity of molecules to fit one another, shape is all-important.

We have learned recently that all the visual pigments known, in both vertebrate and invertebrate eyes, are made with a specifically bent and twisted isomer of retinene. Only this isomer will do because it alone fits the point of attachment on the protein opsin. The intimate union thus made possible between the normally yellow retinene and opsin greatly enhances the color of the retinene, yielding the deep-orange to violet colors of the visual pigments. The only action of light upon a visual pigment is to isomerize—to straighten out—retinene to the all-*trans* configuration. Now it no longer fits opsin, and hence comes away. The deep color of the visual pigment is replaced by the light yellow color of free retinene. This is what is meant by the bleaching of visual pigments by light.

In this succession of processes, however, it is some process associated with the *cis-trans* isomerization that excites vision. The subsequent cleavage of reti-

nene from opsin is much too slow to be responsible for the sensory response. Indeed, in many animals the visual pigments appear hardly to bleach at all. This seems to be the case in all the invertebrate eyes yet examined, in which the entire transformation in light and darkness appears to be restricted to the isomerization of retinene. It seems possible that similar *cis-trans* isomerizations of carotenoid pigments underlie phototropic excitation in plants. Experiments are now in progress in our laboratory to explore this possibility.

Bioluminescence

In addition to responding to light in their various ways, many bacteria, invertebrates and fishes also produce light. All bioluminescent reactions require molecular oxygen; combustions of one kind or another supply the energy that is emitted as light. In photosynthesis light performs organic reductions, releasing oxygen in the process. In bioluminescence the oxidation of organic molecules with molecular oxygen emits light. I used to think that bioluminescence is like vision in reverse; but in fact it is more nearly like photosynthesis in reverse.

What function bioluminescence fulfills in the lives of some of the animals that display it is not yet clear. The flashing of fireflies may act as a signal for integrating their activities, and perhaps as a sexual excitant. What role may be fulfilled by the extraordinary display of red, green and yellow illumination in a railroad worm is altogether conjectural. There is one major situation, however, in which bioluminescence must play an exceedingly important role. This is in the sea, at depths lower than those reached by surface light, and at night at all depths. It would be difficult otherwise to understand how fishes taken from great depths, far below those to which light from the surface can penetrate, frequently have very large eyes. For vision at night or at great depths, it is not necessary that the organisms and objects that are visible themselves be bioluminescent. Bioluminescent bacteria abound in the ocean, and many submerged objects are coated sufficiently with luminous bacteria to be visible to the sensitive eye.

It has lately been discovered that the rod vision of deep-sea fishes is adapted to the wavelengths of surface light that penetrate most deeply into the water: the blue light centered around 475 millimicrons. Furthermore, sensitive new devices for measuring underwater illumination have begun to reveal the remarkable fact that deep-sea bioluminescence may also be most intense at about 475 millimicrons. The same selection of visual pigments that best equips deep-sea fishes to see by light penetrating from the surface seems best adapted to the bioluminescent radiation.

Just as light quanta must be of a certain size to activate or provide the energy for useful chemical reactions, so chemical reactions emit light in the same range of wavelengths. It is for this reason, and no accident, that the range of bioluminescent radiations coincides well with the range of vision and other photobiological processes.

Light and Evolution

The relationship between light and life is in one important sense reciprocal. Over the ages in which sunlight has activated the processes of life, living organisms have modified the terrestrial environment to select those wavelengths of sunlight that are most compatible with those processes. Before life arose, much more of the radiation of the sun reached the surface of the earth than now. We believe this to have been because the atmosphere at that time contained very little oxygen (hence negligible amounts of ozone) and probably very little carbon dioxide. Very much more of the sun's infrared and hard ultraviolet radiation must have reached the surface of the earth then than now.

Some of the short-wave radiation, operating in lower reaches of the atmosphere and also probably in the surface layers of the seas, must have been important in activating the synthesis and interactions of organic molecules which formed the prelude to the eventual emergence of the first living organisms. These organisms, coming into an anaerobic world, surrounded by the organic matter that had accumulated over the previous ages, must have lived by fermentation, and in this process must have produced as a by-product very large quantities of carbon dioxide. Part of this remained dissolved in the oceans; part entered the atmosphere.

Eventually the availability of large amounts of carbon dioxide, much larger than are in the atmosphere today, made possible the development of the process of photosynthesis. This began to remove carbon dioxide from the atmosphere, fixing it in organic form. Simultaneously, through the most prevalent and familiar form of photosynthesis, it began to produce oxygen, and in this way oxygen first became established in our atmosphere. As oxygen accumulated, the layer of ozone that formed high in the atmosphere—itself a photochemical process—prevented the short-wavelength radiation from the sun from reaching the surface of the earth. This relief from antibiotic radiation permitted living organisms to emerge from the water onto the land.

The presence of oxygen also led to the development of the process of cellular respiration, which involves gas exchanges just the reverse of those of photosynthesis. Eventually respiration and photosynthesis came into approximate balance, as they must have been for some ages past.

One may wonder how much of this history could have occurred in darkness,

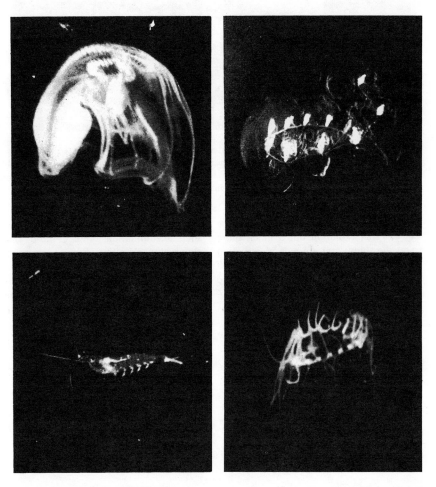

BIOLUMINESCENT CREATURES of the ocean were made to take these photographs of themselves by means of a camera designed by Harold E. Edgerton of Massachusetts Institute of Technology and L. R. Breslau of the Woods Hole Oceanographic Institution. The feeble luminescence of the animals was harnessed to trigger a high-speed electronic flash.

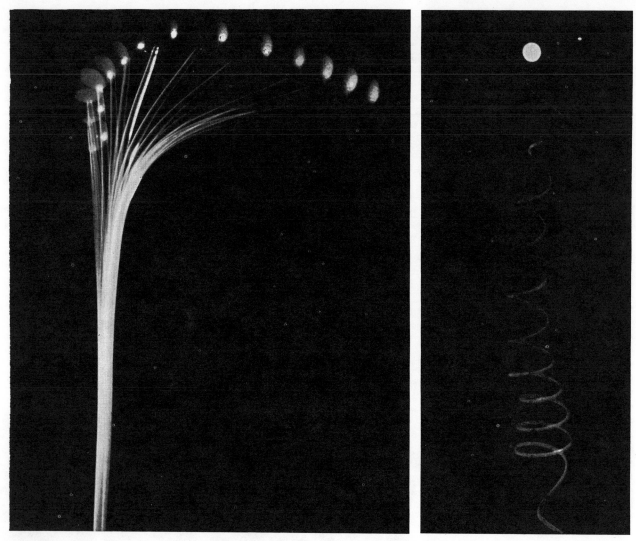

PHOTOTROPISM in the fruiting body of the mold *Phycomyces* is demonstrated in these photographs from the laboratory of Max Delbrück at the California Institute of Technology. The multiple photograph at left, with exposures made at intervals of five minutes, shows the fruiting body growing toward the light source. In the photograph at right, the stalk of the fruiting body has been made to grow in an ascending spiral by placing it on a turntable which revolved once every two hours in the presence of a fixed light-source.

by which I mean not merely the absence of external radiation but a much more specific thing: tne absence of radiation in the range between 300 and 1,100 millimicrons. A planet without this range of radiation would virtually lack photochemistry. It would have a relatively inert surface, upon which organic molecules could accumulate only exceedingly slowly. Granted even enough time for such accumulation, and granted that eventually primitive living organisms might form, what then? They could live for a time on the accumulated organic matter. But without the possibility of photosynthesis how could they ever become independent of this geological heritage and fend for themselves? Inevitably they must eventually consume the organic molecules about them, and with that life must come to an end.

It may form an interesting intellectual exercise to imagine ways in which life might arise, and having arisen might maintain itself, on a dark planet; but I doubt very much that this has ever happened, or that it can happen.

Ultraviolet Radiation and Nucleic Acid

by R. A. Deering
December 1962

*The damaging effects of ultraviolet on living things have
long been known. Now they are being explained in terms
of specific changes in molecules of the genetic material*

Ever since the discovery in 1877 that ultraviolet radiation can kill bacteria, workers in several disciplines have been studying the effects of the radiation on living things. Its actions have turned out to be many and varied. Ultraviolet can temporarily delay cell division and can also delay the synthesis of certain substances by cells; it can change the way in which substances pass across the membranes of the cell; it can cause abnormalities in chromosomes; it can produce mutations. Obviously it is a potent tool for the study of living cells, and it has been extensively employed by experimenters. If its exact modes of action at the molecular level were fully understood, the tool would be even sharper and more useful. This article reports the considerable progress that has been made in the past few years toward understanding the biophysical and biochemical role of ultraviolet.

Most of the recent work has concentrated on the interaction of ultraviolet radiation and the molecule of the genetic material deoxyribonucleic acid (DNA), and that is what I shall discuss. There is no doubt that many of the effects of ultraviolet are exerted solely or chiefly by means of changes in DNA. The fact that DNA strongly absorbs ultraviolet,

and that its absorption spectrum resembles the ultraviolet "action spectrum" for many biological changes (that is, the biological effectiveness of various wavelengths), show that this must be true. Therefore DNA is the logical starting point in the investigation of the biological activity of ultraviolet radiation.

This radiation falls between visible light and X rays in the spectrum of electromagnetic waves, ranging in wavelength from about 4,000 to a few hundred angstrom units. (An angstrom unit is one hundred-millionth of a centimeter.) The important wavelengths for the biologist are those between 2,000 and 3,000 angstroms. The sun is a powerful emitter of ultraviolet, but a layer of ozone in the upper atmosphere absorbs most of the radiation below 2,900 angstroms. Were it not for the ozone, sunlight would damage or kill every exposed cell on earth.

In the laboratory, working with monochromatic ultraviolet radiation at various wavelengths, investigators have established that the region most potent in its effects on living things is near 2,600 angstroms. When DNA was isolated, it was found to absorb most strongly at just these wavelengths. In the past five years workers in several laboratories

have begun to discover what happens to the DNA molecule when it absorbs ultraviolet energy.

Natural DNA, as the readers of this magazine are well aware, normally consists of a double-strand helix. The helices proper—the twin "backbones" of the molecule—consist of an alternation of sugar (deoxyribose) and phosphate groups. Attached to each of the sugars is one of four nitrogenous "bases," generally adenine, guanine, thymine and cytosine. The bases on the two backbones are joined in pairs by hydrogen bonds, the adenine on one chain always being paired with thymine on the other, and the guanine with cytosine. The hydrogen bonds that join the base pairs are weaker than ordinary chemical bonds. Simply heating double-strand DNA breaks the bonds and partially or completely separates the two backbones into two strands of "denatured" DNA.

Ultraviolet radiation falling on DNA is absorbed primarily by the bases, which exhibit about the same absorption peak at 2,600 angstroms as the whole DNA molecule does. This being the case, the first approach was to study the effects of ultraviolet radiation on the isolated bases. It soon turned out that thymine and cytosine, which belong to the class

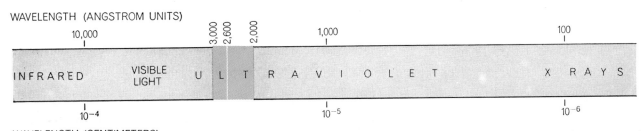

WAVELENGTH (ANGSTROM UNITS)

| 10,000 | | 3,000 2,600 | 2,000 | 1,000 | | 100 |

| INFRARED | VISIBLE LIGHT | U L T R A V I O L E T | | X R A Y S |

| 10^{-4} | | | 10^{-5} | | 10^{-6} |

WAVELENGTH (CENTIMETERS)

ULTRAVIOLET portion of the electromagnetic spectrum lies between visible light and X rays. The wavelengths between 2,000 and 3,000 angstrom units are of primary biological importance. DNA, the genetic material, absorbs most strongly at 2,600 angstroms.

of substances called pyrimidines, are far more sensitive to ultraviolet than are adenine and guanine, which are purines. About one in every 100 quanta of ultraviolet energy absorbed by pyrimidines alters the molecules; for purines the ratio is one in 10,000. (In general only a few of the quanta absorbed by a molecule will be effective in producing permanent changes.) The search was therefore narrowed to the pyrimidines.

The first effect to be discovered was that ultraviolet acts on cytosine molecules or the cytosine units of DNA in water solution, adding a water molecule across a double bond [*see middle illustration, page 122*]. Heating the altered cytosine, even to the temperatures required for biological growth, or acidifying it, partly reverses the reaction. Therefore the hydration of cytosine did not seem likely to be of major biological importance.

For some years, however, this hydration was the only sensitive, ultraviolet-induced change in the bases that could be detected. Heavy doses of radiation did produce complex rearrangements, but these doses were far in excess of the smallest ones known to have biological effects. About three years ago a breakthrough in the photochemistry of DNA came when R. Beukers, J. Ijlstra and W. Berends of the Technological

University of Delft in Holland and Shih Yi Wang, now of Johns Hopkins University, discovered that although in a liquid solution thymine is not particularly sensitive to ultraviolet, in a concentrated, frozen aqueous solution it is extremely sensitive. It developed that irradiation of the frozen solution causes thymine molecules to combine and form two-molecule chains, or dimers. As in the case of the cytosine conversion, a double bond changes to a single, and new bonds between carbon atoms link the two thymines [*see bottom illustration, page 122*]. Unlike the altered cytosine, the thymine dimer is stable to heat and acid. But when the solution is melted, irradiation can convert the dimer back into the two original thymine molecules. What the freezing does is to hold the thymines close together in a crystalline or semicrystalline configuration, making it possible for the dimer bonds to form between two neighboring thymines when they absorb ultraviolet. It seemed likely that such a conversion would also occur in DNA, where thymine units are sometimes adjacent to each other on a helical strand and are held in relatively fixed positions. In 1960 Adolf Wacker and his associates at the University of Frankfort found thymine dimers in DNA extracted from irradiated bacteria.

In order to get more complete information on the formation and splitting

of thymine dimers in polymer chains such as DNA, Richard B. Setlow and I carried out experiments on some model polymers at the Oak Ridge National Laboratory. Similar experiments were performed independently at the California Institute of Technology by Harold Johns and his collaborators. The compounds we used were short polymers—in effect short single strands of DNA in which all the bases were thymine. Some of our test molecules contained only two backbone units and two thymines; others had 12 or more. Since the sugar-phosphate backbone holds the thymines in fairly close proximity, we anticipated that ultraviolet radiation should form dimers between adjacent thymines in a chain even in a liquid solution. And we expected that once the dimers had formed they would be subject to breakage by ultraviolet, as were the isolated thymine dimers. When thymine loses a double bond in changing to a dimer, it also loses its ability to absorb light at 2,600 angstroms. Therefore measuring the change in 2,600-angstrom absorption gives an indication of the ratio between thymine monomers and thymine dimers in the solution.

When we irradiated our polymers, dimers were in fact produced. Since the rate of formation did not vary with thymine concentration, we concluded

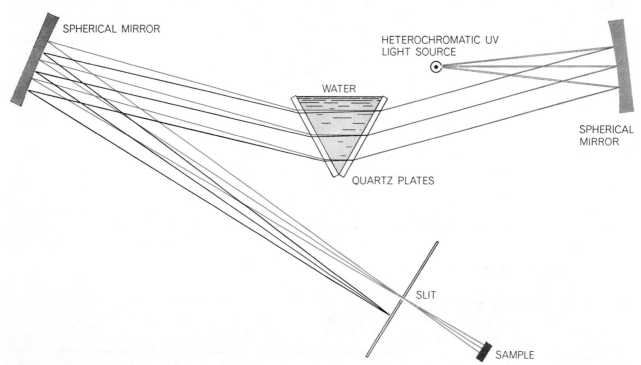

SPHERICAL MIRROR

HETEROCHROMATIC UV LIGHT SOURCE

WATER

SPHERICAL MIRROR

QUARTZ PLATES

SLIT

SAMPLE

MONOCHROMATOR provides ultraviolet light of a single wavelength for experiments. Light of mixed wavelengths is rendered parallel by a spherical mirror and passes through a quartz-and-water prism. (Glass would not transmit the desired wavelengths.) The prism splits the light into many components of different wavelengths, only two of which are indicated here, and the beams are refocused by a second mirror. The sample to be irradiated is positioned behind a slit that excludes all but the desired wavelength.

that they were formed within, rather than between, individual polymers. Dimers also broke up into monomers, but not at the same rate at which they formed. The process is analogous to a reversible chemical reaction in which forward and backward reactions proceed at different rates, with an equilibrium eventually being reached between the reactants and the products. We found that for every wavelength of ultraviolet there is at high doses an equilibrium between the number of dimers being formed and the number being broken [*bottom left illustration, page 123*]. At each wavelength and intensity there is a certain rate for dimer formation and a different one for breakage; the equilibrium level is determined by the relative rates of the forward and backward reactions. At 2,800 angstroms the equilibrium state is on the dimer side: most of the thymines are dimerized. At 2,400 angstroms the opposite is true: most of the thymines are monomers. The relative number of monomers and dimers in the polymer solutions can be controlled by changing the wavelength of the incident ultraviolet.

When the data from a number of experiments are plotted [*see bottom right illustration, page 123*], the resulting curves show the ability of each wavelength to make and break dimers in these model polymers. The curves approximately parallel the absorption spectra of the monomer and dimer respectively, indicating that it is difference in absorption capacity that accounts for the different action of various wavelengths. The "quantum yield," or number of molecules altered by each quantum, does not change greatly with wavelength; for dimer formation in the polymers containing only thymine it is of the order of .01 and for breakage it is near 1.

The next step was to relate molecular changes to changes in the properties of DNA and in its biological activity. Julius Marmur and Lawrence Grossman of Brandeis University have shown recently that when double-strand DNA is exposed to ultraviolet, the two strands become more strongly linked, apparently by chemical bonds rather than by the original weak hydrogen bonds. Marmur and Grossman believe the strong link is the result of interchain dimerization, that is, the formation of dimers between thymine units on opposite strands of the double helix.

At Oak Ridge, Frederick J. Bollum and Setlow found that ultraviolet can induce dimer linkages between adjacent thymine units in single-strand DNA.

They suspect that the same thing can happen between adjacent thymines in natural DNA, but in this case some of the hydrogen bonds in a local region may have to be broken before dimerization is possible. Marmur and Grossman have shown that irradiation does indeed disrupt hydrogen bonding between strands of natural DNA.

Another effect of ultraviolet on isolated DNA that has been clearly identified is a breaking of the sugar-phosphate backbone, but this occurs only at uninterestingly high doses. Among the sensitive reactions only the cytosine and thymine conversions are understood well enough for their biological implications to be assessed. There are surely other important effects, but they remain to be discovered.

Although the biological significance of the cytosine hydration has generally been discounted because it reverses at body temperature or lower, the reversal may be slower in intact DNA than in the isolated base. There is no direct evidence that the hydration product would be detrimental to the biological activity of DNA, but it might affect the hydrogen bonding in a segment of the helix and thereby give rise to the broken bonds observed by Marmur and Grossman.

The formation of thymine dimers should in theory be of great biological significance. When DNA makes a replica of itself, according to the widely accepted hypothesis, the hydrogen bonds break and a new complementary chain forms along each of the old strands. A dimer cross link between strands would interrupt the separation, blocking replication. Dimers between adjacent thymines on the same strand would interfere with proper pairing of the bases. Normally an adenine should come into position opposite each thymine on the parent strand. The joining of two adjacent thymines would probably change matters enough to impair the proper incorporation of adenine; replication might stop short at

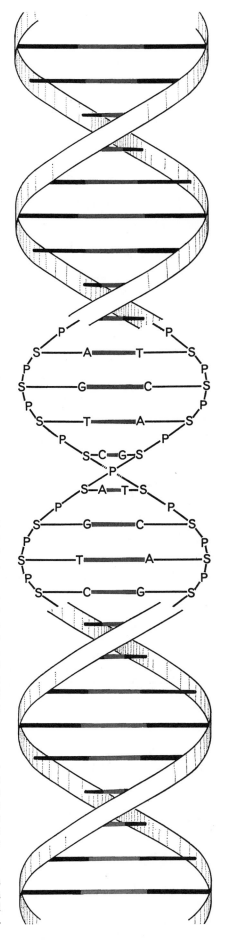

DNA MOLECULE is a double helix, diagramed here schematically. (One strand is actually displaced along the axis of the helix with regard to the other.) The backbone strands are composed of alternating sugar (S) and phosphate (P) groups. Attached to each sugar is one of four bases, usually adenine (A), guanine (G), thymine (T) and cytosine (C). Hydrogen bonds (gray) between bases link the strands. Adenine is always paired with thymine, guanine with cytosine. Genetic information is provided by the sequence of bases along a strand.

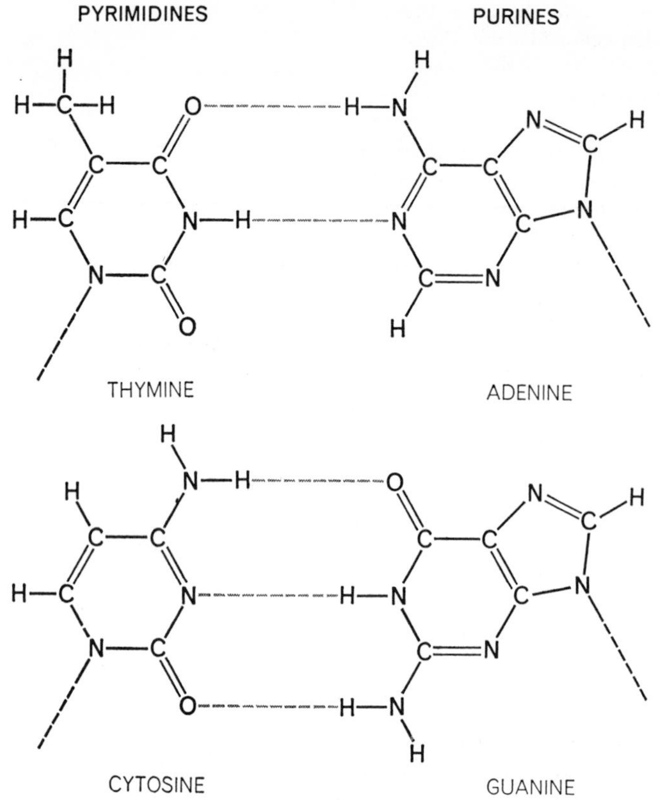

122

PYRIMIDINES   PURINES

THYMINE   ADENINE

CYTOSINE   GUANINE

FOUR BASES are diagrammed as they are paired in DNA. Adenine and guanine, the larger molecules, are purines; thymine and cytosine are pyrimidines. The broken black lines show points of attachment to sugar groups; the broken gray lines are interchain hydrogen bonds.

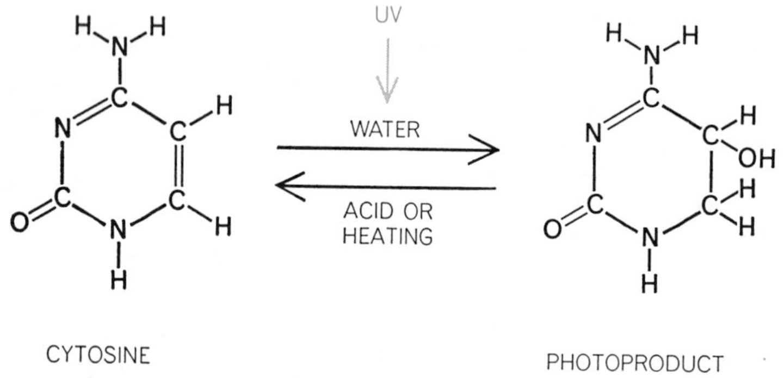

UV

WATER

ACID OR HEATING

CYTOSINE   PHOTOPRODUCT

CYTOSINE in a water solution is altered by irradiation with ultraviolet. A water molecule is added across the double bond between two carbon atoms, the double bond changing to a single bond. When the cytosine solution is heated or acidified, the process is reversed.

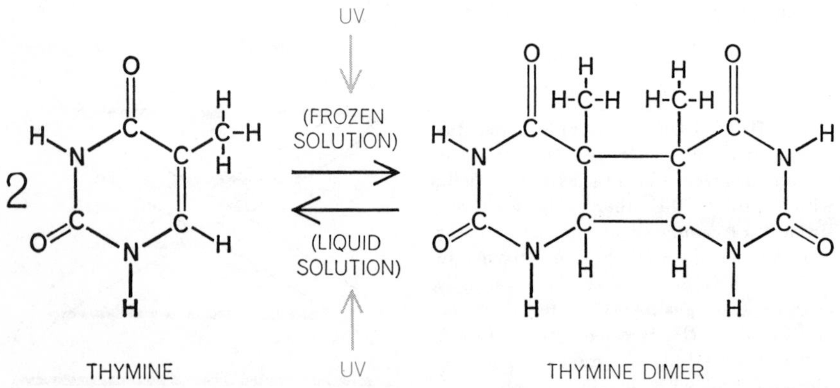

UV

(FROZEN SOLUTION)

(LIQUID SOLUTION)

UV

THYMINE   THYMINE DIMER

THYMINE in a frozen solution undergoes the reaction shown here when it is irradiated. The double bond between carbon atoms changes to single and two thymines are linked in a double molecule, or dimer. When the solution is melted, irradiation breaks the dimer.

that point or might proceed incorrectly, with an altered base sequence on the newly formed chain. On subsequent replication this altered strand would replicate itself, producing a molecule with the wrong base sequence in both strands—in other words, a mutated gene.

Recent work at Oak Ridge has provided direct experimental proof that thymine dimerization is one of the important ways in which the biological activity of DNA is altered by ultraviolet. Setlow and Bollum studied the ability of irradiated single strands of DNA to serve as a template in the manufacture of new DNA in a variety of cell-free test-tube preparations. Irradiation at 2,800 angstroms cut down the priming ability of DNA, the reduction being proportional to the adenine-thymine content of the various preparations. Subsequent irradiation at 2,400 angstroms partially restored template activity. Presumably irradiation at 2,800 angstroms formed dimers between adjacent thymines on the template DNA, blocking or slowing down the normal synthesis of new DNA strands. Irradiation at 2,400 angstroms evidently broke some of the dimers, partially restoring template activity.

In another series of experiments Setlow and his wife Jane K. Setlow worked with a form of DNA called "transforming principle," studying its ability to carry specific bits of genetic information from one cell to another. The measure of the biological activity of the DNA in this case was its effectiveness in transforming a given trait in the new cell. The Setlows found that irradiation at 2,800 angstroms destroyed the transforming ability of a portion of the DNA molecules. Again, when the irradiated DNA was exposed to 2,400-angstrom radiation, some of its molecules regained their transforming ability. The experimenters could account quantitatively for their results by assuming that about 50 per cent of the inactivation of the transforming DNA was due to thymine dimerization. They do not know what changes account for the rest.

When some types of cells that have been damaged by ultraviolet are exposed to ordinary blue light, a great deal of the damage is reversed; even bacteria that appear to have been killed are revived [see "Revival by Light," by Albert Kelner; SCIENTIFIC AMERICAN, May, 1951]. Claud S. Rupert and his associates at Johns Hopkins University had shown that this photoreactivation takes place through light-mediated enzyme reactions, but the details were not known. Recently Daniel L. Wulff and

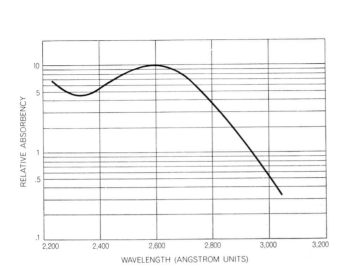

ABSORPTION SPECTRUM of DNA shows its ability to absorb ultraviolet radiation of various wavelengths. The peak is at 2,600 angstrom units, the wavelength known to be most harmful to cells.

MODEL POLYMERS exposed to ultraviolet were synthetic all-thymine DNA strands composed of either two nucleotide (phosphate-sugar-base) units (a) or 12 or more units (b). Irradiation both formed and broke dimers between adjacent thymines (c). Irradiation with high doses of the longer wavelengths led to an equilibrium condition in which most of the thymines were dimerized; exposure to shorter wavelengths tended to break the dimers.

Rupert have identified one mechanism of reactivation: a particular enzyme preparation, in the presence of blue light, can break up to 90 per cent of the thymine dimers in irradiated DNA. Marmur and Grossman have also shown that the enzyme system can break the ultraviolet-induced cross links between two DNA strands, thereby strengthening the idea that these links result from interchain thymine dimers.

Some bacteria apparently can produce enzymes that repair ultraviolet-damaged DNA without the need for visible light. The extreme resistance to ultraviolet displayed by certain bacteria may come from an ability to produce large amounts of these repair enzymes.

To sum up, it is clear that ultraviolet can change DNA in specific ways and can partially reverse those changes. Moreover, both the forward and backward alterations are reflected in DNA function. There is considerable evidence that much of the damage that ultraviolet radiation inflicts on cells and viruses is caused directly by its effects on DNA.

In the case of viruses this may be the whole story. When they infect a cell to

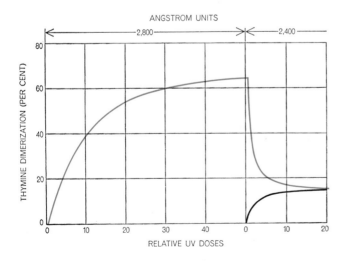

DEGREE OF DIMERIZATION varies with wavelength and dose. Increasing the exposure at 2,800 angstrom units increases the proportion of thymine units that are dimerized until an equilibrium state is attained with 65 per cent of the thymine units as dimers (colored curve). Irradiation at 2,400 angstrom units of the same sample, or of a different sample (black curve), results in a new equilibrium level with about 17 per cent of the thymines dimerized.

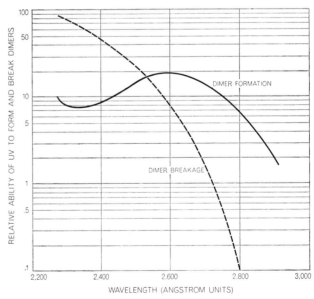

VARIOUS WAVELENGTHS differ in their ability to form and break dimers. Those over 2,540 angstrom units are more able to form dimers; the reverse is true of the shorter wavelengths.

replicate themselves, only their nucleic acid is injected into the cell. Therefore damage to DNA must be directly reflected in the ability of the virus to take over the cell's metabolism and multiply. In cells, however, one cannot assume that the only important effects of ultraviolet are those involving DNA. The ra-

diation is absorbed by proteins, by ribonucleic acid (RNA) outside the nucleus and by other substances that play a part in cell metabolism, and it presumably changes their structures too. The task of identifying all the ultraviolet reactions in living organisms has been well begun, but there is much to learn.

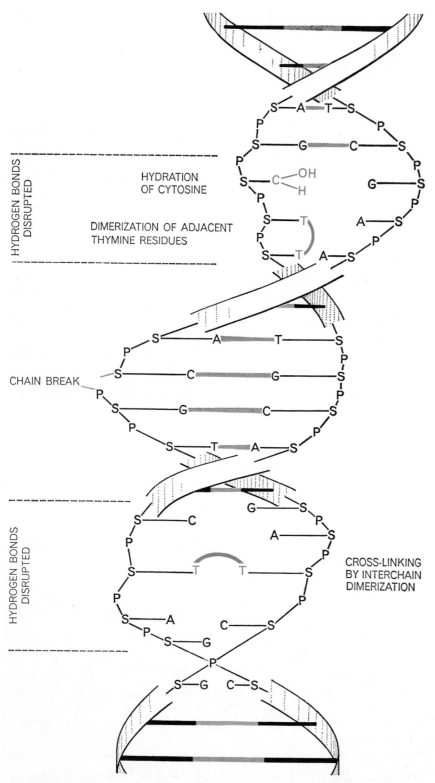

ALTERATIONS IN DNA caused by ultraviolet irradiation are diagramed here. The formation of thymine dimers is the change most likely to do damage to living cells and viruses.

13

The Effects of Light on the Human Body

by Richard J. Wurtman

July 1975

Sunlight tans skin, stimulates the formation of vitamin D and sets biological rhythms. Light is also used in the treatment of disease. Such effects now raise questions about the role of artificial light

Since life evolved under the influence of sunlight, it is not surprising that many animals, including man, have developed a variety of physiological responses to the spectral characteristics of solar radiation and to its daily and seasonal variations. With the coming of summer in the Northern Hemisphere millions of people living in the North Temperate Zone will take the opportunity to darken the shade of their skin, even at the risk of being painfully burned. Coincidentally the sunbathers will replenish their body's store of vitamin D, the vitamin that is essential for the proper metabolism of calcium. Skin-tanning and subcutaneous synthesis of vitamin D from its precursors, however, are only the best-known consequences of exposure to sunlight.

Investigators are slowly uncovering subtler physiological and biochemical responses of the human body to solar radiation or its artificial equivalent. Within the past few years, for example, light has been introduced as the standard method of treatment for neonatal jaundice, a sometimes fatal disease that is common among premature infants. More recently light, in conjunction with a sen-

sitizing drug, has proved highly effective in the treatment of the common skin inflammation psoriasis. It seems safe to predict that other therapeutic uses for light will be found.

At least equally significant for human well-being is the growing evidence that fundamental biochemical and hormonal rhythms of the body are synchronized, directly or indirectly, by the daily cycle of light and dark. For example, my co-workers at the Massachusetts Institute of Technology and I have recently discovered a pronounced daily rhythm in the rate at which normal human subjects excrete melatonin, a hormone synthesized by the pineal organ of the brain. In experimental animals melatonin induces sleep, inhibits ovulation and modifies the secretion of other hormones. In man the amount of the adrenocortical hormone cortisol in the blood varies with a 24-hour rhythm. Although seasonal rhythms associated with changes in the length of the day have not yet been unequivocally demonstrated in human physiology, they are well known in other animals, and it would be surprising if they were absent in man. The findings already in hand suggest that light has an

important influence on human health, and that our exposure to artificial light may have harmful effects of which we are not aware.

The wavelengths of radiation whose physiological effects I shall discuss here are essentially those supplied by the sun after its rays have been filtered by the atmosphere, including the tenuous high-altitude layer of ozone, which removes virtually all ultraviolet radiation with a wavelength shorter than 290 nanometers. The solar radiation that reaches the earth's surface consists chiefly of the ultraviolet (from 290 to 380 nanometers), the visible spectrum (from 380 to 770 nanometers) and the near infrared (from 770 to 1,000 nanometers). About 20 percent of the solar energy that reaches the earth has a wavelength longer than 1,000 nanometers.

The visible spectrum of natural sunlight at sea level is about the same as the spectrum of an ideal incandescent source radiating at a temperature of 5,600 degrees Kelvin (degrees Celsius above absolute zero). The solar spectrum is essentially continuous, lacking only certain narrow wavelengths absorbed by elements in the sun's atmosphere, and at midday it has a peak intensity in the blue-green region from 450 to 500 nanometers [*see upper illustration on next page*]. The amount of ultraviolet radiation that penetrates the atmosphere varies markedly with the season: in the northern third of the U.S. the total amount of erythemal (skin-inflaming) radiation that reaches the ground in December is only about a fifteenth of the amount present in June. Otherwise there is little seasonal change in the spectral composition of the sunlight reaching the ground. The actual number of daylight hours, of course, can vary greatly, depending on the season and the distance north or south of the Equator.

SOME DIRECT AND INDIRECT EFFECTS OF LIGHT on the human body are outlined in the drawing shown on page 127. Indirect effects include the production or entrainment (synchronization) of biological rhythms. Such effects are evidently mediated by photoreceptors in the eye (1) and involve the brain and neuroendocrine organs. For example, excretion of melatonin, a hormone produced by the pineal organ, follows a daily rhythm. In animals melatonin synthesis is regulated by light. The hormone, acting on the pituitary, plays a role in the maturation and the cyclic activity of the sex glands. Ultraviolet radiation acts on the skin to synthesize vitamin D (2). Erythema, or reddening of the skin (3), is caused by ultraviolet wavelengths between 290 and 320 nanometers. In response melanocytes increase their synthesis of melanin (4), a pigment that darkens the skin. Simultaneously the epidermis thickens (5), offering further protection. In some people the interaction of light with photosensitizers circulating in the blood causes a rash (6). In conjunction with selected photosensitizers light can be used to treat psoriasis and other skin disorders (7). In infants with neonatal jaundice light is also used therapeutically to lower the amount of bilirubin circulating in the blood until infant's liver is mature enough to excrete the substance. The therapy prevents the bilirubin from concentrating in the brain and destroying brain tissue.

The most familiar type of artificial light is the incandescent lamp, in which the radiant source is a hot filament of tungsten. The incandescent filament in a typical 100-watt lamp has a temperature of only about 2,850 degrees K., so that its radiation is strongly shifted to the red, or long-wavelength, end of the spec- trum. Indeed, about 90 percent of the total emission of an incandescent lamp lies in the infrared.

Fluorescent lamps, unlike the sun and incandescent lamps, generate visible light by a nonthermal mechanism. With- in the glass tube of a fluorescent lamp ultraviolet photons are generated by a mercury-vapor arc; the inner surface of the tube is coated with phosphors, lumi- nescent compounds that emit visible radi- ations of characteristic colors when they are bombarded with ultraviolet photons. The standard "cool white" fluorescent lamp has been designed to achieve max- imum brightness for a given energy con- sumption. Brightness, of course, is a sub- jective phenomenon that depends on the response of the photoreceptive cells in the retina. Since the photoreceptors are most sensitive to yellow-green light of 555 nanometers, most fluorescent lamps are designed to concentrate much of their output in that wavelength region. It is possible, however, to make fluores- cent lamps whose spectral output close- ly matches that of sunlight [see lower il- lustration on this page].

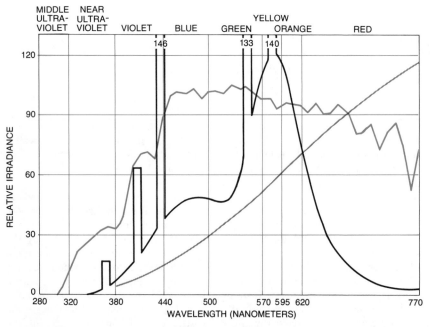

SPECTRUM OF SUN at sea level (color) is compared with the spectra of a typical incandes- cent lamp (gray curve) and of a standard "cool white" fluorescent lamp (black curve). The visible spectrum lies between the wavelengths of 380 and 770 nanometers. The peak of the sun's radiant energy falls in the blue-green region between 450 and 500 nanometers. Cool- white fluorescent lamps are notably deficient precisely where the sun's emission is strongest. Incandescent lamps are extremely weak in the entire blue-green half of the visible spectrum.

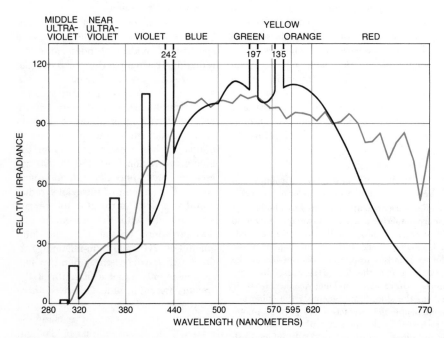

BROAD-SPECTRUM FLUORESCENT LAMP known as Vita-Lite (black curve) closely ap- proximates spectral characteristics of sunlight (color). Wavelengths emitted by fluorescent lamps can be adjusted by selecting phosphors with which inner surface of lamp is coated.

Since fluorescent lamps are the most widely used light source in offices, fac- tories and schools, most people in indus- trial societies spend many of their wak- ing hours bathed in light whose spec- tral characteristics differ markedly from those of sunlight. Architects and lighting engineers tend to assume that the only significant role of light is to provide ad- equate illumination for working and reading. The illumination provided at eye level in artificially lighted rooms is commonly from 50 to 100 footcandles, or less than 10 percent of the light nor- mally available outdoors in the shade of a tree on a sunny day.

The decision that 100 footcandles or less is appropriate for indoor purposes seems to be based on economic and technological considerations rather than on any knowledge of man's biological needs. Fluorescent lamps could provide higher light intensities without exces- sive heat production, but the cost of the electric power needed for substan- tially higher light levels would probably be prohibitive. Nevertheless, the total amount of light to which a resident of Boston, say, is exposed in a convention- ally lighted indoor environment for 16 hours a day is considerably less than would impinge on him if he spent a sin- gle hour each day outdoors. If future studies indicate that significant health benefits (for example better bone min- eralization) might accrue from increas- ing the levels of indoor lighting, our so- ciety might, in a period of energy short- ages, be faced with hard new choices.

Each of the various effects of light on mammalian tissues can be classi- fied as direct or indirect, depending on whether the immediate cause is a photo- chemical reaction within the tissue or a neural or neuroendocrine signal generat-

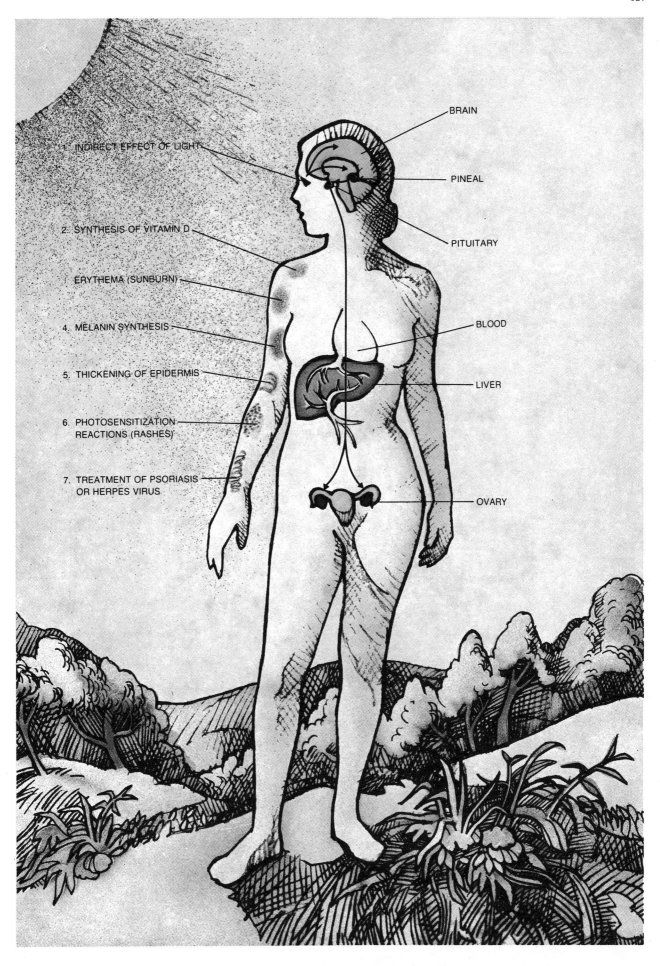

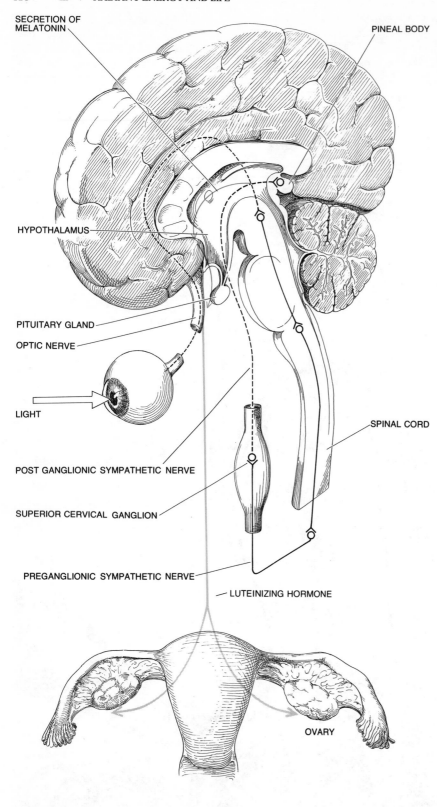

SECRETION OF
MELATONIN

PINEAL BODY

HYPOTHALAMUS

PITUITARY GLAND

OPTIC NERVE

LIGHT

POST GANGLIONIC SYMPATHETIC NERVE

SUPERIOR CERVICAL GANGLION

SPINAL CORD

PREGANGLIONIC SYMPATHETIC NERVE

LUTEINIZING HORMONE

OVARY

INDIRECT EFFECT OF LIGHT ON OVARIES OF RATS is shown schematically. Light activates receptors in the retina, giving rise to nerve impulses that travel via a chain of synapses through the brain, the brain stem and the spinal cord, ultimately decreasing the activity of neurons running to the superior cervical ganglion (in the neck) and of the sympathetic nerves that reenter the cranium and travel to the pineal organ. There the decrease in activity reduces both the synthesis and the secretion of melatonin. With less melatonin in blood or cerebrospinal fluid, less reaches brain centers (probably in hypothalamus) on which melatonin acts to suppress secretion of luteinizing hormone from anterior pituitary. Thus more hormone is released, facilitating ovarian growth and presumably ovulation.

ed by a photoreceptor cell. When the effect is direct, the molecule that changes may or may not be the one that actually absorbs the photon. For example, certain molecules can act as photosensitizers: when they are raised to transient high-energy states by the absorption of radiation, they are able to catalyze the oxidation of numerous other compounds before they return to the ground state. Photosensitizers sometimes present in human tissues include constituents of foods and drugs and of toxins produced in excess by some diseases.

In order to prove that a particular chemical change in a tissue is a direct response to light one must show that light energy of the required wavelength does in fact penetrate the body to reach the affected tissue. In addition the photoenergetic and chemical characteristics of the reaction must be fully specified, first in the test tube, then in experimental animals or human beings, by charting the reaction's "action spectrum" (the relative effectiveness of different spectral bands in producing the reaction) and by identifying all its chemical intermediates and products. Visible light is apparently able to penetrate all mammalian tissues to a considerable depth; it has even been detected within the brain of a living sheep.

Ultraviolet radiation, which is far more energetic than visible wavelengths, penetrates tissues less effectively, so that erythemal radiations barely reach the capillaries in the skin. The identification of action spectra for the effects of light on entire organisms presents major technical problems: few action spectra have been defined for chemical responses in tissues other than the skin and the eyes.

The indirect responses of a tissue to light result not from the absorption of light within the tissue but from the actions of chemical signals liberated by neurons or the actions of chemical messengers (hormones) delivered by circulation of the blood. These signals in turn are ultimately the result of the same process as the one that initiates vision: the activation by light of specialized photoreceptive cells. The photoreceptor transduces the incident-light energy to a neural signal, which is then transmitted over neural, or combined neural-endocrine, pathways to the tissue in which the indirect effect is observed. For example, when young rats are kept continuously under light, photoreceptive cells in their retina release neurotransmitters that activate brain neurons; these neurons in turn transmit signals over complex neuroendocrine pathways that reach the anterior pituitary gland, where they stimulate the secretion of the gonadotropic hormones that accelerate the maturation of the ovaries [*see illustration on opposite page*].

That the ovaries are not responding directly to light can be shown by removing the eyes or the pituitary gland of the rat before exposing it to continuous light. After either procedure light no longer has any influence on ovarian growth or function. Various studies confirm that the effect of light on the ovaries is mediated by photoreceptive cells in the retina. It has not been possible to show, however, which of the photoreceptors in the eye release the neurotransmitters that ultimately affect the pituitary gland.

Natural sunlight acts directly on the cells of the skin and subcutaneous tissues to generate both pathological and protective responses. The most familiar example of a pathological response is sunburn; in susceptible individuals exposed over many years sunlight also causes a particular variety of skin cancer. The chief protective response is tanning. Ultraviolet wavelengths in the narrow band from 290 to 320 nanometers cause the skin to redden within a few hours of exposure. Investigators generally agree that the inflammatory reaction, which may persist for several days, results either from a direct action of ultraviolet photons on small blood vessels or from the release of toxic compounds from damaged epidermal cells. The toxins presumably diffuse into the dermis, where they damage the capillaries and cause reddening, heat, swelling and pain. A number of compounds have been proposed as the offending toxins, including serotonin, histamine and bradykinin. Sunburn is largely an affliction of industrial civilization. If people were to expose themselves to sunlight for one or two hours every day, weather permitting, their skin's reaction to the gradual increase in erythemal solar radiation that occurs during late winter and spring would provide them with a protective layer of pigmentation for withstanding ultraviolet radiation of summer intensities.

Immediately after exposure to sunlight the amount of pigment in the skin increases, and the skin remains darker for a few hours. The immediate darkening probably results from the photooxidation of a colorless melanin precursor and is evidently caused by all the wavelengths in sunlight. After a day or two, when the initial response to sunlight has subsided, melanocytes in the epidermis begin to divide and to increase their synthesis of melanin granules, which are then extruded and taken up into the adjacent keratinocytes, or skin cells [*see illustration at left*]. Concurrently accelerated cell division thickens the ultraviolet-absorbing layers of the epidermis. The skin remains tan for several weeks and offers considerable protection against further tissue damage by sunlight. Eventually the keratinocytes slough off and the tan slowly fades. (In the U.S.S.R. coal miners are given suberythemal doses of ultraviolet light every day on the theory that the radiation provides protection against the development of black-lung disease. The mechanism of the supposed protective effect is not known.)

In addition to causing sunburn and tanning, sunlight or its equivalent initiates photochemical and photosensitization reactions that affect compounds present in the blood, in the fluid space between the cells or in the cells themselves. A number of widely prescribed drugs (such as the tetracyclines) and constituents of foods (such as riboflavins) are potential photosensitizers. When they are activated within the body by light, they may produce transient intermediates that can damage the tissues in sensitive individuals. A typical response is the appearance of a rash on the parts of the body that are exposed to the sun.

In individuals with the congenital disease known as erythropoietic protoporphyria unusually large amounts of porphyrins (a family of photosensitizing chemicals) are released into the bloodstream as a result of a biochemical ab-

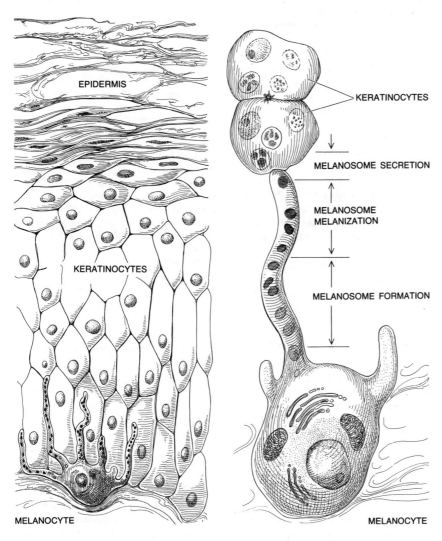

EPIDERMIS

KERATINOCYTES

KERATINOCYTES

MELANOSOME SECRETION

MELANOSOME MELANIZATION

MELANOSOME FORMATION

MELANOCYTE

MELANOCYTE

MECHANISM OF SUN-TANNING is an extension of the mechanism responsible for skin pigmentation. After exposure to the sun melanocytes begin to divide and increase their output of melanin granules, produced in the tiny intracellular bodies called melanosomes. The melanosomes are secreted into the adjacent keratinocytes, or skin cells, where the melanin causes the skin to take on a darker appearance. The tan fades as the keratinocytes slough off.

normality. The porphyrins absorb visible radiations and give rise to intermediates that are toxic to tissues. Patients with the disease complain at first of a burning sensation in areas of the skin that are exposed to sunlight; reddening and swelling soon follow.

Investigators can easily induce these typical symptoms without serious consequences in patients suffering from mild forms of erythropoietic protoporphyria, so that the disease is one of the few of its kind where the action spectrum for a direct effect of light has been studied in detail. The skin damage is caused by a fairly narrow band of wavelengths in the region of 400 nanometers. This band has also been shown to coincide with one of the absorption peaks of abnormal porphyrins. The symptoms of the disease can be ameliorated by administering photoprotective agents such as carotenoids, which quench the excited states of oxygen produced as intermediates in the photosensitization reactions.

In the past few years physicians have treated several skin diseases by deliberately inducing photosensitization reactions on the surface of the body or within particular tissues. The intent is to cause selective damage to invading organisms (such as the herpes virus), to excessively proliferating cells (as in psoriasis) or to certain types of malignant cells. The activated photosensitizers appear to be capable of inactivating the DNA in the viruses or in the unwanted cells. In treating herpes infections the photosensitizer (usually a dye, neutral red) is applied directly to the skin or to the mucous membrane under the ruptured blister; the area is then exposed to low-intensity white fluorescent light.

The treatment for psoriasis was devised by John A. Parrish, Thomas B. Fitzpatrick and their colleagues at the Massachusetts General Hospital. They administer a special photosensitizer (8-methoxypsoralen, or methoxalen) by mouth and two hours later expose the afflicted skin areas for about 10 minutes to the radiation from special lamps that emit strongly in the long-wave ultraviolet at about 365 nanometers. The sensitizing agent is present in small amounts in carrots, parsley and limes. It is derived commercially from an Egyptian plant (*Ammi majus* Linn.) that was used in ancient times to treat skin ailments. Scores of patients have responded successfully to the new light treatment, which will soon be generally available.

The formation of vitamin D_3, or cholecalciferol, in the skin and subcutaneous tissue is the most important of the beneficial effects known to follow exposure to sunlight. Vitamin D_3 is formed when ultraviolet radiation is absorbed by a precursor, 7-dehydrocholesterol. A related biologically active compound, vitamin D_2, can be obtained by consuming milk and other foods in which ergosterol, a natural plant sterol, has been converted to vitamin D_2 by exposure to ultraviolet radiation. Although vitamin D_2 can cure rickets in children who are deficient in vitamin D_3, it has not been demonstrated that vitamin D_2 is biologically as effective as the vitamin D_3 formed in the skin.

In a population of normal white adults living in St. Louis, studied by John G. Haddad, Jr., and Theodore J. Hahn of the Washington University School of Medicine, some 70 to 90 percent of the vitamin D activity in blood samples was found to be accountable to vitamin D_3 or its derivatives. The investigators concluded that sunlight was vastly more important than food as a source of vitamin D. (Although vitamin D_3 is also found in fish, seafood is not an important source in most diets.) In Britain and several other European countries the fortification of foods with vitamin D_2 has now been sharply curtailed because of evidence that in large amounts vitamin D_2 can be toxic, causing general weakness, kidney damage and elevated blood levels of calcium and cholesterol.

A direct study of the influence of light on the human body's ability to absorb calcium was undertaken a few years ago by Robert Neer and me and our coworkers. The study, conducted among elderly, apparently normal men at the Chelsea Soldiers' Home near Boston, suggests that a lack of adequate exposure to ultraviolet radiation during the long winter months significantly impairs the body's utilization of calcium, even when there is an adequate supply in the diet. The calcium absorption of a control group and an experimental group was followed for 11 consecutive weeks from the onset of winter to mid-March.

During the first period of seven weeks, representing the severest part of the winter, all the subjects agreed to remain indoors during the hours of daylight. Thus both groups were exposed more or less equally to a typical low level of mixed incandescent and fluorescent lighting (from 10 to 50 footcandles). At the end of the seven weeks the men in both groups were found to absorb only about 40 percent of the calcium they ingested. During the next four-week period, from mid-February to mid-March, the lighting was left unchanged for the control subjects, and their ability to ab-

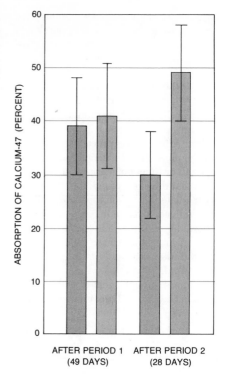

CALCIUM ABSORPTION was increased by a daily eight-hour exposure to broad-spectrum artificial light in a study made by the author and his colleagues at a veterans' home. During the first seven weeks after the beginning of winter, control subjects (*gray bars*) and experimental subjects (*colored bars*) were equally exposed to the same low levels of typical indoor lighting. The bars at the left show their ability to absorb calcium at the end of the initial period. During the next four-week period conditions for the control subjects were unchanged; their ability to absorb calcium fell about 25 percent. The experimental subjects, who were exposed to 500 footcandles of broad-spectrum fluorescent light for eight hours per day for four weeks, showed an average increase of about 15 percent in their calcium absorption.

sorb calcium fell by about 25 percent. The men in the experimental group, however, were exposed for eight hours per day to 500 footcandles of light from special fluorescent (Vita-Lite) lamps, which simulate the solar spectrum in the visible and near-ultraviolet regions. In contrast with the control subjects' loss of 25 percent of their capacity to absorb calcium, the experimental group exhibited an increase of about 15 percent [*see illustration above*]. The additional amount of ultraviolet radiation received by the experimental subects was actually quite small: roughly equivalent to what they would get during a 15-minute lunchtime walk in the summer.

Our study indicates that a certain amount of ultraviolet radiation, whether it is from the sun or from an artificial source, is necessary for adequate calcium metabolism. This hypothesis receives support from a recent study conducted by Jean Aaron of the Mineral Metabolism Unit at the General Infirmary at Leeds in England, who found that undermineralization (osteomalacia) is far more prevalent in autopsy samples collected in England during the winter months than it is in samples collected during the summer. Thus it seems likely that properly designed indoor lighting environments could serve as an important public-health measure to prevent the undermineralization of bones among the elderly and others with limited access to natural sunlight.

Perhaps 25,000 premature American infants were sucessfully treated with light last year as the sole therapy for neonatal jaundice. The rationale for this remarkable treatment is as follows. When red blood cells die, they release hemoglobin, which soon degrades into the yellow compound bilirubin. An increase in the concentration of bilirubin in the blood, due to excessive production of the compound or to failure of the liver to remove it, gives the skin its characteristic jaundiced color.

A potentially dangerous form of hyperbilirubinemia afflicts from 15 to 20 percent of premature infants because their liver is physiologically immature; in some cases the amounts of bilirubin released into the bloodstream are also increased as a result of blood-type incompatibility or concurrent infections. In such infants the bilirubin, which is soluble in fat, becomes concentrated in certain parts of the brain, where it can destroy neurons, producing the clinical syndrome kernicterus (yellow nuclei). The toxicity of bilirubin is aggravated by other factors, such as anoxia, acidosis, low body temperature, low blood sugar, low blood protein and infection. The brain damage resulting from kernicterus is often irreversible; it can cause various degrees of motor and mental retardation, leading to cerebral palsy and even death.

All current therapies for neonatal hyperbilirubinemia are based on the hope that if the level of bilirubin in the blood plasma can be kept from reaching between 10 and 15 milligrams per 100 milliliters until the maturing liver is able to remove the offending substance, there will be no brain damage. One widely used therapy involves exchange transfusions, in which jaundiced blood from the infant is completely replaced with normal blood from the donor.

Some years ago it was discovered that bilirubin in solution could be bleached by light and thus destroyed; the nature of the photodecomposition products remains unknown. This observation prompted R. J. Cremer, P. W. Perryman and D. H. Richards, who were then working at the General Hospital at Rockford in England, to see if light might be effective in lowering the plasma bilirubin in infants suffering from hyperbilirubinemia. That possibility was supported by informal observations that newborn infants whose crib had been placed near an open window tended to show less evidence of jaundice than infants whose crib was less exposed to light. Perhaps sunlight was accelerating the destruction of bilirubin. If it was, it should be possible to reproduce the effect with artificial light.

The efficacy of light therapy was fully confirmed in a controlled study conducted by Jerold F. Lucey of the University of Vermont College of Medicine. The treatment consists in exposing jaundiced infants to light for three or four days, or until their liver is able to metabolize bilirubin. Although it was initially assumed that the light converted the bilirubin into nontoxic products that could be excreted, it now turns out that a major fraction of the excreted material is unchanged bilirubin itself. Hence it is at least conceivable that phototherapy has a direct beneficial effect on the liver and the kidneys.

Many questions remain concerning the mechanism of phototherapy for hyperbilirubinemia and the long-term effectiveness of that therapy in protecting infants against brain damage. Blue light is the most effective in decomposing pure solutions of bilirubin. In clinical tests, however, full-spectrum white light in almost any reasonable dosage (continuous, intermittent or in brief strong pulses) has proved effective in lowering plasma-bilirubin levels, regardless of the fraction of the radiant energy that falls in the blue region of the spectrum. Thus the mechanism by which light destroys bilirubin in infants may differ from the simple photochemical reaction that takes place in a test tube. For example, a photosensitization reaction, perhaps mediated by circulating riboflavin, may underlie the desirable effect. Another possibility is that the light may act on the

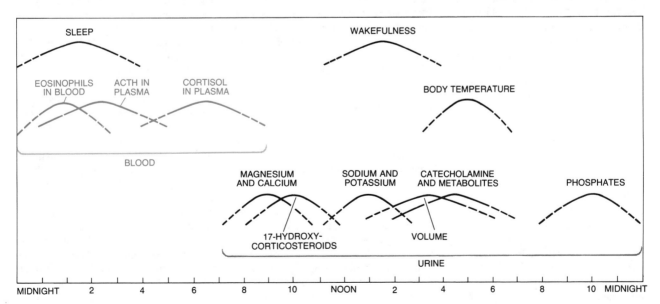

DAILY RHYTHMS are characteristic of many human physiological functions. Whether these 24-hour rhythms are produced by the daily light-dark cycle or are simply entrained by that cycle remains to be unequivocally established. Each curve represents the typical daily peak for a physiological state or for the levels of particular substances that circulate in the blood or are excreted in the urine.

plasma albumin to which most of the circulating bilirubin is bound. Alternatively the physiologically effective wavelength may be not in the blue region at all but in some other region of the spectrum, still to be identified, that is present in all the white-light sources.

The observation that ordinary sunlight or artificial light sources can drastically alter the plasma level of even one body compound (in this case bilirubin) opens a Pandora's box for the student of human biology. It presents the strong possibility that the plasma or tissue levels of many additional compounds are similarly affected by light. Some such responses must be physiologically advantageous, but some may not be.

Let us turn now to the indirect effects of light, those associated in one way or another with biological rhythms. The amount of time that all mammals are exposed to light varies with two cycles: the 24-hour cycle of day and night and the annual cycle of changing day length. (Even at the Equator there are small seasonal variations in the light-dark cycle.) These light cycles appear to be associated with many rhythmic changes in mammalian biological functions. Physical activity, sleep, food consumption, water intake, body temperature and the rates at which many glands secrete hormones all vary with periods that approximate 24 hours.

In human beings, for example, the concentration of cortisol, one of the principal hormones produced by the adrenal cortex, varies with a 24-hour rhythm [see illustration on opposite page]. The level is at a maximum in the morning hours, soon after waking, and drops to a minimum in the evening. When people reverse their activity cycle, by working at night and sleeping during the day, the plasma-cortisol rhythm takes from five to 10 days to adapt to the new conditions. When the cortisol level is studied in rats, it is found that the rhythm persists in animals that are blinded but not in animals kept under continuous illumination. Blindness in human beings seems to upset the rhythm, so that the times of the daily peaks and valleys are out of phase with the normal pattern and may even vary from day to day.

Among the rhythmic functions that can be closely studied in one and the same animal (specifically rhythms in sleep, physical activity and food consumption) it has been shown that in the absence of cyclic exposure to light the rhythms become "circadian" (that is, their periods become approximately 24

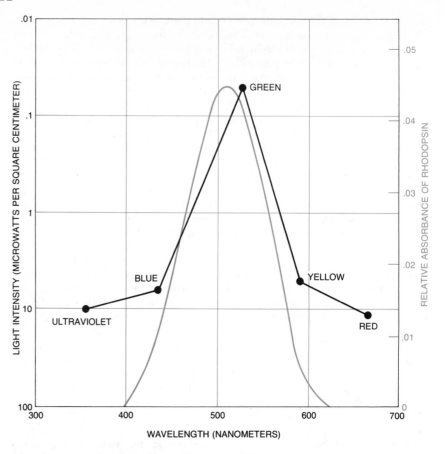

BODY-TEMPERATURE RHYTHM IN RATS, which follows a 24-hour cycle, can be altered by shifting the hours of the light-dark cycle. The author and his colleagues have found that green light is much more effective in establishing a new rhythm than radiation of other wavelengths (red, yellow, blue or ultraviolet). Black curve estimates the wavelength and intensity needed to establish a new 24-hour rhythm in half of an experimental group of rats. It closely follows the relative sensitivity of rhodopsin from rat retinas (curve in color).

hours in length rather than exactly). The fact that such rhythms can "free-run" suggests that they are not simply reflex responses to 24-hour cycles of light or some other environmental component. The factors responsible for the rhythms are not yet known; they might include other cyclic inputs such as the consumption of food or the intake of water, which also free-run in the absence of light. Some investigators are convinced that the rhythms are generated by intrinsic oscillators, commonly called biological clocks.

Little is known about the action spectra or the light intensities needed either to generate or to "entrain" (synchronize) daily rhythms in mammals. There is strong presumptive evidence, however, that in most mammals light exerts its effects indirectly through photoreceptors in the eye. It is not known whether the photoreceptors are the same ones (the rods and the cones) that mediate vision, discharging into nonvisual pathways, or

whether they are a distinct family of photoreceptors with their own neural network.

In our laboratory at M.I.T. we have investigated the daily rhythmicity in the body temperature of rats to see what colors of light are most effective in inducing a change in rhythm to a new light-dark cycle and what intensities are needed. The body temperature of rats normally rises by one or two degrees C. at the onset of darkness and falls again at daybreak. We found that green light is the most potent in changing the phase of the temperature cycle and that ultraviolet and red wavelengths are the least potent. The action spectrum plotted from these results closely follows the absorption spectrum for rhodopsin, the photosensitive pigment in the rods of the retina [see illustration above]. In separate studies a similar action spectrum, peaking in the green, was found for the wavelengths of light that are most effective in inhibiting the function of the pineal gland of rats [*see top illustration at right*].

Cycles in environmental lighting can interact with biological rhythms in two ways. The light cycle may directly induce the rhythm, in which case either continuous light or darkness should rapidly abolish it, or the cycle may simply entrain the biological rhythm so that all animals of a given species exhibit maximums or minimums at about the same time of day or night. In the latter case the rhythmicity itself may be generated by a cyclic input other than light, either exogenous (for example food intake) or endogenous (a biological clock). If the cycle is simply entrained by light, an environment of continuous light or darkness might not extinguish it. In human beings psychosocial factors are probably of greater importance than light cycles in generating or synchronizing biological rhythms. The biological utility of even so dramatic a rhythm as that of sleep and wakefulness, for example, remains to be discovered.

Annual rhythms in sexual activity, hibernation and migratory behavior are widespread among animals. The rhythms enable members of a species to synchronize their activities with respect to one another and to the exigencies of the environment. For example, sheep ovulate and can be fertilized only in the fall, thus anticipating the spring by many months, when food will be available to the mother for nursing the newborn. In man no annual rhythms have been firmly established, except, of course, those (such as in sun-tanning and vitamin D_3 levels) that are directly correlated with exposure to summer sunlight.

The best-characterized indirect effect of light on any process other than vision is probably the inhibition of melatonin synthesis by the pineal organ of mammals. Although melatonin seems to be the major pineal hormone, its precise role has not yet been established. When melatonin is administered experimentally, it has several effects on the brain: it induces sleep, modifies the electroencephalogram and raises the levels of serotonin, a neurotransmitter. In addition melatonin inhibits ovulation and modifies the secretion of other hormones from such organs as the pituitary, the gonads and the adrenals, probably by acting on neuroendocrine control centers in the brain.

Experiments performed on rats and other small mammals during the past decade provide compelling evidence that the synthesis of melatonin is suppressed by nerve impulses that reach the pineal

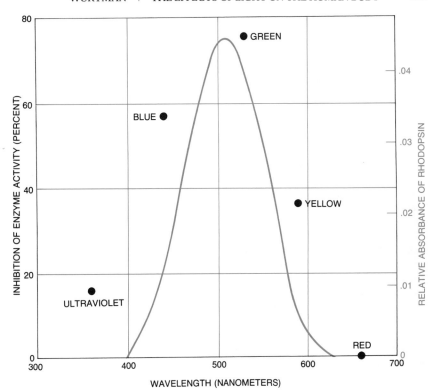

PINEAL ACTIVITY OF RATS can be suppressed by exposing the animals continuously to light. As in the case of the daily temperature rhythm, green light is more effective than light of other spectral colors in suppressing the organ's enzyme activity, as is shown by the labeled dots. The enzyme that is measured is the melatonin-forming enzyme hydroxyindole-O-methyltransferase. Presumably the suppression is mediated by rhodopsin (*curve in color*).

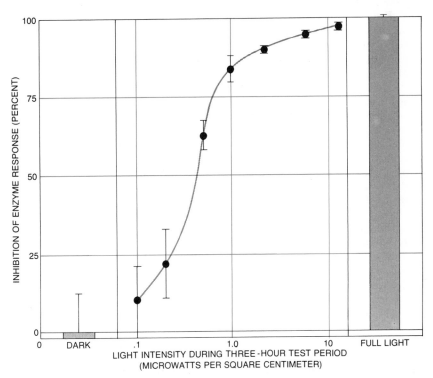

GRADATION IN LIGHT INTENSITY leads to proportional inhibition in the activity of the rat's pineal gland, indicating that light controls synthesis of the pineal hormone, melatonin. Rats that had been kept in constant light for 48 hours were exposed to various light intensities for the next three hours. Their pineals were then analyzed for serotonin-N-acetyltransferase, an enzyme that participates in melatonin synthesis, with the results plotted here.

over pathways of the sympathetic nervous system. These impulses in turn vary inversely with the amount of visible light impinging on the retina. In rats the pineal function is depressed to half its maximum level when the animals are subjected to an amount of white light only slightly greater than that shed by the full moon on a clear night [*see bottom illustration on preceding page*]. A multisynaptic neuronal system mediates the effects of light on the pineal. The pathway involved, which is apparently unique to mammals, differs from the route taken by the nerve impulses responsible for vision.

Quite recently Harry Lynch, Michael Moskowitz and I have found a daily rhythm in the rate at which normal human subjects excrete melatonin. During the third of the day corresponding to the bedtime hours, 11:00 P.M. to 7:00 A.M., the level of melatonin in the urine is much higher than it is in any other eight-hour period [*see illustration at right*]. It remains to be determined whether the rhythm in melatonin excretion in humans is induced by light or is simply entrained by it.

In some birds and reptiles the pineal responds directly to light, thereby serving as a photoreceptive "third eye" that sends messages about light levels to the brain. In the pineal organ of mammals any trace of a direct response to light is lost. Evidently photoreceptors in the retina mediate the control of the pineal by light. Since, as I have noted, the function of the pineal in rats is influenced most strongly by green light, corresponding to the peak sensitivity of the rod pigment rhodopsin, the retinal photoreceptor would seem to be a rod cell, at least in this species.

Light levels and rhythms influence the maturation and subsequent cyclic activity in the gonads of all mammals and birds examined so far. The particular response of each species to light seems to depend on whether the species is monestrous or polyestrous, that is, on whether it normally ovulates once a year (in the spring or fall) or at regular intervals throughout the year. Examples of polyestrous species are rats (ovulation every four or five days), guinea pigs (every 12 to 14 days) and humans (every 21 to 40 days). The gonadal responses also seem to depend on whether the members of the species are physically active during the daylight hours or during the night. Recently Leona Zacharias and I had the opportunity to examine more than a score of girls and women (members of a diurnally active polyestrous species) who had become blind in the

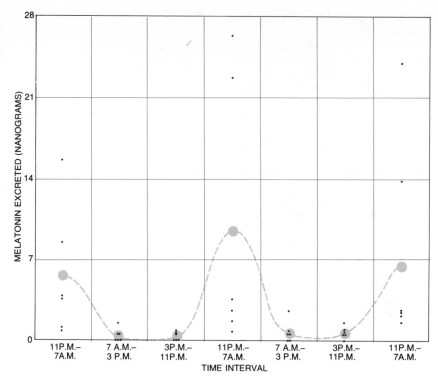

RHYTHM IN MELATONIN SECRETION in human beings has been found by the author and his colleagues. The black dots show the melatonin content of urine samples from six subjects during consecutive eight-hour periods. The colored circles and broken curve correspond to the mean values. High values that were recorded for the 11:00-P.M-to-7:00-A.M. samples suggest that synthesis of melatonin in man, as in rats, increases with onset of darkness.

first year of life. We observed that gonadal maturation had in general occurred earlier in this group than in normal girls. In contrast, in rats (a polyestrous species that is active at night) blindness delays maturation, and continuous illumination accelerates the maturation of weanlings with normal vision.

The gonads of most birds and of most diurnally active, monestrous animals (the ferret, for instance) mature in the spring, in response to the gradual increase in day length. Ovulation can be accelerated in such animals by exposing them to artificially long days. The annual gonadal activity in domestic sheep, on the other hand, occurs in the fall, in response to the decrease in day length. The mechanisms that cause some species to be monestrous and others polyestrous, or that cause some animals to sleep by day and others by night, are entirely unknown, as are the factors that cause the gonadal responses of various species to light to vary as widely as they do.

The multiple and disparate effects of light I have described support the view that the design of light environments should incorporate considerations of human health as well as visual and aesthetic concerns. We have learned that the chemical constituents of the environ-ment in the form of food, drugs and pollutants must be monitored and regulated by agencies with suitable powers of enforcement. A major part of their responsibility is to see that nothing harmful is put into food or drugs and that nothing essential is left out of food. The food and drug industries, for their part, look to public and private research organizations, including their own laboratories, for intellectual guidance in creating wholesome and beneficial (as well as profitable) products.

In contrast, only minuscule sums have been expended to characterize and exploit the biological effects of light, and very little has been done to protect citizens against potentially harmful or biologically inadequate lighting environments. Both government and industry have been satisfied to allow people who buy electric lamps—first the incandescent ones and now the fluorescent—to serve as the unwitting subjects in a long-term experiment on the effects of artificial lighting environments on human health. We have been lucky, perhaps, in that so far the experiment has had no demonstrably baneful effects. One hopes that this casual attitude will change. Light is potentially too useful an agency of human health not to be more effectively examined and exploited.

Light and Plant Development

14

by W. L. Butler and Robert J. Downs
December 1960

*Events in the life of a plant such as flowering are responses
to a change in the length of the night. The light-sensitive
enzyme that mediates these responses
has now been discovered*

Various kinds of plant germinate, grow, flower and fruit at different times in the year, each in its own season. Thus some plants flower in the spring, others in the summer and still others in the autumn. And in the autumn, trees and shrubs stop growing in apparent anticipation of winter, usually well before the weather turns cold. What is the nature of the clock or calendar that regulates these cycles in the diverse life histories of plants? Some 40 years ago it was discovered that the regulator is the seasonal variation in the length of the day and night. Since this is the one factor in the environment that changes at a constant rate with the change of the seasons, in retrospect the discovery does not seem so surprising. But how do plants detect the change in the ratio of daylight to darkness? The answer to this question is just now becoming clear. It appears that a single light-sensitive pigment, common to all plants, triggers one or another of the crises in plant growth, from the sprouting of the seed to the onset of dormancy, depending upon the plant species. This discovery is a major breakthrough toward a more complete understanding of the life processes of plants, and it places within reach a means for the artificial regulation of these processes.

The pigment has been called phytochrome by the investigators who discovered it at the Plant Industry Station of the U. S. Department of Agriculture in Beltsville, Md. It has been partially isolated, and it has been made to perform in the test tube what seems to be its critical photosensitive reaction: changing back and forth from one of its two forms to the other upon exposure to one or the other of two wavelengths of light that differ by 75 millimicrons. (A millimicron is a ten thousandth of a centimeter.) Phytochrome appears to be chemically active in one of its forms and inactive in the other. In the tissues of the plant it functions as an enzyme and probably catalyzes a biochemical reaction that is crucial to many metabolic processes.

The first step toward the discovery of phytochrome was taken in the 1920's, when W. W. Garner and H. A. Allard of the Department of Agriculture recognized "photoperiodism" [see "The Control of Flowering," by Aubrey W. Naylor; SCIENTIFIC AMERICAN Offprint 113]. They showed that many plants will not flower unless the days are of the right length—some species flowering when the days are short, some when the days are long. Indeed, some plants seem to react not simply to seasonal changes but to changes in the length of the day from one week to the next. Photoperiodism explained why plants of one type, even though planted at different times, always flower together, and why some plants do not fruit or flower in certain latitudes.

Other investigators soon reasoned that if a plant needs a certain length of day to flower, then keeping it in the dark for part of the day would inhibit its flowering. They tried to demonstrate such an effect in the laboratory. Nothing happened: the plants always bloomed in the proper season. The riddle was solved when the reverse experiment was tried, that is, when the night was interrupted with a brief interval of light. Chrysanthemums, poinsettias, soybeans, cockleburs and other plants that flower during the short days and long nights of autumn and early winter remained vegetative (*i.e.*, nonflowering). Moreover, they could be made to bloom out of season, when the night was lengthened by keeping them in the dark at the beginning or the end of a long summer day.

Conversely, a brief interval of light interrupting the long winter night induced flowering in petunias, barley, spinach and other plants that normally bloom in the short-night summer season. Artificial lengthening of the short summer night kept these plants vegetative. Interrupting or prolonging the nighttime darkness correspondingly affected stem growth and other processes as well as flowering in many plants.

It was evident that light must act upon a photoreceptive compound, or compounds, to set some mechanism that runs to completion in darkness. As a first step toward elucidating the chemistry of the process H. A. Borthwick, Marion W. Parker and Sterling B. Hendricks of the Department of Agriculture in 1944 set out to determine the wavelength or color of light that is most effective in inhibiting flowering in long-night plants. They exposed each of a series of Biloxi soybean plants, from which they had stripped all but one leaf, to different wavelengths of light from a large spectrograph [*as in middle illustration on following page*]. Several days after the treatment the plants were examined for the effect of this exposure upon the formation of buds. Red light with a wavelength of 660 millimicrons proved to be by far the most effective inhibitor of flowering. The cocklebur and other long-night plants gave the same response. By plotting on a graph the energy of light required to inhibit flowering at various wavelengths, the investigators obtained the "action spectrum" of the mechanism that inhibits flowering. This showed that to interfere with flowering, much more light energy is required at, for example, 520 or 700 millimicrons than at (or very near) 660 millimicrons [*see illustration at bottom of page 141*]. The wavelength at which the unknown substance absorbs light

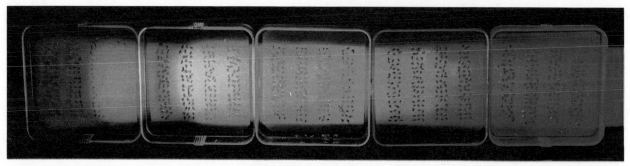

LIGHT FROM SPECTROGRAPH exposes lettuce seeds to different wavelengths. Only 2 per cent of those in far-red light at far left will sprout, while 90 per cent of those in red will germinate. In this photograph only a few of seeds in red region are visible.

CATALPA TREE SEEDLINGS, kept on short days and long nights, are exposed to this spectrum in the middle of the night. Those seen in red will grow. Seedlings in far-red light at far left and all of the others will stop growing and become dormant.

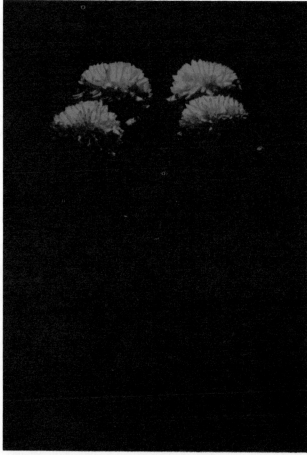

CHRYSANTHEMUM PLANTS at left have had long nights interrupted by period of red light. This divides night into two short dark periods, and plants will not bloom. Chrysanthemums at right also had nights interrupted by red light, but they were irradiated with far-red, as seen here. Far-red light reversed effects of red light, and the plants bloomed just as if nights had been uninterrupted.

most efficiently was thus shown to be 660 millimicrons.

Borthwick and his associates then turned to short-night plants. The same spectrographic experiments yielded exactly the same action spectrum. But in this case the effect of the exposure was to promote—not inhibit—flowering! Since the same wavelength of light caused the greatest response in both cases, the investigators could only conclude that a single photoreceptive substance was involved in these two diametrically opposed responses.

Subsequent experiments implicated the same compound in the control of stem elongation and leaf growth. Recent work has shown that light at 660 millimicrons also acts upon mature apples to turn them red by enabling them to make the pigment anthocyanin. The same red light controls the production of anthocyanin in a number of seedling plants.

With the collaboration of a research group headed by Eben H. Toole, Borthwick and his associates next set out to determine which wavelengths of light trigger germination in those seeds that must be exposed to light in order to grow. Many weed and crop seeds are of this type. The action spectrum for the promotion of seed germination turned out to be essentially the same as that established for other plant responses. Whereas about 20 per cent of the seeds germinated in the dark or when they were exposed to green, blue and other shorter-wavelength colors, more than 90 per cent sprouted after irradiation by red light at 660 millimicrons.

This finding led the two groups of investigators to the study of an entirely different effect of light upon germination. It had been observed in the late 1930's that germination was inhibited when seeds were exposed to the longer wavelengths of far-red light which are invisible to the human eye. That observation was speedily confirmed. In fact, seeds that had been pushed to maximum germinative capacity by exposure to red light failed to germinate when they were subsequently irradiated with far-red light. The plotting of the action spectrum for this effect showed that far-red light at 735 millimicrons wavelength is the most potent in inhibiting the germination of seeds.

Still more interesting was the discovery that the diametrically opposed effects of red and far-red light upon germination are fully reversible. After a series of alternate exposures to light of 660 and 735 millimicrons, the seeds re-

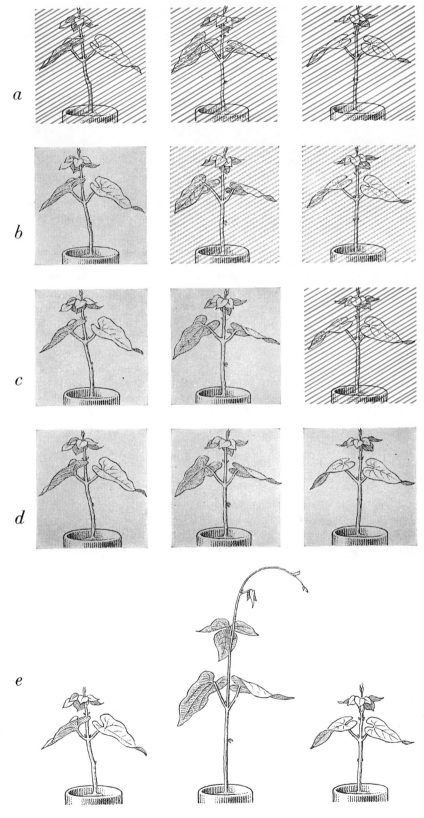

a

b

c

d

e

STEM ELONGATION in pinto-bean seedlings is promoted by exposure to far-red light on four successive evenings. All three plants are in red-irradiated condition at end of each day (*a*). Second and third plants are exposed briefly to far-red light (*b*); third plant is given dose of red light, which reverses effects of far red (*c*). Then all three have normal nights (*d*). Some days later first and third plants are still short, but center plant is tall (*e*).

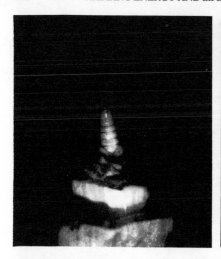

GROWING TIPS OF BARLEY, photographed through dissecting microscope at magnification of 20 diameters, show effects of long and short nights. Tip at left, dissected out from leaves, displays only leaf buds and little growth after being kept on long nights for two weeks. Plant at right had short nights; its tip has grown long and produced many tiny flower buds.

sponded to the light by which they were last irradiated. If it was red, they germinated; if far-red, they remained dormant [see illustrations at top of pages 141 and 142]. Now all of the phenomena growth and flowering had to be re-examined for the effect of far-red as well as of red light. In each case the experiments demonstrated that irradiation by far-red light reversed the effects obtained by irradiation with red light.

The reversibility of these reactions and the clear definition of their action spectra strongly suggested that a single light-sensitive substance is at work in every case, and that this substance exists in two forms. One form, which was designated phytochrome 660, or P_{660}, absorbs red light in the region of 660 millimicrons. When P_{660} is irradiated at this wavelength, it is transformed into phytochrome 735 (P_{735}), which absorbs far-red light at 735 millimicrons. When P_{735} is irradiated with light at 735 millimicrons, it reverts in turn to the P_{660} form.

In order to substantiate these deductions phytochrome had to be extracted from the plant. This required a method for detecting the presence of the compound other than the responses of a living plant. Since those responses occur at sharply defined wavelengths, there was reason to expect that phytochrome itself would prove to be more opaque, or "dense," to light at the wavelengths of 660 and 735 millimicrons when examined in a spectrophotometer—an instrument that measures the intensity of transmitted light at discrete wavelengths. The transformation of phytochrome from one form to the other would also show up well.

Measuring the very small amounts of phytochrome present in plants was not a simple task. K. H. Norris and one of the authors (Butler), respectively an engineer and a biophysicist in the Department of Agriculture, had been studying the pigment composition of intact plant-tissue, and had developed some sensitive spectrophotometers that measured the absorption of light by leaves and other plant parts. With Hendricks, a chemist, and H. W. Siegelman, a plant physiologist who had been investigating the chemistry of phytochrome, they formed a research group to look for the reversible pigment. Initially the absorption of light by plant parts failed to reveal the presence of phytochrome. The plant tissue, however, contained large amounts of chlorophyll, which absorbs strongly at 675 millimicrons. Apparently the absorption of light by chlorophyll—at a wavelength so near the phytochrome absorption peak of 660 millimicrons—was masking the absorption by phytochrome.

It was no great problem, however, to get around this obstacle. Seedlings can be sprouted in the dark and can grow for a while on the food energy stored in the seed; they do not begin to synthesize chlorophyll until they are exposed to the light. Corn seedlings were accordingly grown for several days in complete darkness. They were then chopped up and exposed to red light to put the phytochrome into the P_{735} form in which it absorbs far-red light. In the spectrophotometer a weak beam of light at 735 millimicrons was projected through the sample, weak light being used so that the phytochrome would not change

form. The absorption spectrum showed that the phytochrome was indeed absorbing light at 735 millimicrons; the same sample passed light at 660 millimicrons. On the other hand, after exposure to relatively bright far-red light the chopped seedlings were found to absorb more light at 660 millimicrons and less light at 735 millimicrons. Repeated demonstration of this reversibility fully confirmed all that had been predicted from the responses of whole, growing plants. No doubt remained that a single compound was responsible for the reversible changes in growing plants.

The spectrophotometer measures the changes in "optical density" with such high sensitivity that it can be used to assay the amount of phytochrome in plant tissue. Thus with the help of this instrument a search was instituted for a plant that would supply phytochrome in sufficient abundance for chemical separation. Certain plant tissues, such as the flesh and seed of the avocado and the head of the cauliflower, showed a relatively high concentration of phytochrome. The cotyledons (the first leaves,

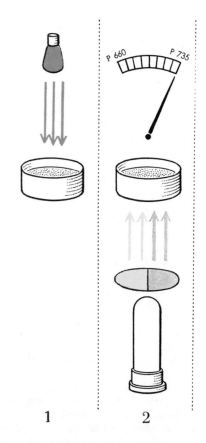

SPECTROPHOTOMETER TESTS, diagrammed here, show effects of red and far-red light on chopped corn seedlings in glass-bottomed dish. Dish is exposed to bright

which feed the seedling) of most leg-umes synthesize phytochrome, and the concentration reaches its maximum about five days after the seeds start soaking up water. The growing shoot, or hypocotyl, as well as the first leaves of the legumes, contain less phytochrome than the cotyledons. In cabbage and turnip seedlings that have been sprouted in the dark the cotyledons are also a good source of phytochrome. However, five-day-old, dark-grown corn seedlings proved to be the best source, because they develop a high phytochrome con-tent, have large stems and are easy to grow and harvest.

The preliminary and partial chemical isolation of phytochrome was easily ac-complished. Corn shoots were ground up in a blender along with water and a mild alkaline buffer, and a clear solution was separated from the solid material by fil-tration. This extract exhibits exactly the same reversible optical-density changes at 660 and 735 millimicrons as the chopped seedlings themselves. Thor-ough study of the partially purified ma-terial has developed no evidence that any compound other than phytochrome participates in the photoreaction.

Though phytochrome has not yet been isolated in pure form, the outlook is favorable. The compound shows all the properties of a relatively stable soluble protein, and Hendricks and Siegelman are now using the techniques of protein chemistry to purify it. They have sub-jected it to dialysis (diffusion through a porous membrane), and it has retained its photochemical activity. Oxidizing and reducing agents do not affect it. More-over, the photoconversion occurs at zero degrees centigrade as readily as at 35 degrees, as would be expected of a strictly photochemical reaction. Higher temperatures denature the protein and destroy its photochemical activity.

Measurements of the amount of red and far-red light necessary to bring about the conversion show that P_{660} consumes only one third as much energy in being transformed to P_{735} as P_{735} con-sumes in being changed into P_{660}. In both cases, however, the energy con-sumption is relatively small, indicating that the phytochrome absorbs light ef-ficiently. The pigment should turn out to be a blue or blue-green, the colors com-plementary to red, but the concentration achieved so far has been too low to make the color visible.

Experiments with growing plants had indicated that P_{735} slowly changes back into P_{660} in darkness, whereas red-ab-sorbing P_{660} is stable. Direct measure-ment of the changes in the form of phy-tochrome in intact corn seedlings have confirmed these indications. In seedlings that have never been exposed to light, phytochrome occurs entirely in the red-absorbing, or P_{660}, form. These seed-lings are exposed briefly to red light to convert the P_{660} to P_{735}. They are then returned to the darkroom and are exam-ined with the spectrophotometer at in-tervals thereafter. Such measurements show that it takes about four hours at room temperature for the P_{735} to change back into P_{660}.

This conversion in the absence of light is apparently mediated by enzymes. It is markedly retarded by lowering the temperature, and it does not occur at

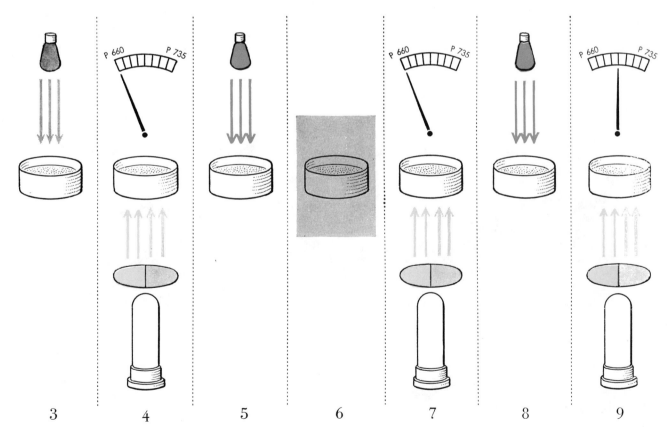

3   4   5   6   7   8   9

red light (1). Filters are used in spectrophotometer to project weak red and far-red light through dish (2); dial shows absorption at wavelength of 735 millimicrons. Now dish is exposed to bright far-red light (3). Next test (4) indicates phytochrome has changed to form that absorbs light at 660 millimicrons. After second ex-posure to bright red light (5) seedlings are placed in dark for four hours (6), and P_{735} again changes to P_{660} (7). Exposure to red light (8) shows that phytochrome lost half its activity in dark (9).

ROOTS OF AMERICAN HOLLY show response to light. Both cuttings were rooted during winter. That at right was kept on natu- ral days and nights. Long nights were interrupted with 30 foot- candles of light for plant at left; its roots grew prodigiously.

MATURE APPLES do not turn red if they are kept in the dark while ripening (right). Ethyl alcohol collects instead of red pig- ment. The apples can manufacture anthocyanin, the red pigment, only if they are exposed to light when they are mature (left).

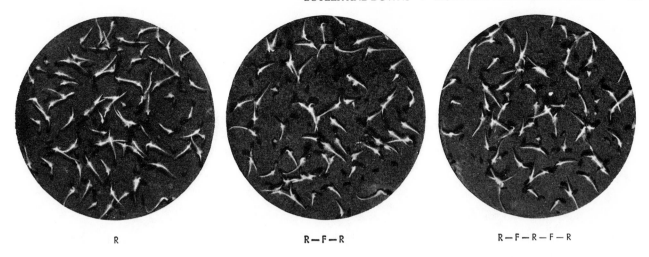

R   R—F—R   R—F—R—F—R

GERMINATION OF LETTUCE SEEDS placed on moist disks of blotting paper is promoted by exposure to red light. "R" indicates an exposure to red light; "F," to far-red light. The last type of light given in the series determines whether the seeds sprout.

all in the absence of oxygen. In the partially purified clear liquid extracts of seedlings, however, P_{735} is stable in the dark, indicating that the dark-conversion enzyme system has been removed. Half the phytochrome activity is lost in intact seedlings during the dark-conversion of P_{735} to P_{660}. After a second illumination with red light, total phytochrome activity declines still more. In continuous light, phytochrome activity is quite low, but is still detectable. It was fortunate that the presence of chlorophyll made it necessary to grow seedlings in darkness for the early experiments. If the seedlings had received even small amounts of light, the unstable P_{735} would have formed and would have soon lost its activity to such an extent that phytochrome might never have been detected.

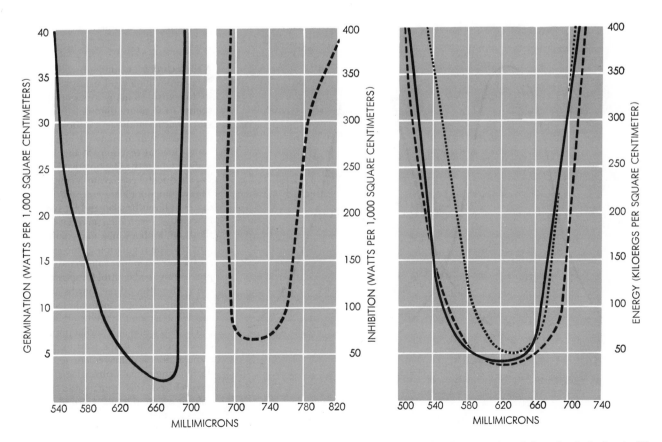

ACTION SPECTRA for the promotion (left) and inhibition (middle) of germination in lettuce seeds show energy of light required (vertical scale) at each wavelength (horizontal scale) to produce the desired effect in 50 per cent of the seeds. Curves at right are spectra for the promotion of flowering in barley (solid line), and for the inhibition of flowering in soybeans (long dashes) and cockleburs (short dashes). Barley flowers during short nights, and the other two flower when the nights are long.

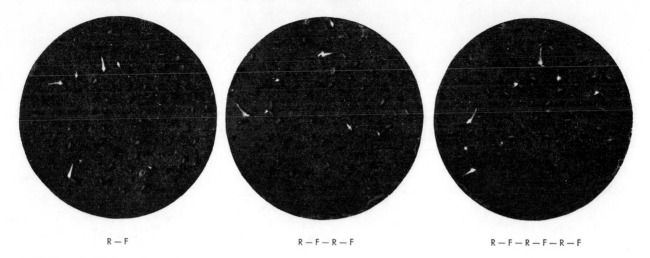

R — F   R — F — R — F   R — F — R — F — R — F

GERMINATION IS INHIBITED by irradiation with far-red light. While the great majority of seeds germinated after final exposure to red light (*opposite page*), practically none of the lettuce seeds irradiated last with far-red light will ever be able to germinate.

How phytochrome exerts its manifold influences on plant growth is still unknown. P_{735} seems to be the active form, while P_{660} appears to be a quiescent form in which the plant can store the potentially active compound. At the close of any period of exposure to light, phytochrome is predominantly in the far-red-absorbing form. The rate at which P_{735} is then carried through the dark conversion back to P_{660} provides the plant with a "clock" for measuring the duration of the dark period. The rate of conversion probably varies from one plant to another, and must depend in part upon such factors as temperature. The effective dark period might be the time required for the complete conversion, or it might be the time in darkness after the conversion is finished. In either case a brief interval of light in the middle of the dark period would cause a plant to respond as though the dark period were short.

Phytochrome is undoubtedly an enzyme—a biological catalyst. Its ability to control so many kinds of plant response in so many different tissues suggests that it catalyzes a critical reaction that is common to many metabolic pathways. Several reactions of this kind are known. One is the reaction that forms the so-called acetyl coenzyme-A compounds. These compounds are essential intermediates in fat utilization and fat synthesis, in cellular respiration and in the synthesis of anthocyanin and sterol compounds. The regulation of the supply of acetyl coenzyme-A compounds would provide an ideal control for growth processes. More than three fourths of all the carbon in a plant is incorporated in this coenzyme at some stage or other.

The extraction and partial purification of phytochrome is the starting point of a major forward movement in the understanding of plant physiology. Further research on this remarkable protein, ubiquitous in the plant world, should answer many questions concerning germination, growth, flowering, dormancy and coloring. It should also provide a means to control all of these plant processes to the great benefit of agriculture.

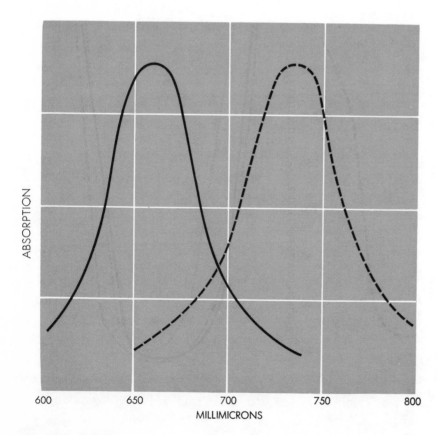

ABSORPTION

600   650   700   750   800
MILLIMICRONS

ABSORPTION SPECTRA for the two forms of phytochrome are shown here. The form known as P_{660} (*solid line*) absorbs the most light at a wavelength of 660 millimicrons, while P_{735} (*broken line*) is far more absorbent, or opaque, to light at a wavelength of 735 millimicrons. The reactions of plants to these wavelengths indicate that P_{735} is the active form.

The Infrared Receptors of Snakes

by R. Igor Gamow and John F. Harris
May 1973

The snakes of two large families have sensitive organs that can detect the heat radiation emitted by their prey. The performance of these detectors is investigated with the aid of an infrared laser

A boa constrictor will respond in 35 milliseconds to diffuse infrared radiation from a carbon dioxide laser. A sensitive man-made instrument requires nearly a minute to make what is essentially the same measurement. Reflecting on this comparison, one wonders what feat of bioengineering nature has performed to make the snake's sensor so efficient. One also wonders if a better understanding of the animal's heat-sensing apparatus would provide a basis for improving the man-made ones. It was the pursuit of these questions that gave rise to the somewhat unusual situation in which a group of workers in an aerospace engineering laboratory (our laboratory at the University of Colorado) was investigating snakes.

Snakes belong to one of the four large orders that comprise the living members of the class Reptilia. The order Testudinata contains such members as the turtles, the tortoises and the terrapins. The order Crocodilia contains the crocodiles and the alligators. The third order, the Rhynchocephalia, has only one member: the tuatara of New Zealand. In the fourth order, the Squamata, are the lizards and the snakes.

On the basis of outward appearance one might suppose that the lizards are more closely related to the crocodiles and the alligators than to the snake. Evolutionary evidence, however, clearly indicates that the snakes arose from the lizard line. Although the lizard is therefore the snake's closest relative, the two animals have developed pronounced differences during the course of evolution. Most lizards have limbs and no snakes have limbs, although vestigial ones are found in certain snakes. Most lizards have two functional lungs, whereas most snakes have only one. Again a few snakes have a small second lung, which

is another indication of the direction of evolution from the lizard to the snake.

Today most herpetologists would agree that the first step in the evolution of the snake occurred when the animal's ancestral form became a blind subterranean burrower. In evolving from their lizard-like form the ancestral snakes lost their limbs, their eyesight and their hearing as well as their ability to change coloration. Later, when the animals reappeared on the surface, they reevolved an entire new visual system but never regained their limbs or their sense of hearing.

Today the snakes constitute one of the most successful of living groups, being found in almost every conceivable habitat except polar regions and certain islands. They live in deep forests and in watery swamps. Some are nocturnal, others diurnal. Some occupy freshwater habitats, others marine habitats. Certain snakes are arboreal and survive by snatching bats from the air, others live in the inhospitable environment of the desert. Their success is indicated by the fact that their species, distributed among 14 families, number more than 2,700.

Two of the 14 families are distinguished by the fact that all their members have heat sensors that respond to minute changes of temperature in the snake's environment. The snake employs these sensors mainly to seek out and capture warm-bodied prey in the dark. It seems probable that the snake also uses the sensors to find places where it can maintain itself comfortably. Although snakes, like all reptiles, are cold-blooded, they are adept at regulating their body temperature by moving from place to place. Indeed, a snake functions well only within a rather narrow range of temperatures and must actively seek environments of the proper temperature.

A case in point is the common sidewinder, which maintains its body temperature in the range between 31 and 32 degrees Celsius (87.8 and 89.6 degrees Fahrenheit). One advantage of a heat sensor is that it enables the snake to scan the temperature of the terrain around it to find the proper environment.

One of the families with heat receptors is the Crotalidae: the pit vipers, including such well-known snakes as the rattlesnake, the water moccasin and the copperhead. The other family is the Boidae, which includes such snakes as the boa constrictor, the python and the anaconda. Although all members of both families have these heat receptors, the anatomy of the receptors differs so much between the families as to make it seem likely that the two types evolved independently.

In the pit vipers the sensor is housed in the pit organ, for which these snakes are named. There are two pits; they are located between the eye and the nostril and are always facing forward. In a grown snake the pit is about five millimeters deep and several millimeters in diameter. The inner cavity of the pit is larger than the external opening.

The inner cavity itself is divided into an inner chamber and an outer one, separated by a thin membrane. A duct between the inner chamber and the skin of the snake may prevent differential changes in pressure from arising between the two chambers. Within the membrane separating the chambers two large branches of the trigeminal nerve (one of the cranial nerves) terminate. In both snake families this nerve is primarily responsible for the input from the heat sensor to the brain. Near the terminus the nerve fibers lose their sheath of myelin and fan out into a broad, flat,

GREEN PYTHON of New Guinea (*Chondropython viridis*) is a member of the family Boidae that has visible pits housing its infra-red detectors. The pits extend along the jaws. Photograph was made by Richard G. Zweifel of the American Museum of Natural History.

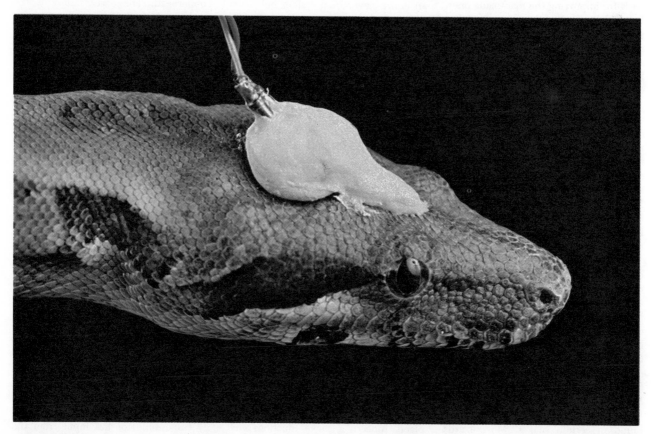

BOA CONSTRICTOR is a boid snake with infrared detectors that are not visible externally, although they are in the same location as the green python's. This boa wears an apparatus with which the authors recorded responses of the brain to infrared stimuli.

palmate structure. In this structure the nerve endings are packed full of the small intracellular bodies known as mitochondria. Evidence obtained recently by Richard M. Meszler of the University of Maryland with the electron microscope strongly suggests that the mitochondria change morphologically just after receiving a heat stimulus. This finding has led to the suggestion that the mitochondria themselves may constitute the primary heat receptor.

In the family Boidae there are no pit organs of this type, although somewhat different pits are often found along the snakes' upper and lower lips. Indeed, it was once thought that only the boid snakes with labial pits had heat sensors. An extensive study by Theodore H. Bullock and Robert Barrett at the University of California at San Diego has shown, however, that boid snakes without labial pits nonetheless have sensitive heat receptors. One such snake is the boa constrictor.

For experimental purposes the boid snakes are preferable to the pit vipers because the viper is certain to bite sooner or later, and the bite can be deadly. The boids, in contrast, can be described as friendly, and they get along well in a laboratory. When our laboratory became interested several years ago in the possibility of using an infrared laser as a tool to help unravel the secrets of the mode of operation of the snake's heat sensor, we chose to work with boid snakes.

Bullock and his collaborators have done most of the pioneering work on the heat receptors of snakes. In their original experiments, using the rattlesnake, they first anesthetized the animal and then dissected out the bundle of large nerves that constitute the main branches of the trigeminal nerve. It is these branches that receive the sensory information from the receptor.

Bullock and his colleagues found by means of electrical recording that the frequency of nerve impulses increased as the receptor was warmed up and decreased as it was cooled. The changes were independent of the snake's body temperature; they were related only to changes of temperature in the environment. The Bullock group also determined that the operation of the sensor is phasic, meaning that the receptor gives a maximum response when the stimulus is initiated and that the response quickly subsides even if the stimulus is continued. (Many human receptors, such as the ones that sense pressure on the skin, are phasic; if they were not, one would be constantly conscious of such things as a wristwatch or a shirt.)

Our work was built on the foundations laid by Bullock and his associates. In addition we had in mind certain considerations about electromagnetic receptors in general. Biological systems utilize electromagnetic radiation both as a source of information and as a primary source of energy. Vision is an example of electromagnetic radiation as a source of information, and photosynthesis is a process that relies on electromagnetic radiation for energy.

All green plants utilize light as the source of the energy with which they build molecules of carbohydrate from carbon dioxide and water. To collect this energy the plants have a series of pigments (the various species of chlorophyll molecules) that absorb certain frequencies of electromagnetic radiation. Indeed, green plants are green because they absorb the red part of the spectrum and reflect the green part. Because the chlorophyll molecule absorbs only a rather narrow spectral frequency, it can

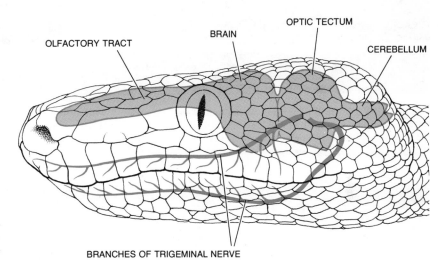

ANATOMY OF RECEPTOR in a boa is indicated. The scales along the upper and lower jaws have behind them an elaborate network of nerves, which lead into the two branches of the trigeminal nerve shown here. When the system detects an infrared stimulus, the trigeminal nerve carries a signal to the brain. A response can be recorded from the brain within 35 milliseconds after a boid snake receives a brief pulse of infrared radiation.

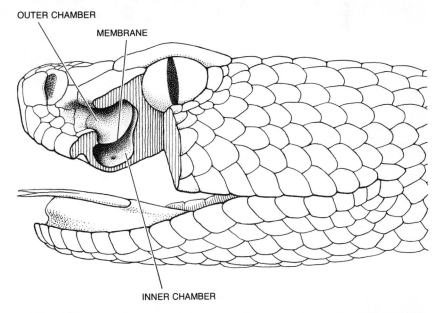

STRUCTURE OF PIT in a pit viper, the rattlesnake *Crotalus viridis*, differs from the anatomy of the infrared receptor in boid snakes. A pit viper has two pits, located between the eye and the nostril and facing forward. Each pit is about five millimeters deep, with the opening narrower than the interior. The elaborate branching of the trigeminal nerve is in the thin membrane that separates the inner and outer chambers of the pit organ.

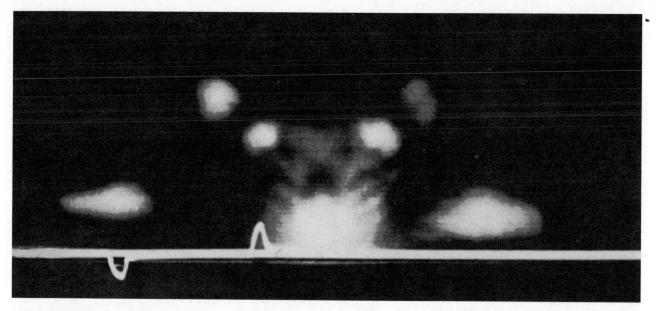

INFRARED VIEW OF RAT suggests what a snake "sees" through its infrared detectors when it is stalking prey. Snakes with such detectors prey on birds and small mammals. This view was obtained with a Barnes thermograph, which detects infrared radiation. In a thermogram the coolest areas have the darkest appearance and the warmest areas, such as the nose of the rat, appear as white spots.

BOA IN LASER BEAM was tested in the authors' laboratory at the University of Colorado for responses to infrared radiation. The carbon dioxide laser, emitting in the infrared at 10.6 microns, appears as a glowing area in the upper right background. Its beam is spread by a lens so that the snake, even when moving about in the cage, is doused in infrared radiation delivered in occasional pulses lasting eight milliseconds each. A brain signal recorded by the electrode assembly on the snake's head goes to a preamplifier and then to an oscilloscope and to a signal-averager. Electroencephalograms recorded in this way appear in the illustration, page 149.

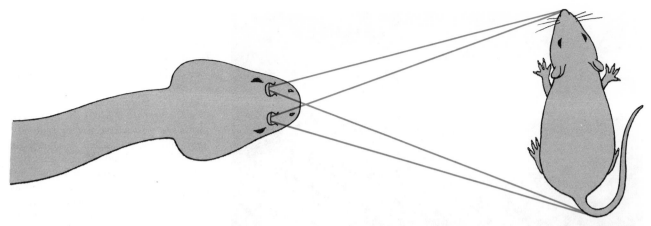

DIRECTIONALITY OF SENSOR in a pit viper is indicated by the location of the two pits on the snake's head and by the geometry of the pits. It appears certain that in stalking its prey, which include birds and small mammals, such a snake can establish the direction in which the prey lies by shifting its head as it does in using eyesight. Rattlesnake and copperhead are among the pit vipers.

be called a frequency (or wavelength) detector.

The eye is also a frequency detector, but it does not use radiation as an energy source. The incoming radiation triggers the release of energy that has been stored in the nerves previously, having been produced by normal metabolism. The eye, like other frequency detectors, operates within a narrow band of the electromagnetic spectrum, namely at wavelengths from about 300 to about 1,000 nanometers (billionths of a meter). One can see how narrow the band is by recalling that man-made instruments can detect electromagnetic wavelengths from 10^{-20} meter to 10^5 meters, a full 25 orders of magnitude, whereas the range of the human eye is from $10^{-6.4}$ to about $10^{-6.1}$ meter. Within this range the eye can resolve thousands of different combinations of wavelengths, which are the number of shades of color one can recognize. Although the eye is a good frequency detector, it is a poor energy detector: a dim bulb appears as bright as a bright one to the dark-adapted eye, which is to say that the eye adjusts its sensitivity according to the conditions to which it has become adapted.

Why has nature chosen this frequency range for its photobiology? From an evolutionary point of view the answers seem clear. One reason is that 83 percent of the sunlight that reaches the surface of the earth is in that frequency range. Moreover, it is difficult to imagine a biological sensor that would detect X rays or hard ultraviolet radiation, because the energy of the photons would be higher than the bonding energy of the receiving molecules. The photons would destroy or at least badly disrupt the structure of the sensor. Low-frequency radiation presents just the opposite difficulty. The energy of long-wavelength infrared radiation and of microwaves is so low that the photons cannot bring about specific changes in a molecule of pigment. Hence the sensor must operate in a frequency range that provides enough energy to reliably change biological pigment molecules from one state to another (from a "ground" state to a transitional state) but not so much energy as to destroy the sensor.

Early workers on the heat detectors of snakes had determined that the receptor responded to energy sources in the near-infrared region of the spectrum. The work left unanswered the question of whether the sensor contained a pigment molecule that trapped this long-wave radiation, thus acting as a kind of eye, or whether the sensor merely trapped energy in proportion to the ability of the tissue to absorb a given frequency and was thus acting as an energy detector. We therefore directed our experiments toward trying to resolve this issue.

To make sure that the response we obtained was maximal, we wanted to work with snakes that were functioning as close to their normal physiological level as possible. First we studied the normal feeding behavior of boa constrictors that were healthy and appeared to be well adjusted. The work entailed seeing how the snake sensed, stalked and captured prey animals such as mice and birds. Since the snake can capture prey in complete darkness as well as in light, it is clear that the heat receptors play a crucial role.

Barrett, while he was a graduate student working with Bullock, went further with this type of behavioral study. He found that the snake would strike at a warm sandbag but not at a cold, dead mouse. On the other hand, the snake would swallow the cold mouse (after a great deal of tongue-flicking and examination) if the mouse was put near the snake's mouth, but it never tried to swallow a sandbag. Barrett concluded that the snake has a strike reflex that is triggered by the firing of the heat receptors, whereas another set of sensory inputs determines whether or not the snake will swallow the object.

In searching for a reliable index that would tell us whether or not the heat receptor was responding to an infrared stimulus, we first tried measuring with an electrocardiograph the change in heartbeat after the snake received a stimulus. This venture ran afoul of the difficulty of finding the heart in such a long animal. (It is about a third of the way along the body from the head.) A more serious difficulty was our discovery that a number of outside influences would change the rate of the heartbeat, so that it was hard to establish a definite stimulus-response relation.

We next turned to a method that proved to be much more successful. It entailed monitoring the electrical activity of the snake's brain with an electroencephalograph. A consistent change in the pattern of an electroencephalogram after a stimulus has been received by the peripheral nervous system is called the evoked potential. When a neural signal from a sensory receptor arrives at the cortex of the brain, there is a small perturbation in the brain's electrical activity. When the signal is small, as is usually the case, it must be extracted from the electrical background noise. The process is best accomplished by averaging a substantial number of evoked potentials. This procedure results in a highly sensi-

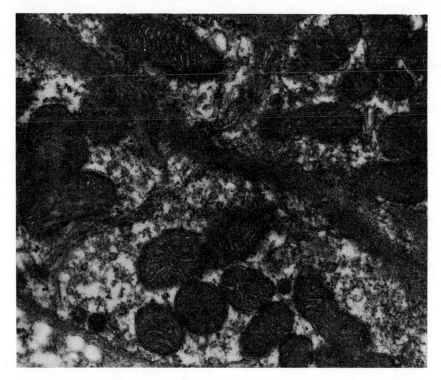

APPEARANCE OF MITOCHONDRIA in the nerve endings of the infrared receptor of a cottonmouth moccasin (*Agkistrodon piscivorus piscivorus*) after exposure of the receptor to an infrared stimulus is shown in this electron micrograph made by Richard M. Meszler of the University of Maryland. The enlargement is 34,000 diameters. In contrast to the mitochondria in the micrograph below, which was made when the receptor was exposed to a cold body, these mitochondria are condensed, as shown by the dense matrix and the organization of the inner membrane. Change in morphology of the mitochondria after a heat stimulus has led to the suggestion that they constitute the primary receptors in the detector.

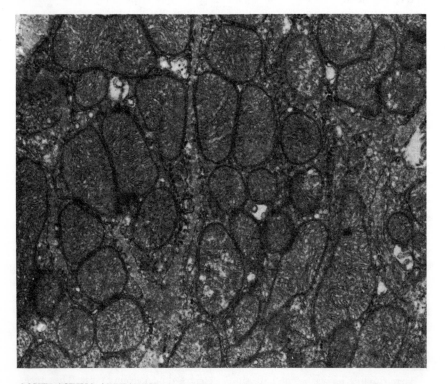

CONTRASTING APPEARANCE of the mitochondria in the infrared detector of a cottonmouth moccasin when the receptor was exposed to a cold body is evident in this electron micrograph made by Meszler. The enlargement is 27,000 diameters. A cold body, in contrast to a warm one, is known to reduce the firing of discharges by the heat-sensitive receptor.

tive measure of a physiological response.

The boa constrictors used in our study ranged from 75 to 145 centimeters in length and from 320 to 1,200 grams in weight. For several weeks before we involved them in experiments they lived under normal conditions in our laboratory. To prepare a snake for the experiments we anesthetized it with pentobarbital and then installed an electrode assembly on its head. After a postoperative recovery period the animal appeared to behave in the same way as snakes that had not been operated on.

A brain signal recorded by this apparatus went to a preamplifier and then to an oscilloscope and to a signal-averager. The signal-averager, which is in essence a small computer, is the workhorse of our system. By averaging the electroencephalogram just before and just after a stimulus it extracts the evoked potential, which would otherwise be buried in the background noise of the brain. In general we average the evoked potentials from about 20 consecutive stimuli.

The birds and mammals that the boa constrictor hunts emit infrared radiation most strongly at wavelengths around 10 microns. A carbon dioxide laser is ideal for our experiments because it produces a monochromatic output at a wavelength of 10.6 microns. We pulse the laser by means of a calibrated camera shutter so that it will deliver a stimulus lasting for eight milliseconds. The opening of the shutter also triggers the signal-averager, thus establishing precisely the time when the stimulus is delivered.

After the beam passes through the shutter it is spread by a special infrared-transmitting lens, so that the entire snake is doused in the radiation. The intensity of the radiation is measured by a sensitive colorimeter placed near the snake's head. This is the instrument we mentioned at the outset that takes nearly a minute to measure the power, whereas the snake gives a maximal response within 35 milliseconds after a single eight-millisecond pulse. Another indication of the sensitivity of the snake's receptor can be obtained by putting one's hand in the diffused laser beam; one feels no heat, even over a considerable period of time.

In order to verify that the responses of the snake resulted directly from stimulation of the heat receptor, we repeated the entire procedure with a common garter snake, which has no heat sensor. Even at laser powers far exceeding the stimulus given to the boas, we found no response in recordings from the garter snake. On the other hand, both species showed clear responses to visible light.

Our data strongly suggested an answer to the question of whether the receptor is a photochemical frequency detector like an eye or is an energy detector. The answer is that the receptor is an energy detector. One argument supporting this conclusion is that the stimulus is so far out in the low-energy infrared region of the spectrum (10.6 microns) that it would not provide enough power to activate an eyelike frequency detector, and yet the snake shows a full response. Another argument has to do with the 35-millisecond interval between the stimulus and the response. Photochemical reactions are quite fast, occurring in periods of less than one millisecond. Although the time a nerve impulse from the eye takes to reach the cortex is about the same (35 milliseconds) as the time the nerve impulse from the heat sensor takes, the neural geometry of the two systems is quite different. The visual pathway incorporates a large number of synapses (connections between neurons), which account for most of the delay. In the trigeminal pathway no synapses are encountered until the signal reaches the brain. We therefore believe the delay found in the heat-receptor response is largely a result of the time required to heat the sensor to its threshold.

We also tested the snake's receptor in the microwave region of the spectrum, where the signals have longer wavelength, lower frequency and lower energy than in the infrared. The reason was that in view of the many problems that have arisen in contemporary society about exposure to radiation we wanted to see whether an organism experienced physiological or psychological effects after being exposed for various periods of time to low-energy, long-wavelength radiation. There is no question that high-intensity microwave radiation can be detected not only by snakes but also probably by all animals; after all, a microwave oven can cook a hamburger in a matter of seconds. Our concern was with the kind of exposure arising from leaky microwave appliances such as ovens and from the increasing use of radar.

Testing the snakes with microwave radiation as we did with infrared, we obtained a clear-cut response [see illustration on this page]. Our result provides what we believe is the first unambiguous physiological demonstration that a biological system can indeed be influenced by such low-energy microwave radiation. Our conviction that the snake's heat receptor functions entirely as an energy detector is therefore reinforced.

The question of how much energy is required to activate the detector can be answered with certain reservations: it is approximately .00002 (2×10^{-5}) calorie per square centimeter. The reason for the reservations is that it is difficult to obtain an absolute threshold of sensitivity for any biological phenomenon. For one thing, a biological system shows considerable variability at or near its threshold of response. Moreover, there is always a certain amount of variation in the amount of energy put out by our sources of energy. With these reservations we have determined that the snake can easily and reliably detect power densities from the carbon dioxide laser ranging from .0019 to .0034 calorie per square centimeter per second. Since this density is administered in a short time period (eight milliseconds), the total energy that the snake is responding to is about .00002 calorie per square centimeter. The density of microwave power that is needed for a reliable response from the snake is about the same as the amount of laser power.

Our studies have shown that the heat-sensing snakes have evolved an extremely sensitive energy-detecting device giving responses that are proportional to the absorbed energy. It will be interesting to see whether the growing understanding of the snake's heat sensor will point the way toward an improvement in man-made sensors.

a

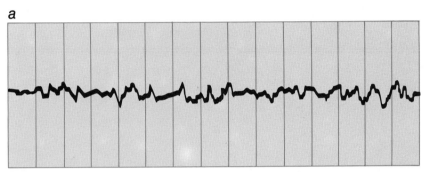

b   **c**

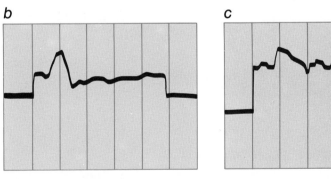

d   **e**

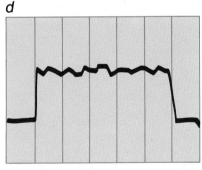
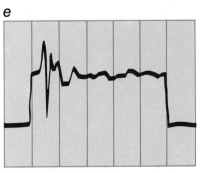

ELECTROENCEPHALOGRAMS of a boa constrictor were recorded under various conditions. The normal activity of a boa's brain (a) was traced directly on a strip recorder; the interval of time between each pair of colored vertical lines was 100 milliseconds. The remaining electroencephalograms were recorded through a signal-averager that reflected both the stimulus and the response. In each case the first rise shows the time of the stimulus and the next rise, if any, shows the response. The time interval is 100 milliseconds in every tracing but b, where it is 50 milliseconds. Traces show the averaged evoked response after an infrared stimulus (b) and a microwave stimulus (c), in the control situation in which the snake was shielded from the stimulus (d) and after a series of visible-light flashes (e).

IV

TEMPERATURE AND LIFE

IV

TEMPERATURE AND
LIFE

INTRODUCTION

All life processes that we know of occur in aqueous solution. This fact defines the temperature range in which organisms can survive. At sea level, water boils at 100°C. At 0°C it freezes to ice. We can thus reasonably consider these two temperatures as the thermal limits for biological activity. These limits are not quite as sharp as we might suppose, however. We know of many organisms that thrive at temperatures considerably below 0°C, owing to the presence of solutes such as glycerol or sucrose in their cells or in the immediate environment of the cells. The solutes act as an "antifreeze," preventing ice crystals from forming. On the other hand, some organisms are known to survive temperatures above 100°C if kept under pressures sufficiently high to prevent the water from boiling.

Although some organisms are adapted to survive or even thrive at the extremes of this temperature range, most are adapted to a very narrow temperature range. In this section we will consider some of the effects of temperature on life processes. Among the clearest examples of these effects are the seasonal variations in the tempo of life so obvious at polar latitudes of the earth and at high altitudes.

We should distinguish between the many different aspects of temperature effects on living organisms. We readily comprehend that boiling water or ice crystals can damage a living cell, but why do some organisms succumb to a change of temperature, say, from 25°C to 30°C? What are the characteristics of some strains of organisms that enable them to grow at temperatures at which closely related strains cannot survive? What about the microorganisms adapted to live at high temperatures, the thermophiles—why do they stop growing at normal temperatures?

In his article, "Heat and Life" (page 157), Frank H. Johnson presents a brief review of the range of temperatures suitable for biological activity and discusses some of the reasons for the limits. Basically, the upper limit is imposed by the nature of the chemistry of living matter. We have seen earlier that the delicate supramolecular structures of which living cells are built are held together by relatively weak bonds, and some of the structures are more readily disrupted than others.

Enzymes, the all-important catalysts of biochemical reactions, are protein molecules. To maintain its specific function, an enzyme must retain its unique three-dimensional configuration, which is determined by the weak bonds that cause the folding of the polypeptide chain. Because of differences in their amino acid composition, proteins differ in the ease with which their structures can be disrupted, or denatured. Among a multitude of enzymes some will be much more easily denatured than others.

A complex, multienzyme life process will be stopped at the temperature that denatures the most sensitive of the enzymes. The sensitivity of proteins to heat denaturation is affected by the presence of solutes in their vicinity. At high concentrations, alcohol and certain salts can cause the denaturation of some proteins even at room temperature. On the other hand, salts containing ions like Ca^{2+} and Mg^{2+} can function to protect biological structures from heat denaturation.

Another point worth remembering is that the rate of an enzyme-catalyzed reaction is affected by temperature in at least two ways. One is the direct effect on the kinetic energy of the reacting molecules. The other is the effect on the state of the enzyme itself. The change in the overall reaction rate is the net result of the two effects. Disruptions of intricate metabolic networks because of differences in the sensitivity of the enzymes explains why so many animals and plants do not survive relatively small deviations from their optimal temperatures.

It has been demonstrated that the addition of nutrients such as vitamins or amino acids to the culture medium will permit some bacteria and protozoa to survive at temperatures that they do not otherwise tolerate. This indicates that the specific metabolic reactions for the synthesis of the nutrient in question are the ones that are affected by the unfavorable temperature. The extraneous supply of the metabolite obviates the need for that metabolic pathway and enables the organism to grow anyway.

Not only proteins but other very important components of the cell are affected by changes in temperature. In his article, "Heat Death" (page 161), the late L. V. Heilbrunn emphasizes the role of the biological membrane, especially the lipids of this structure, in the organism's responses to temperature stress. Although at the time the article was written we knew considerably less than we do now about the architecture of a biological membrane, the significance of the membrane lipids was already well-established.

The state of the lipids in the cell has a direct effect on all membranous structures. As you recall from Fox's article, which dealt with the properties of membranes, lipids constitute the essential matrix of membranes. The organization of different proteins on and within the lipid layer determines the biological properties of the membrane. It is readily apparent, therefore, that an increase or decrease in the fluidity of the lipids due to temperature changes can cause changes in the distribution of membrane proteins and alter their activity.

As an example of how changes in the properties of the membrane affect other cellular functions, Heilbrunn discusses the release of calcium ions from the membrane or cortex of the cell into the interior. The released ions could cause the coagulation of other proteins in the cytoplasm.

Living organisms have devised various strategies for coping with a cold environment. One is to avoid the cold altogether, as many migratory animals do. Another is to hibernate: the organism becomes inactive during the coldest months. Most plants and many animals use this strategy.

The evolutionary development of constant-temperature bodies in mammals and birds has enabled these organisms to adapt to a wide range of ambient temperatures. We have discussed some of the effects of temperature on living cells, with particular reference to enzymes and membranes. In "Adaptations to Cold" (page 165), Laurence Irving discusses the adaptation of higher, warm-blooded animals to low ambient temperatures that are potentially lethal to their cells. Efficient insulation and modified circulation patterns are evolutionary refinements that apparently enable warm-blooded animals to survive exposure to severe cold.

An impressive aspect of adaptation to cold in the Arctic regions is the extremely rapid tempo of life processes during the short summer season. Plants and insects complete their life cycles in several weeks and then return to a dormant winter state to await the next summer. An analogous strategy of life in the adaptation of desert plants to short, widely separated periods of rainfall was mentioned in Part I.

We have considered the damage that can be done to living organisms when the temperature deviates slightly from the optimum for their life. In his interesting article, "The Freezing of Living Cells" (page 171), A. S. Parkes considers the biological consequences not just of chilling but of actually freezing living organisms. New types of damage result from freezing. The formation of ice crystals causes mechanical tearing of biological structures. As the water freezes, the solute concentration in the remaining liquid increases. Most living cells, especially those of higher organisms, are very sensitive to the total solute content of the solution in which they are immersed.

It is apparent from Parkes's article that the primary reason for research on the survival of living cells in the frozen state is the practical applications of such knowledge. Since the article was written this field has advanced greatly, and at present many kinds of organisms can be maintained frozen in storage in a state of suspended animation. By controlling the rate of freezing and the composition of the liquid in which the cells are suspended, it is possible to freeze many bacteria, protozoa, and even mammalian cell suspensions. After freezing it is possible to remove most of the ice from the cells by sublimation under high vacuum and then store the freeze-dried material until required. The rehydration and revival of activity of such cells also require careful control of the conditions. These successes apply only to cell suspensions; the nondestructive freezing of whole, complex animals is not much further advanced than it was when the article was written.

Maintaining living matter in a freeze-dried state has interesting implications for theories of the origin of life on earth. One such theory is called the "pan-spermia" theory. It suggests that living matter in a state of suspended animation, such as seeds or spores, can be transported from one solar system to another. When such spores reach an appropriate environment they germinate, thus spreading life throughout the universe. This theory does not attempt to explain the origin of life *per se*—only its arrival on the earth. The demonstration that living matter can withstand freezing and dehydration under appropriate conditions is encouraging. However, to be transported through space, living matter must also be able to withstand exposure to highly energetic radiation.

We have discussed some of the adaptations of animals for coping with changing climatic conditions. Plants have developed other strategies for coping with such changes. One of the strategies available to animals, migration, is obviously not available to plants.

Means of conserving heat or of disposing of excess heat have evolved in plants and enable them to withstand relatively mild temperature fluctuations of short duration. Dormancy is the most common strategy used by plants to cope with longer and more severe periods of harsh conditions.

The handling of heat loads by some plants is the subject of the article, "Heat Transfer in Plants" (page 176), by David M. Gates. Water loss by transpiration appears to be the most common method devised by plants for unloading excess heat. When water conservation becomes a vital necessity, however, plants resort to different methods. The size and shape of the leaves appear to be of the utmost importance, as is the nature of the cuticle on the leaf surface. The placement and control of the stomatal opening are also evolutionary developments for handling the heat load by controlling water loss. Movement of leaves to expose smaller areas to direct sunlight is also accomplished by some plants.

We close this section with the article, "Thermal Pollution and Aquatic Life" (page 184), by John R. Clark. Human activities have a pronounced effect on the ecology of large land masses. Agriculture is an enterprise for increasing the biological productivity of the land. Needless to say, we are absolutely committed to such manipulations of our ecosystem in order to maintain our species at even its present population level. If we are to allow the population to increase, it is inevitable that our degree of control and manipulation of the environment must also increase in direct proportion.

We should evaluate the problems of thermal pollution from this point of view. In his article, Clark presents the consequences of unloading large amounts of excess heat from industrial sources into relatively small bodies of water. The expected increase in temperature can cause ecological havoc among the native populations of those waters.

The hydrosphere is, in general, much more stable than the lithosphere. Organisms living in cold waters will not tolerate a temperature increase of more than several degrees. Similarly, warm-water organisms will die if the temperature decreases significantly. It is thus to be expected that the heating of a river by thermal pollution will cause a change in the flora and fauna of the river. Heating the water to the extent indicated in the article will not eradicate life from the affected habitats, however. Some living organisms exist even in hot springs where the temperature approaches the boiling point. Our desire to maintain the existing biota of streams and rivers must be weighed against the economic needs of our society.

LUMINESCENT BACTERIA *Photobacterium phosphoreum,* suspended in saline solution, are photographed by their own light.
The flask at upper left, at 5 degrees centigrade, is one fifth as luminous as the flask at upper right, which is at 22 degrees C. This
is roughly the optimum temperature for luminosity in this species
(*see graph on the opposite page*). The flask at lower left, at 32
degrees C., has gone past the optimum point and has returned to
about the same intensity as the first flask. The flask at lower right
has been heated to 32 degrees C. and then cooled to 22 degrees C.,
its equal intensity indicating the reversibility of the reaction.

Heat and Life

by Frank H. Johnson
September 1957

*Between the lowest temperatures and the highest is a
narrow zone in which living organisms have evolved. This
is the range where enzymes can exist and speed the
chemical reactions of metabolism*

Without heat, all life processes cease. With a little too much heat, they cease just as surely. And the difference between too little and too much is only a hair's-breadth in the broad scale of temperature known to modern science. Life is pretty much confined to the range between about 0 and 50 degrees centigrade, or 32 to 122 degrees Fahrenheit. Very few organisms can live long at temperatures above or below those limits. It has been estimated that, if the average daily temperature on the face of the earth as a whole were suddenly raised or lowered by only 20 degrees, all life would perish. As a matter of fact, life could not adjust to any great change in temperature even with unlimited time for evolutionary adaptation, for the chemistry of organisms is subject to unchangeable limitations.

We might consider first some of the exceptional situations. At the cold end of the biological scale, there is, for example, the Alaskan stonefly, which apparently lives a normal life at freezing temperatures; it has been observed to mate on the frozen ceiling of ice caverns. Some species of bacteria and molds are able to grow slowly at temperatures several degrees below freezing. The polar codfish remains active at below-zero temperatures. Yet these few extremes only emphasize the central principle. As a general rule, at freezing or near-freezing temperatures living organisms can exist, if they survive at all, only because their normal living activities, including metabolism, come to a virtual standstill. Some species of bacteria can survive being kept for months at a time at temperatures around 300 degrees below zero F., and a remarkable array of living forms, including luminous bacteria, moss spores, pollen grains, seeds of higher plants and some lower animals,

have recovered their normal activity after exposure to the temperature of liquid helium, within a couple of degrees of absolute zero. Among the more complex animals, a number can keep alive at low temperatures by hibernation, which is a state of suspended activity. Even among the higher mammals some can survive a drastic lowering of the body temperature; for example, young white rats have recovered after being chilled to 37 degrees F., when all perceptible signs of life, including the heartbeat and circulation of the blood, disappeared. The rats in effect underwent a general anesthesia. The anesthetic effects of refrigeration of course are well known; it has been used by surgeons for hundreds of years, and Napoleon's surgeon-in-chief, Larrey, is said to have performed 200 amputations in a single day with the assistance of freezing temperatures on the battlefield.

At the hot end of the biological scale, where the chemistry of life is speeded up rather than stalled, organisms are far more vulnerable to temperatures exceeding the limit. Here too, however, there are freaks. Perhaps the hottest of all animals is the "cold-blooded" fish *Barbus thermalis*, which lives in hot springs of Ceylon at a temperature of 122 degrees F. The albino mouse lives at a body temperature of 102.2 degrees, and the songbirds have a normal temperature of nearly 113 degrees. To find really warm organisms, however, we must turn to the world of primitive

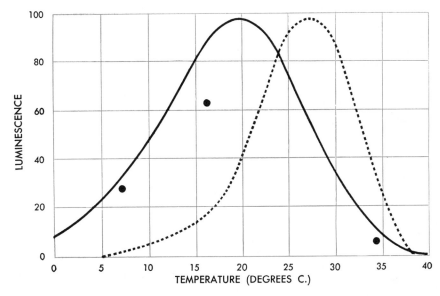

LUMINESCENCE VARIATION occurs in two species of bacteria as the temperature of their environment changes. The solid line shows the increase and decrease of intensity of luminescence for *Photobacterium phosphoreum* as the temperature is increased. The dots show the reversibility of this reaction by readings taken after heating these bacteria to 40 degrees, then cooling. The dashed line shows the response of *Achromobacter fischeri*.

158

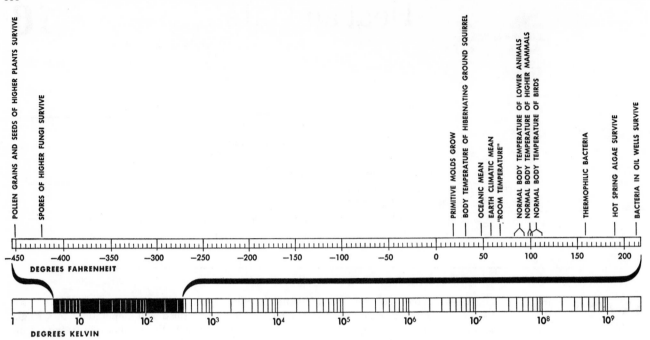

LIFE TEMPERATURES of organisms are largely confined to a narrow band on the thermometer. Such organisms as luminous bacteria and pollen grains can survive to about 450 degrees below zero by virtually suspending metabolic activity. Primitive plants, such as bacteria in oil wells, mark the upper end of the scale at 212 degrees. Most life processes occur between 32 and 122 degrees.

plants. Some species of bacteria thrive at 158 degrees or above. In some of the hot pools of the Yellowstone a blue-green alga carries on a simmering existence at 185 degrees. Certain bacteria indigenous to deep oil-well brines are believed to exist at a temperature of 212 degrees F. or higher.

We are concerned here with what temperatures are most favorable to life. The optimal temperature varies, of course, not only with the organism but also with the state of the organism. In a human being a rise in temperature, for instance, may help to fight an infection; fever therapy of syphilitic paresis depends on the fact that spirochetes cannot withstand as high a temperature as human cells can.

Growth and size have their optimal temperature, but here the relationship is not at all simple. Although cold retards

EDITOR'S NOTE

The subject of this article is more fully discussed in *The Kinetic Basis of Molecular Biology*, by Frank H. Johnson, Henry Eyring and Milton J. Polissar. This book was published by John Wiley & Sons, Inc., in 1954.

biological processes, among lower animals it is not unusual to find larger body sizes associated with cold habitats. Many arctic species grow to a larger size than their near relatives in warm climates, and flies are larger in winter than in summer. On the other hand, relatively high temperatures may greatly increase the growth of cells. When cultures of the vinegar-forming bacterium *Acetobacter* are incubated at 104 degrees F., which exceeds the normal optimum for the rate of reproduction, cell division almost ceases but the cells themselves continue to grow—sometimes to as much as 150 times the normal size.

Even longevity has an optimum temperature. A study of the water flea *Moina* shows that it lives longest at a temperature of about 60 degrees F. No doubt there is a similar optimum for longevity among the warm-blooded animals, including man, but obviously it would be more difficult to determine.

Biologists would like to develop a precise theory of optimal temperatures for man and other animals, but the temperature relationships are many and complex. In living organisms so many distinct reactions take part in an end result—growth, cell division, respiration, reproduction, longevity and so on—that it is difficult, indeed, to identify which one or which ones may be most important in determining the observed response to temperature.

The crux of the whole problem of temperature and life lies in the effects of heat upon enzymes, the body's catalysts. Enzymes lower the temperature required for reactions to proceed at a given rate. Their activity increases with rise in temperature, just as the rates of chemical reactions in general do. But because they are large, complex proteins, they are unstable to any excess of heat. This dual effect of heat on enzymes accounts primarily for the narrow limits of temperature within which life is possible, and also for the varieties of temperatures at which different life processes and different organisms do best. In each organism there is a multiplicity of enzymes, each catalyzing a specific step or type of reaction, and heat affects both the activity and the stability of some enzymes more than of others. The net effect of temperature on the total process is conditioned by the specific enzymes involved and the characteristics of their response to heat; these characteristics in turn depend on the biological source of the enzymes and on the nature of the chemical environment in which they act.

Life is sensitive to many other environmental variables besides heat—salt concentration, acidity, alkalinity and so on. Ultimately, however, heat is still the fundamental factor; the optimal concentrations of salt, acid and other agents depend in part upon the temperature. Both the activity and the stability of enzymes and other essential constituents of living cells are influenced by the concentration of such agents, but temperature is still basically involved. An interesting illustration lies in the fact that when an egg is placed in alcohol or acid, it will coagulate at room temperature. Heat makes the reaction go, just as when the egg is boiled in water, but the catalytic action of the alcohol or acid lowers the required temperature. (In living organisms alcohol produces fundamentally similar effects, though its concentration is not usually high enough to cause coagulation; the alcohol reduces, by less drastic and usually reversible reactions, the activity of the sensitive enzymes.)

In view of the large number of different reactions that are responsible for the end result of a complex biological process, and in view of the equally large number of enzymes involved in these reactions, it would appear to be a well-nigh hopeless undertaking to attempt to analyze the total rate according to laws governing simple reactions. Surprisingly, however, some of the most complicated living processes respond to temperature as a whole in precisely the manner of a simple chemical reaction. Thus at temperatures below the normal optimum of the system the rate of a single enzyme reaction, or of the heartbeat, or of mental processes, may increase with temperature in the manner that the Swedish chemist Svante Arrhenius discovered is true of many simple reactions; *i.e.*, the logarithm of the rate is a linear function of the reciprocal of the absolute temperature. The reason is that the total rate is limited primarily by the slowest step in the group of reactions, and this slowest reaction, as well as all the others, obeys the same physicochemical laws that hold for all reactions.

At a given temperature above the normal optimum, the rate of the process often decreases in a manner also in accord with the Arrhenius relation. The decrease suggests irreversible destruction of one or more of the essential enzymes. However, if the exposure to the above-optimum temperature is short, the rate in some cases recovers at once on cooling. This effect, which was first quantitatively demonstrated in living cells with bacterial luminescence, indicates an equilibrium change between catalytically active and inactive states of an essential enzyme. We are probably touching here on a general mechanism

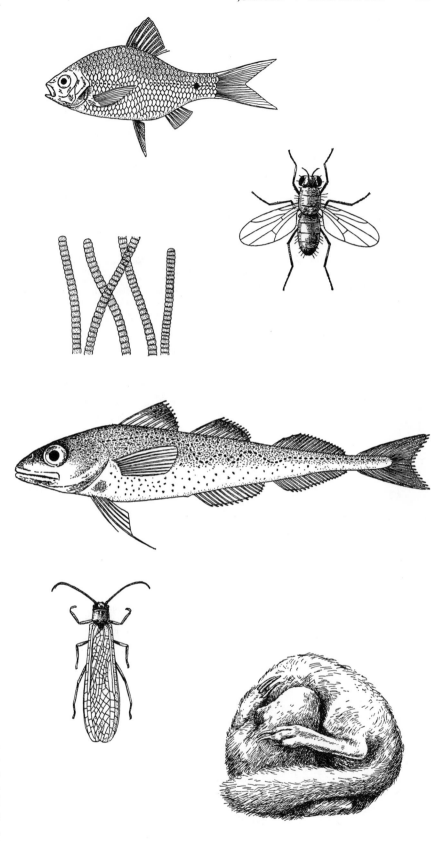

EXTREME DWELLERS in the realm of life temperatures are illustrated here. The fish *Barbus thermalis* (*top*) lives in the hot springs of Ceylon at 122 degrees F. The brine fly *Ephydra* can be found in hot salt pools at 115 degrees. The blue-green algae *Oscillatoria* are quite adaptable and live up to 190 degrees. The polar codfish *Boreogadus* is one of the coldest creatures at 29 degrees. The Alaskan stonefly *Nemoura columbiana* mates at 32 degrees. The ground squirrel *Citellus tridecemlineatus* hibernates in winter at 32 degrees.

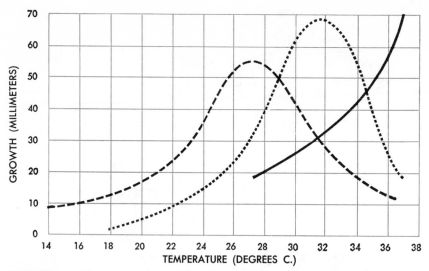

PLANTS respond to temperatures differently according to species variations. Increments of growth for 48 hours are shown for a temperate climate plant, *Lupinus* (*long-dashed line*) and two plants of warmer climes, *Zea* (*short-dashed line*) and *Cucumis melo* (*solid*).

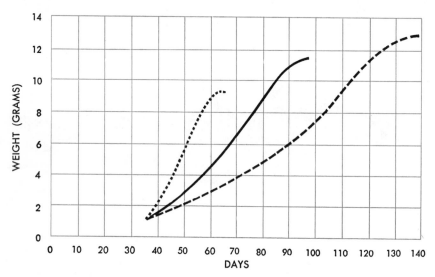

FISH EMBRYOS indicate a varying response of growth to temperature. One hundred trout larvae were raised at three temperatures: 16 degrees C. (*short-dashed line*), 10 degrees (*solid line*) and 5 degrees (*long-dashed line*). Weight increase was greatest for the coldest.

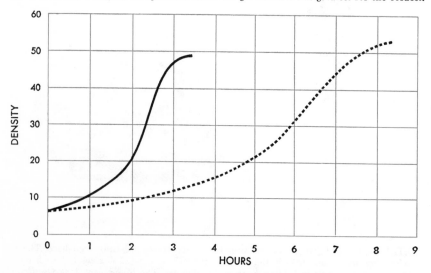

BACTERIA indicate a differential response of growth to temperature. Two cultures of *Escherichia coli* were kept at 37 degrees C. (*solid line*) and 23 degrees. The higher temperature caused an increase of rate in growth at the expense of total bacterial population.

in control of physiological rates, for this equilibrium is sensitive not only to temperature but also to many drugs, including alcohol, which lower the temperature at which the change in equilibrium occurs.

We suppose that the intensity of light emitted by luminescent bacteria depends on an oxidative enzyme, whose catalytic activity is reversibly increased with rise in temperature, but whose stability is reversibly decreased, to a different extent. On this assumption it is possible to account with considerable accuracy for the relation between temperature and light intensity over a broad range in temperature. Although many reactions are involved, the net result, as measured by luminescence, responds to temperature as if the whole process depended upon the reactivity of a single species of molecules, namely, the enzyme that limits the process of light emission. There are many other instances wherein a complex biological process as a whole responds to heat in the manner of a relatively simple chemical reaction.

No one knows, and no one is ever likely to know for certain, the temperature of the earth when life first appeared. There is reason to believe that the temperature was quite a bit hotter than it is today. Life probably arose from a single molecule or at most from a very few molecules having the same type of optical activity. It arose when this molecule, or these few molecules, reproduced themselves, possibly by acting as templates for the formation of their successors. The fact that left-handed (levo) optical forms predominate over right-handed (dextro) forms in the living world today, though the two forms have exactly the same chemical reactivity, is difficult to account for on any other hypothesis. The original molecules, with which life began, may have set the left-handed pattern, and the few dextro isomers can be explained as inversions without survival value.

The first self-duplicating molecules must have been relatively small and simple. Because of their simplicity they were more stable to heat. With the cooling of the earth it became possible for more complex molecules to survive, and evolution proceeded in every feasible direction. Out of it has emerged present-day life in all its magnificent array. Heat makes it go and heat makes it stop going. Whatever the future holds in store for us rests ultimately on heat— and on the compromise between too little and too much.

Heat Death

by L. V. Heilbrunn
April 1954

*A brook trout perishes at room temperature; a frog, when
one of its limbs is placed in warm water. Every animal has
its own fatal temperature, determined by the stuff of its cells*

What is it that makes a man die of the heat—or a mouse, a fish or an insect? They all do die when the temperature of their bodies rises only a few degrees above normal. Of course there is a difference between a man and a mouse on the one hand and a fish and an insect on the other. Men and mice keep their body temperature reasonably constant. This is somewhat easier to accomplish for a man than it is for a mouse. The mouse, being much smaller, loses heat more rapidly and on a cold day has rather a hard time keeping its body temperature normal; in spite of its protecting fur, it doesn't always succeed. Both animals produce heat as a result of muscular activity, for when muscles contract they give off heat. When it is hot, our bodies must lose some heat if we are to survive. Sweating is a help. The mouse does not sweat, but it is cooled by evaporation from its moist mouth and lung surfaces.

The insect and the fish have no protection against the heat of their environment. Some fish will die at temperatures that seem cool to us. The heat any animal can tolerate depends on the temperature it normally lives at. Tropical fish can live in rather warmish water; a brook trout, which normally lives in cold mountain streams, cannot survive at ordinary room temperature. Marine fish accustomed to life in the cool ocean soon die in warmer water. On the other hand, fishes of shallow warm ponds and tropical fish can stand more heat. In some cases animals live their lives through in water just a degree or two cooler than the temperature which kills them. This is true of polyps which build coral reefs. Our own normal temperature is 98.6 degrees Fahrenheit. If a violent fever increases our temperature 9 or 10 degrees, we are sure to die.

Why is this true? Why is the stuff we are made of so sensitive to heat? The heat death of an animal, be it man or mouse, fish or insect, is due to the heat death of cells. What is it that makes a cell die when its temperature goes up a few degrees?

Typically a cell is a tiny fluid droplet consisting of a nucleus, a surrounding fluid in which are suspended numerous granules, and a firmer outer rim called the cortex. The chemical components of the cell are water, proteins, fats or fatlike substances, carbohydrates and salts.

In considering how heat kills a cell, we can well exclude the nucleus, for cells can live for quite a while even if the nucleus is removed from them. The main effect of killing heat on the cytoplasm of the cell is a marked stiffening of the fluid. This can be measured with a centrifuge. The most useful centrifuge for such tests is a hand-operated one whose tubes are whirled by means of a handle like that on an old-fashioned coffee grinder. A convenient cell on which to experiment is the egg of the sea urchin. A sea urchin egg has a great many granules evenly distributed throughout its protoplasm. In its normal condition, when the egg is subjected to a centrifugal force several thousand times gravity almost all its granules, being heavier than the rest of the protoplasm, move toward one end of the cell. But if sea urchin eggs are first heated to a temperature of 93 degrees F. for a few minutes, centrifuging will not move the granules at all. The main mass of the protoplasm has lost its fluidity and the granules are stuck in the stiffened mass.

What causes this stiffening? The simplest explanation would be that the proteins of the cell have coagulated, exactly like the white of a boiled egg.

But the proteins in egg white harden only at rather high temperatures—not at 93 degrees but at something closer to 212 degrees. Some proteins, it is true, may coagulate at lower temperatures, but there are cold-climate animals that die at 60 degrees, and one could scarcely talk about heat coagulation of proteins at that temperature.

If heat death is not due to coagulation of proteins, is it possible that other substances are primarily affected—and what other substances? There is a strong suspicion that fats and fatlike substances may be involved. Here is some of the evidence: There is a correlation between the melting points of the fats of various animals and the temperatures at which they die. Cold-blooded animals such as fishes ordinarily die at 70 to 80 degrees F. Birds and mammals do not die at temperatures below 100 degrees. Correlated with this is the fact that at room temperature the fats of fishes ordinarily are fluid (for example, cod-liver oil) while those of birds and mammals are solid (*e.g.*, chicken or beef fat). The same relation holds in plants. Those plants that live in more northern latitudes and are more sensitive to heat have fats which ordinarily are fluid at room temperature (for example, linseed oil), while the fats of tropical plants are often solid (for instance, cocoa butter). In general the melting points of the fats of northern plants are lower than the melting points of the fats of southern plants. This is true even of similar species growing in the north and in the south.

There is an even more striking fact than this. When an animal of a particular species lives at lower temperatures, its fats have lower melting points. Thus salmon living in northern waters have more fluid fats than those that live farther south. And this sort of thing can also

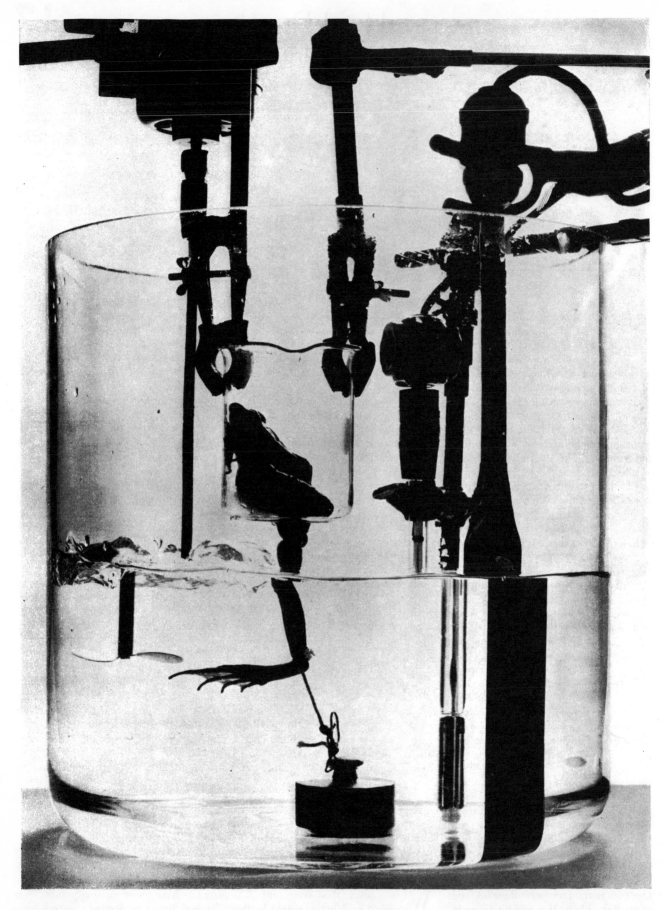

FROG'S LEG is immersed in water at 110 degrees Fahrenheit. Here the leg is tied off by a ligature and the frog shows no ill effects. When the ligature is removed, a substance produced in the leg by heat circulates through the rest of the body and the frog dies. Immersed in the water at the far right is a heating element. Next to it is a temperature-regulating device. At the far left is a stirrer.

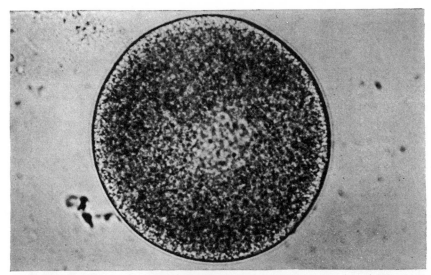

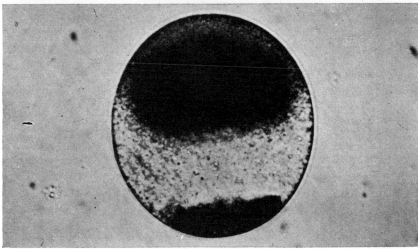

EGGS OF CHAETOPTERUS, a marine worm, are photographed under the microscope. At the top is a normal egg; fine granules are distributed through its protoplasm. At the bottom is an egg that has been spun in a centrifuge; the granules have moved toward the periphery of the cell. If before it is centrifuged the egg is placed in oxalate or citrate solution, which removes the calcium in its outer layer, the granules pass out of it entirely.

be demonstrated in laboratory experiments. For example, if flies are raised at different temperatures, the warmer flies have more solid fats than those raised in the cold. An amusing experiment of this sort was done many years ago in Denmark. Two Danish physiologists kept pigs in woolen underwear, so that their skin temperatures were higher than those of their naked brethren. (Ordinarily the skin temperature of an animal is not constant and is lower than the body temperature.) When the cutaneous fats of the pigs in underwear were examined, they were found to have higher melting points.

This introduces an interesting possibility. Couldn't we perhaps change the constitution of the fats of an animal by changing its diet? Then if its fats became more solid, it should be able to resist heat more readily. It has long been known that if a pig is fed more solid fats, the fats in its body become more solid—this is a way of producing better bacon. Years ago a student of mine discovered that frog tadpoles which were fed beef had fats of higher melting points and were better able to resist heat than tadpoles which grew up on a fish diet. Later some Canadian biologists did more extensive experiments on goldfish. Goldfish that were fed lard were much more tolerant to heat than others raised on fish oil. Another way of doing the same experiment is to feed animals a diet rich in carbohydrates. An animal converts carbohydrates into fats with relatively high melting points. The Canadian investigators Louis-Paul Dugal and Mercedes Thérien demonstrated that rats fed a diet rich in carbohydrates were more resistant to heat. I have heard it said that people who work in the hot sun in the South like to drink sugar water. The

French army used to issue sugar to its troops in the tropics.

Actually little is known about what might be done to improve the ability of man to withstand tropical heat. The problem is an old one. The Romans wondered why their soldiers and colonists in Africa could not stand heat as well as the natives. With the world becoming ever more crowded, it certainly would be advantageous for the white race to discover ways of living comfortably in hot climates. Much has been done in the way of air conditioning, but one can hardly expect to live at all times in air-conditioned rooms. The late explorer Ellsworth Huntington once wrote: "The climate of many countries seems to be one of the great reasons why idleness, dishonesty, immorality, stupidity and weakness of will prevail. If we can conquer climate, the whole world will become stronger and nobler." Whether this is true or not, it is rather surprising that so little scientific effort has been expended to improve man's resistance to heat.

Let us return to our inquiry as to why heat kills cells. How does the fact that its fats become more fluid affect the life of a cell? The answer seems to lie in the cortex of the cell. The cortex contains fats and proteins, and bound to them chemically is some calcium. Apparently anything that tends to liquefy or dissolve the fats will loosen their bonds to calcium, with the result that calcium is released from the cortex into the interior of the cell.

Calcium is an extremely important element in the chemistry of the body [see "Calcium and Life," by L. V. Heilbrunn; SCIENTIFIC AMERICAN, June, 1951]. One of its vital functions is to act as a clotting agent. The stiffness of the cell cortex depends on calcium. This can be demonstrated by an experiment on the egg of the marine worm Chaetopterus. If the egg is placed in a solution of oxalate or citrate, which removes calcium, the cortex becomes so fluid that granules can readily be displaced from it. Calcium is necessary for the clotting of blood and of protoplasm. The normal activity of protoplasm depends on this clotting; for instance, if you inject a little calcium into muscle cells it produces a gentle clotting which causes the cells to contract. But the release of a large amount of calcium into a cell will cause severe clotting and death.

When a marine worm egg is exposed, even for a short time, to temperatures above those to which it is accustomed, the cortex loses its rigidity and the interior protoplasm of the cell becomes

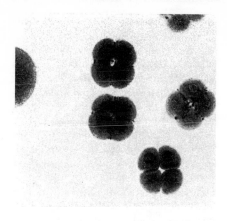

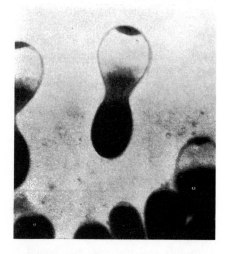

EGGS OF THE SEA URCHIN are also photographed. At the top are normal eggs, some of which are dividing. Like the eggs of *Chaetopterus*, they contain evenly distributed granules. At the bottom is a photomicrograph of sea urchin eggs in a centrifuge turning at 10,000 revolutions per minute. Here again the granules move toward the periphery of the cell. If the cell is heated, the granules do not move at all. The bottom photograph was made by Ethel Browne Harvey of Princeton University.

immersed in water at that temperature as it does if its whole body is exposed to 110-degree air. The same is true of a frog. It seems that the cells killed by the heat produce a poison which affects the whole animal. If the circulation from the legs is tied off so that blood from the heated legs cannot go to the rest of the animal, the animal survives.

What is the poison produced by scalded or burned cells? This question has been asked by many scientific workers, but there is at present no definite answer. In my own work I have come to the conclusion that when the protoplasm clots, it produces a clotting enzyme, just as clotted blood produces the enzyme thrombin. The thrombin-like substance produced when cells are killed by heat (or other agents) passes by way of the blood to all parts of the body and can, I believe, cause the clotting or death of cells. But if this view is to be proved correct, much more experimental work needs to be done.

Surprisingly enough, the effect of heat on protoplasm has not been given the attention it deserves. A great deal remains to be learned. Perhaps work along these lines in the future will not only help us to understand the nature of living substance but may also be of help in solving problems that concern the welfare of the human race.

clotted. Presumably the heat releases calcium from its bonds to fat in the cortex and allows it to enter the interior protoplasm. Exposure of cells to a substance that dissolves fat, such as ether or alcohol, has the same effect. Indeed, we can fortify the effect of heat by adding one of these solvents to the treatment. This may be one reason why it is unwise for people in the tropics to drink alcohol to excess.

In an animal as complex as a mammal the situation is complicated by the fact that some types of cells are more sensitive to heat than others. And there are general interactions among the cells. Consider the rat. A rat cannot live at a temperature of 110 degrees F. The interesting point is that it dies just as fast if only its legs below the knee joint are

Adaptations to Cold

by Laurence Irving
January 1966

*One mechanism is increased generation of heat by a rise
in the rate of metabolism, but this process has its limits.
The alternatives are insulation and changes in the
circulation of heat by the blood*

All living organisms abhor cold. For many susceptible forms of life a temperature difference of a few degrees means the difference between life and death. Everyone knows how critical temperature is for the growth of plants. Insects and fishes are similarly sensitive; a drop of two degrees in temperature when the sun goes behind a cloud, for instance, can convert a fly from a swift flier to a slow walker. In view of the general hostility of cold to life and activity, the ability of mammals and birds to survive and flourish in all climates is altogether remarkable.

It is not that these animals are basically more tolerant of cold. We know from our own reactions how sensitive the human body is to chilling. A naked, inactive human being soon becomes miserable in air colder than 28 degrees centigrade (about 82 degrees Fahrenheit), only 10 degrees C. below his body temperature. Even in the Tropics the coolness of night can make a person uncomfortable. The discomfort of cold is one of the most vivid of experiences; it stands out as a persistent memory in a soldier's recollections of the unpleasantness of his episodes in the field. The coming of winter in temperate climates has a profound effect on human well-being and activity. Cold weather, or cold living quarters, compounds the misery of illness or poverty. Over the entire planet a large proportion of man's efforts, culture and economy is devoted to the simple necessity of protection against cold.

Yet strangely enough neither man nor other mammals have consistently avoided cold climates. Indeed, the venturesome human species often goes out of its way to seek a cold environment, for sport or for the adventure of living in a challenging situation. One of the marvels of man's history is the endurance and stability of the human settlements that have been established in arctic latitudes.

The Norse colonists who settled in Greenland 1,000 years ago found Eskimos already living there. Archaeologists today are finding many sites and relics of earlier ancestors of the Eskimos who occupied arctic North America as long as 6,000 years ago. In the middens left by these ancient inhabitants are bones and hunting implements that indicate man was accompanied in the cold north by many other warm-blooded animals: caribou, moose, bison, bears, hares, seals, walruses and whales. All the species, including man, seem to have been well adapted to arctic life for thousands of years.

It is therefore a matter of more than idle interest to look closely into how mammals adapt to cold. In all climates and everywhere on the earth mammals maintain a body temperature of about 38 degrees C. It looks as if evolution has settled on this temperature as an optimum for the mammalian class. (In birds the standard body temperature is a few degrees higher.) To keep their internal temperature at a viable level the mammals must be capable of adjusting to a wide range of environmental temperatures. In tropical air at 30 degrees C. (86 degrees F.), for example, the environment is only eight degrees cooler than the body temperature; in arctic air at −50 degrees C. it is 88 degrees colder. A man or other mammal in the Arctic must adjust to both extremes as seasons change.

The mechanisms available for making the adjustments are (1) the generation of body heat by the metabolic burning of food as fuel and (2) the use of insulation and other devices to retain body heat. The requirements can be expressed quantitatively in a Newtonian formula concerning the cooling of warm bodies. A calculation based on the formula shows that to maintain the necessary warmth of its body a mammal must generate 10 times more heat in the Arctic than in the Tropics or clothe itself in 10 times more effective insulation or employ some intermediate combination of the two mechanisms.

We need not dwell on the metabolic requirement; it is rarely a major factor. An animal can increase its food intake and generation of heat to only a very modest degree. Moreover, even if metabolic capacity and the food supply were unlimited, no animal could spend all its time eating. Like man, nearly all other mammals spend a great deal of time in curious exploration of their surroundings, in play and in family and social activities. In the arctic winter a herd of caribou often rests and ruminates while the young engage in aimless play. I have seen caribou resting calmly with wolves lying asleep in the snow in plain view only a few hundred yards away. There is a common impression that life in the cold climates is more active than in the Tropics, but the fact is that for the natural populations of mammals, including man, life goes on at the same leisurely pace in the Arctic as it does in warmer regions; in all climates there is the same requirement of rest and social activities.

The decisive difference in resisting cold, then, lies in the mechanisms for conserving body heat. In the Institute of Arctic Biology at the University of Alaska we are continuing studies that have been in progress there and elsewhere for 18 years to compare the

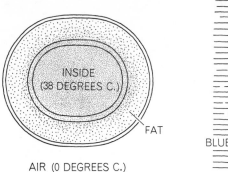

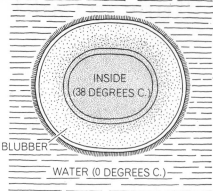

AIR (0 DEGREES C.)

WATER (0 DEGREES C.)

TEMPERATURE GRADIENTS in the outer parts of the body of a pig (*left*) and of a seal (*right*) result from two effects: the insulation provided by fat and the exchange of heat between arterial and venous blood, which produces lower temperatures near the surface.

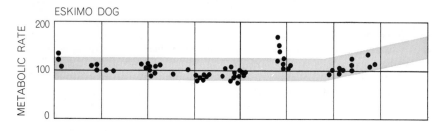

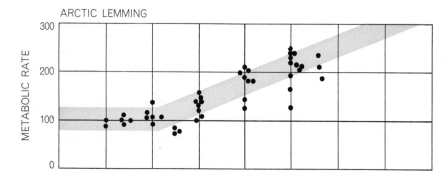

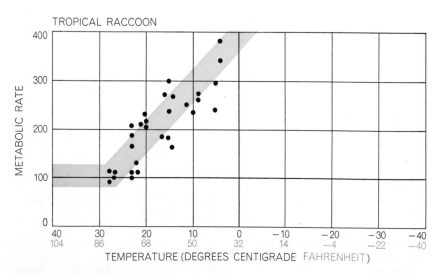

RATE OF METABOLISM provides a limited means of adaptation to cold. The effect of declining temperatures on the metabolic rate is shown for an Eskimo dog (*top*), an arctic lemming (*middle*) and a tropical raccoon (*bottom*). Animals in warmer climates tend to increase metabolism more rapidly than arctic animals do when the temperature declines.

mechanisms for conservation of heat in arctic and tropical animals. The investigations have covered a wide variety of mammals and birds and have yielded conclusions of general physiological interest.

The studies began with an examination of body insulation. The fur of arctic animals is considerably thicker, of course, than that of tropical animals. Actual measurements showed that its insulating power is many times greater. An arctic fox clothed in its winter fur can rest comfortably at a temperature of −50 degrees C. without increasing its resting rate of metabolism. On the other hand, a tropical animal of the same size (a coati, related to the raccoon) must increase its metabolic effort when the temperature drops to 20 degrees C. That is to say, the fox's insulation is so far superior that the animal can withstand air 88 degrees C. colder than its body at resting metabolism, whereas the coati can withstand a difference of only 18 degrees C. Naked man is less well protected by natural insulation than the coati; if unclothed, he begins shivering and raising his metabolic rate when the air temperature falls to 28 degrees C.

Obviously as animals decrease in size they become less able to carry a thick fur. The arctic hare is about the smallest mammal with enough fur to enable it to endure continual exposure to winter cold. The smaller animals take shelter under the snow in winter. Weasels, for example, venture out of their burrows only for short periods; mice spend the winter in nests and sheltered runways under the snow and rarely come to the surface.

No animal, large or small, can cover all of its body with insulating fur. Organs such as the feet, legs and nose must be left unencumbered if they are to be functional. Yet if these extremities allowed the escape of body heat, neither mammals nor birds could survive in cold climates. A gull or duck swimming in icy water would lose heat through its webbed feet faster than the bird could generate it. Warm feet standing on snow or ice would melt it and soon be frozen solidly to the place where they stood. For the unprotected extremities, therefore, nature has evolved a simple but effective mechanism to reduce the loss of heat: the warm outgoing blood in the arteries heats the cool blood returning in the veins from the extremities. This exchange occurs in the *rete mirabile* (wonderful net), a network of small arteries and veins near the junc-

ARCTIC ZONE (20 TO −60 DEGREES C.)

TEMPERATE ZONE (20 TO −20 DEGREES C.)

TROPICAL ZONE (35 TO 25 DEGREES C.)

RANGE OF TEMPERATURES to which warm-blooded animals must adapt is indicated. All the animals shown have a body temperature close to 100 degrees Fahrenheit, yet they survive at outside temperatures that, for the arctic animals, can be more than 100 degrees cooler. Insulation by fur is a major means of adaptation to cold. Man is insulated by clothing; some other relatively hairless animals, by fat. Some animals have a mechanism for conserving heat internally so that it is not dissipated at the extremities.

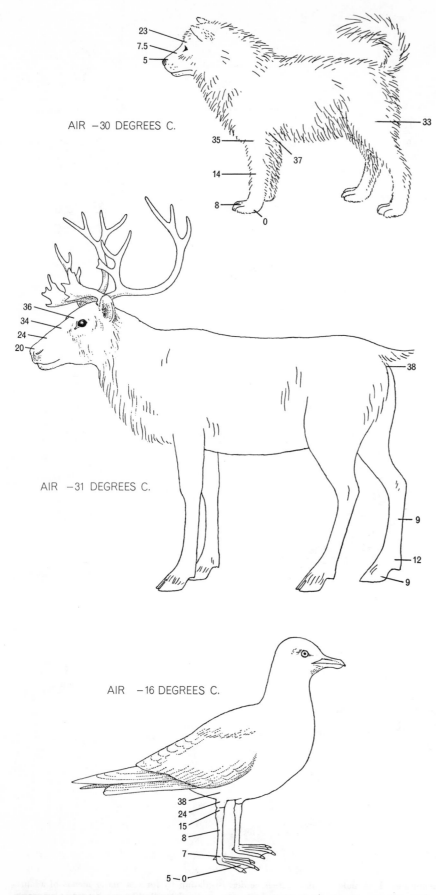

AIR −30 DEGREES C.

AIR −31 DEGREES C.

AIR   −16 DEGREES C.

TEMPERATURES AT EXTREMITIES of arctic animals are far lower than the internal body temperature of about 38 degrees centigrade, as shown by measurements made on Eskimo dogs, caribou and sea gulls. Some extremities approach the outside temperature.

tion between the trunk of the animal and the extremity [see "'The Wonderful Net,'" by P. F. Scholander; SCIENTIFIC AMERICAN, April, 1957]. Hence the extremities can become much colder than the body without either draining off body heat or losing their ability to function.

This mechanism serves a dual purpose. When necessary, the thickly furred animals can use their bare extremities to release excess heat from the body. A heavily insulated animal would soon be overheated by running or other active exercise were it not for these outlets. The generation of heat by exercise turns on the flow of blood to the extremities so that they radiate heat. The large, bare flippers of a resting fur seal are normally cold, but we have found that when these animals on the Pribilof Islands are driven overland at their laborious gait, the flippers become warm. In contrast to the warm flippers, the rest of the fur seal's body surface feels cold, because very little heat escapes through the animal's dense fur. Heat can also be dissipated by evaporation from the mouth and tongue. Thus a dog or a caribou begins to pant, as a means of evaporative cooling, as soon as it starts to run.

In the pig the adaptation to cold by means of a variable circulation of heat in the blood achieves a high degree of refinement. The pig, with its skin only thinly covered with bristles, is as naked as a man. Yet it does well in the Alaskan winter without clothing. We can read the animal's response to cold by its expressions of comfort or discomfort, and we have measured its physiological reactions. In cold air the circulation of heat in the blood of swine is shunted away from the entire body surface, so that the surface becomes an effective insulator against loss of body heat. The pig can withstand considerable cooling of its body surface. Although a man is highly uncomfortable when his skin is cooled to 7 degrees C. below the internal temperature, a pig can be comfortable with its skin 30 degrees C. colder than the interior, that is, at a temperature of 8 degrees C. (about 46 degrees F.). Not until the air temperature drops below the freezing point (0 degrees C.) does the pig increase its rate of metabolism; in contrast a man, as I have mentioned, must do so at an air temperature of 28 degrees C.

With thermocouples in the form of needles we have probed the tissues of pigs below the skin surface. (Some pigs, like some people, will accept a little pain to win a reward.) We found that with the air temperature at −12 degrees C. the cooling of the pig's tissues extended as deep as 100 millimeters (about four inches) into its body. In warmer air the thermal gradient through the tissues was shorter and less steep. In short, the insulating mechanism of the hog involves a considerable depth of the animal's fatty mantle.

Even more striking examples of this kind of mechanism are to be found in whales, walruses and hair seals that dwell in the icy arctic seas. The whale and the walrus are completely bare; the hair seal is covered only with thin, short hair that provides almost no insulation when it is sleeked down in the water. Yet these animals remain comfortable in water around the freezing point although water, with a much greater heat capacity than air, can extract a great deal more heat from a warm body.

Examining hair seals from cold waters of the North Atlantic, we found that even in ice water these animals did not raise their rate of metabolism. Their skin was only one degree or so warmer than the water, and the cooling effect extended deep into the tissues—as much as a quarter of the distance through the thick part of the body. Hour after hour the animal's flippers all the way through would remain only a few degrees above freezing without the seals' showing any sign of discomfort. When the seals were moved into warmer water, their outer tissues rapidly warmed up. They would accept a transfer from warm water to ice water with equanimity and with no diminution of their characteristic liveliness.

How are the chilled tissues of all these animals able to function normally at temperatures close to freezing? There is first of all the puzzle of the response of fatty tissue. Animal fat usually becomes hard and brittle when it is cooled to low temperatures. This is true even of the land mammals of the Arctic, as far as their internal fats are concerned. If it were also true of extremities such as their feet, however, in cold weather their feet would become too inflexible to be useful. Actually it turns out that the fats in these organs behave differently from those in the warm internal tissues. Farmers have known for a long time that neat's-foot oil, extracted from the feet of cattle, can be used to keep leather boots and harness flexible in cold weather. By laboratory examination we have found that the fats in the bones of the lower leg and foot of the caribou remain soft even at 0 degrees C. The melting point of the fats in the leg steadily goes up in the higher portions of the leg. Eskimos have long been aware that fat from a caribou's foot will serve as a fluid lubricant in the cold, whereas the marrow fat from the upper leg is a solid food even at room temperature.

About the nonfatty substances in tissues we have little information; I have seen no reports by biochemists on the effects of temperature on their properties. It is known, however, that many of the organic substances of animal tissues are highly sensitive to temperature. We must therefore wonder how the tissues can maintain their serviceability over the very wide range of temperatures that the body surface experiences in the arctic climate.

We have approached this question by studies of the behavior of tissues at various temperatures. Nature offers many illustrations of the slowing of tissue functions by cold. Fishes, frogs and water insects are noticeably slowed down by cool water. Cooling by 10 degrees

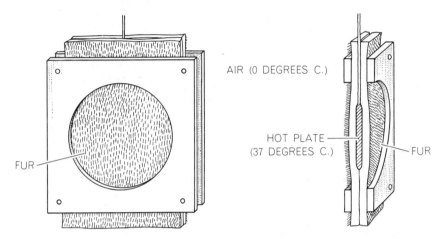

INSULATION BY FUR was tested in this apparatus, shown in a front view at left and a side view at right. The battery-operated heating unit provided the equivalent of body temperature on one side of the fur; outdoor temperatures were approximated on the other side.

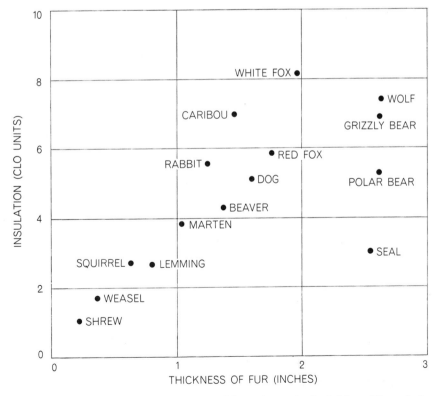

INSULATING CAPACITY of fur is compared for various animals. A "clo unit" equals the amount of insulation provided by the clothing a man usually wears at room temperature.

C. will immobilize most insects. A grass-hopper in the warm noonday sun can be caught only by a swift bird, but in the chill of early morning it is so sluggish that anyone can seize it. I had a vivid demonstration of the temperature effect one summer day when I went hunting on the arctic tundra near Point Barrow for flies to use in experiments. When the sun was behind clouds, I had no trouble picking up the flies as they crawled about in the sparse vegetation, but as soon as the sun came out the flies took off and were uncatchable. Measuring the temperature of flies on the ground, I ascertained that the difference between the flying and the slow-crawling state was a matter of only 2 degrees C.

Sea gulls walking barefoot on the ice in the Arctic are just as nimble as gulls on the warm beaches of California. We know from our own sensations that our fingers and hands are numbed by cold. I have used a simple test to measure the amount of this desensitization. After cooling the skin on my fingertips to about 20 degrees C. (68 degrees F.) by keeping them on ice-filled bags, I tested their sensitivity by dropping a light ball (weighing about one milligram) on them from a measured height. The weight multiplied by the distance of fall gave me a measure of the impact on the skin. I found that the skin at a temperature of 20 degrees C. was only a sixth as sensitive as at 35 degrees C. (95 degrees F.); that is, the impact had to be six times greater to be felt.

We know that even the human body surface has some adaptability to cold. Men who make their living by fishing can handle their nets and fish with wet hands in cold that other people cannot endure. The hands of fishermen, Eskimos and Indians have been found to be capable of maintaining an exceptionally vigorous blood circulation in the cold. This is possible, however, only at the cost of a higher metabolic production of body heat, and the production in any case has a limit. What must arouse our wonder is the extraordinary adaptability of an animal such as the hair seal. It swims in icy waters with its flippers and the skin over its body at close to the freezing temperature, and yet under the ice in the dark arctic sea it remains sensitive enough to capture moving prey and find its way to breathing holes.

Here lies an inviting challenge for all biologists. By what devices is an animal able to preserve nervous sensitivity in tissues cooled to low temperatures? Beyond this is a more universal and more

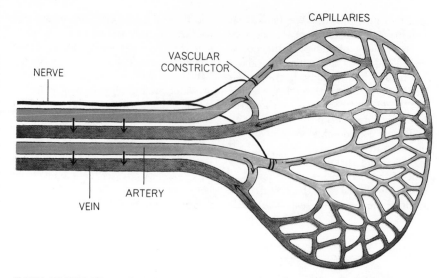

ROLE OF BLOOD in adaptation to cold is depicted schematically. One mechanism, indicated by the vertical arrows, is an exchange of heat between arterial and venous blood. The cold venous blood returning from an extremity acquires heat from an arterial network. The outgoing arterial blood is thus cooled. Hence the exchange helps to keep heat in the body and away from the extremities when the extremities are exposed to low temperatures. The effect is enhanced by the fact that blood vessels near the surface constrict in cold.

interesting question: How do the warm-blooded animals preserve their overall stability in the varying environments to which they are exposed? Adjustment to changes in temperature requires them to make a variety of adaptations in the various tissues of the body. Yet these changes must be harmonized to maintain the integration of the organism as a whole. I predict that further studies of the mechanisms involved in adaptation to cold will yield exciting new insights into the processes that sustain the integrity of warm-blooded animals.

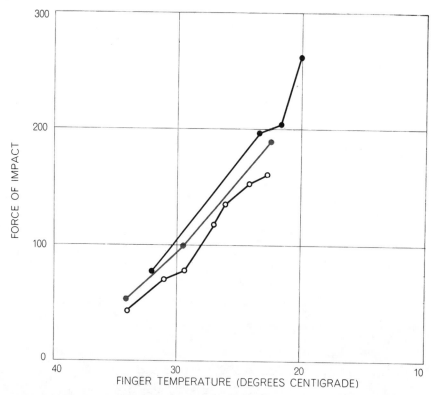

FINGER EXPERIMENT performed by the author showed that the more a finger was chilled, the farther a one-milligram ball had to be dropped for its impact to be felt on the finger. The vertical scale is arbitrary but reflects the relative increase in the force of impact.

The Freezing of Living Cells

by A. S. Parkes
June 1956

*Human blood and cattle sperm can now be stored at low
temperature for more than a year. It has also been shown
that a whole hamster can be frozen for an hour and thawed
out none the worse for wear*

Suspended animation in the frozen state—conferring potential immortality—has often figured in legend and literature, but it is less known as a theme of laboratory experiment. Indeed, it is only within the last two or three centuries that any accurate study of the effects of cold on living organisms has been possible. In 1683 the chemist Robert Boyle, writing of his observations on the subject, remarked:

"It may to most men appear a work of needless Curiosity, or superfluous diligence, to examine sollicitously by what Criterion or way of estimate the Coldness of Bodies and the degrees of it are to be judg'd; Since Coldness being a Tactile Quality, it seems impertinent to seek for any other judges of It than the Organs of that sense, whose proper object it is."

When, in 1714, Gabriel Fahrenheit invented the first accurate thermometer, objective measurement of the "coldness of bodies" became not only pertinent but a subject of interest to biologists. They turned eagerly to testing the effects of low temperatures upon plants and animals. Outstanding among these early students was Lazzaro Spallanzani, the 18th-century Italian naturalist, who performed many experiments on cells and tissues and made some remarkably shrewd and penetrating observations. In our own century there has been a great extension of such investigations. With the aid of liquefied gases (*e.g.*, air) living cells and tissues have been subjected to temperatures far below those ever recorded in nature.

Scientists engaged in this kind of work no doubt have been activated in part by "needless curiosity," but the work also has a practical side: namely, the hope of finding ways to preserve cells and tissues (*e.g.*, blood) for long periods. Progress was not very rapid until less than a decade ago. Experimenters

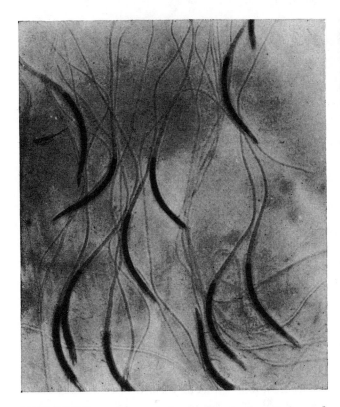 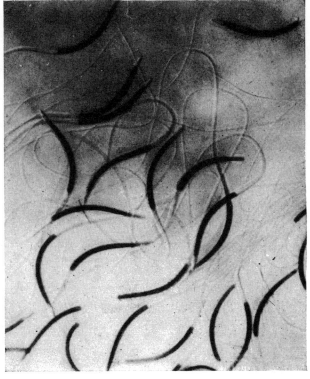

FOWL SPERM are photographed under the microscope before and after freezing. After the photomicrograph at left had been made, glycerol was added to the sperm and their temperature lowered to −79 degrees centigrade. At right they are shown after thawing.

WHOLE BLOOD is stored in a deep-freeze unit in the laboratory of Henry A. Sloviter in the Harrison Department of Surgical Research at the University of Pennsylvania School of Medicine. After glycerol solution is added, the blood is stored at –20 degrees C.

had some success in freezing and reviving lower organisms, but the cells of vertebrates, including mammals, were found to be very sensitive to freezing and thawing. It was thought that they were irreparably damaged by the formation of ice crystals during freezing. Hoping to avoid this damage, some investigators, notably B. J. Luyet and his co-workers at the St. Louis University Institute of Biophysics, experimented with stratagems such as partial dehydration of the cells and ultrarapid freezing. The drying method gave encouraging results. For example, some experimenters found that the spermatozoa of fowl occasional-

ly survived freezing and thawing when they were treated with a strong sugar solution as a dehydrating agent. However, the treatment succeeded in protecting only a very small proportion of the sperm. Much of this early work is summarized in the classic monograph, *Life and Death at Low Temperatures*, published by Luyet and his associate P. M. Gehenio in 1940.

In the fall of 1948 a curious chance discovery made possible a fresh onslaught on the whole matter. At the National Institute for Medical Research in London, Audrey U. Smith and Chris-

topher Polge were re-investigating the sugar dehydration technique. They used solutions of levulose, the fruit sugar, and froze fowl sperm with dry ice to minus 79 degrees centigrade (112 degrees below zero Fahrenheit). The experiments were unsuccessful: hardly any of the sperm survived after thawing. Pending some new inspiration, the experimenters put away some of their levulose solution in the refrigerator. Some months later they went back to the experiment, using what they took to be the stored levulose solution. This time the solution proved to be almost completely successful in protecting frozen fowl sperm: the sperm not only regained motility after thawing, but what was more, at least some also retained fertilizing power.

This very curious change in the solution's effect naturally aroused our keenest interest. The miraculous solution was analyzed, and it was then discovered that what had been taken from the refrigerator shelf was not the levulose solution but a mislabeled bottle of Mayer's albumen—a mixture of albumen (egg white) and glycerol (or glycerin)!

Tests soon showed that the albumen

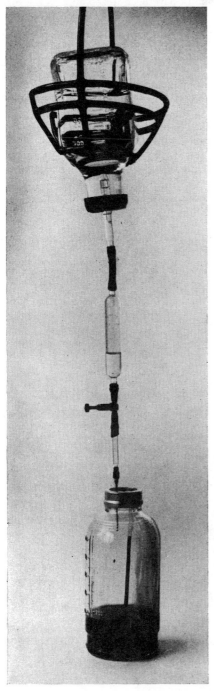

GLYCEROL SOLUTION is held in a vessel suspended above the blood bottle. The solution is added to the blood through a needle inserted into the cap of the bottle.

had nothing to do with the protection of frozen sperm, and our attention therefore focused on the glycerol—a syrupy, sweetish alcohol. It was only after our chance observation of glycerol's protective effect that we learned of an important earlier observation along the same lines. In 1946 Jean Rostand of Paris had observed that frog spermatozoa survived a slight degree of freezing when he added glycerol to the medium.

I record this story in some detail partly to add it to the literature of chance discoveries, partly to provide an authoritative account of the incident, and partly to correct any notion that our work with glycerol was based—without acknowledgment—on Rostand's observation, of which we had no knowledge at the time.

It was fortunate that the accident of the switched bottles occurred while fowl spermatozoa were being investigated. Glycerol would not have protected spermatozoa of mammals under the procedures we were using at the time, which included rapid freezing. Even with the fowl sperm, one snag after another cropped up. In the first place, it was found that the glycerol treatment, even without freezing, deprived most of the sperm of their fertilizing power. Considerably later we realized that this was probably due to the fact that the glycerol caused osmotic damage to the sperm after they were transferred to a normal environment in the oviduct of the hen. Experiments then established that the sperm would recover their normal fertilizing power if the glycerol was removed from the sperm suspension by a gradual process before the insemination.

The clues leading to this discovery stemmed from work by Smith on human red blood cells. She found that the red cells, like fowl sperm, would resist freezing and thawing without destruction if they were immersed in a saline solution with 15 per cent of glycerol. However, they disintegrated when they were returned to a normal medium—in this case the blood. Examination under the microscope showed that when the red cells were immersed in the glycerol solution, they first shrank, as water was drawn out of them by the glycerol, and then swelled back to their normal size and shape as glycerol slowly infiltrated the cell. When the thawed cells were returned to serum, they burst, because water passed into them more rapidly than glycerol passed out.

It was Henry A. Sloviter of the University of Pennsylvania School of Medi-

cine, a visiting worker in our laboratory, who developed the technique for gradual removal of glycerol from cell suspensions; he used the process known as dialysis. Red cells so treated survived for their normal lifetime when they were returned to the bloodstream.

A second snag in the protection of frozen sperm was even more illuminating and reorientated the whole course of our freezing research. It had been tacitly assumed that cells kept at minus 79 degrees C. would not deteriorate and might be held in a state of suspended animation indefinitely. Work with the frozen fowl spermatozoa showed that this assumption was fallacious. When the cells were kept at minus 79 degrees, most of them lost their ability to recover motility within a few weeks. The recovery rate declined even when the sperm were held at the extremely low temperature of minus 190 degrees C. in liquefied air. Since it could not be supposed that biochemical processes were going on at this temperature, we had to conclude that the cells were damaged in some way by physical causes. At this stage our team was joined by a physical chemist, J. E. Lovelock. He was able to show that the losses during storage were greatly influenced by the nature of the medium in which the cells were frozen.

Red blood cells, like fowl sperm, deteriorated at minus 79 degrees C., but more slowly. Lovelock proceeded to develop a modification of the glycerol-saline medium in which red cells could be preserved for a year or more with negligible loss. The cells apparently do not age during storage, for P. L. Mollison and his co-workers found that such cells have a normal lifetime after later transfusion.

This work has made possible the long-term storage of blood. So far the method has been used in blood banks only for preservation of rare types of blood, because of the expense and complication of removing the glycerol before use, but it is applicable on a large scale if necessary. The fact that studies of frozen fowl spermatozoa, prompted in part by "needless curiosity," led directly to a new and potentially important method of storing human blood is a good example of the value of basic research.

We turned next to experiments in freezing the sperm of cattle. The first results were disconcerting: bull sperm, when treated by the technique that had been successful with fowl sperm, failed to revive after thawing. We then realized that the reason for the failure might be the fact that we were using the method of rapid cooling. Mammalian spermatozoa are subject to shock when cooled suddenly. A more gradual cooling method was therefore adopted. The bull semen was cooled slowly, at the rate of one degree centigrade per minute, down to minus 15 degrees C., and then more rapidly to minus 79 degrees. This change made a great difference, and with other improvements in technique we were able to raise the survival rate of the bull sperm to about 70 per cent. Moreover, it appeared that in the case of bull sperm it was unnecessary to remove the glycerol after thawing: the glycerolized sperm survived return to a normal medium without damage.

We moved on at once to field tests in which cows were inseminated with bull semen that had been glycerolized, frozen and thawed. The first season's tests failed, but in the second season, after a slight modification of technique, preserved semen produced as high a rate of pregnancies as does normal fresh semen. Frozen bull semen has now been stored successfully without deterioration for several years. A bull that has been dead nearly three years is still getting cows in calf. In cattle, at least, telegenesis in time as well as space has arrived.

Experiments with other kinds of cells, including human sperm, naturally followed. Successful human pregnancies from sperm preserved by freezing have been reported. Varieties of herring that spawn at different times of year (spring and autumn) have been crossed for the first time by the use of sperm kept in cold storage. And there have been promising results with mammalian eggs, protozoa and various other types of cells.

Further, the work has now gone beyond single cells to tissues and whole organs. Substantial success has been achieved in preserving for long periods the adrenals and gonads of rats, the skin of rabbits and the uterus of the guinea pig. Long-term storage of human tissue for grafting is an obvious possible development. It has already been achieved with the cornea of the eye, for which storage banks kept at minus 79 degrees C. are being established. The technique should be useful for other tissues which, like the cornea, can be transplanted from one person to another. It will be most intriguing to see to what extent the provision of spare parts for human surgery can be developed.

Inevitably we were drawn to a still more fascinating question: Could a whole animal survive freezing? There are, of course, many stories of fish and frogs reviving after being found frozen solid in blocks of ice, but these stories must be apocryphal. Any that revived must have been frozen only on the surface, for no vertebrate, even of the cold-blooded variety, will survive conversion of all its body water to ice under any conditions so far known.

A warm-blooded animal is, of course, vulnerable to cooling of its body temperature at levels far short of freezing. In the rat, for example, the heartbeat and breathing stop when the internal temperature falls to about 15 degrees C. (59 degrees F.). Hibernating animals may maintain the heartbeat and other vital functions, at a reduced rate, at a body temperature as low as 2.5 degrees

BOTTLES OF BLOOD are shown before (*left*) and after (*right*) freezing. After the blood is thawed out, and before it can be used for transfusion, it must be treated to dilute or remove the glycerol. The red blood cells apparently do not age during cold storage.

C., but what would happen if the body temperature fell below zero?

In short, the reduction of a whole mammal to a state of suspended animation by freezing seemed an unpromising project. In considering the matter we saw at least two major problems: firstly, how to stop the heartbeat and breathing in such a way that they could be restored by rewarming, and secondly, how to avoid damage to the animal from freezing of its body water. At this point in our cogitations we were fortunate in making contact with Radoslav Andjus of the University of Belgrade, who seemed to have solved the first of our two problems. Andjus had succeeded in reanimating rats after chilling them almost to freezing temperature. He first placed the rats in cold air in a closed vessel, where they rebreathed their own expired air so that the proportion of oxygen decreased and that of carbon dioxide increased. When their body temperature fell to about 15 degrees C., he put the animals in iced water. Their respiration and heartbeat soon stopped, and their body temperature dropped to about one degree above freezing. Andjus found that he could revive some of the rats so treat-

ed by placing a hot spatula over the heart and giving artificial respiration.

Andjus came to our laboratory as a visiting worker, and his method was developed further so that rats were revived in almost every case after the body temperature had been reduced nearly to the freezing point of water. The way was thus cleared to study the cooling of mammals below zero. For this purpose it was decided to use the golden hamster, which in nature is a hibernating animal.

Audrey Smith of our group first cooled the hamster to about 3 degrees C. by Andjus' method. At this point, when its heartbeat and breathing were about to cease, the animal was transferred to a very cold bath kept liquid by an antifreeze solution of 50 per cent propylene glycol. It was kept at about 5 degrees below zero centigrade in a deep-freeze cabinet. The hamster's deep body temperature, as measured by a thermocouple inserted into the colon, soon went below zero. At this stage one of two things happened. The animal might start to freeze, at first on the outside and then in the deeper layers of the body. If taken out within one hour, when about 50 per cent of its total body water had been

converted to ice, the hamster could be revived by rewarming and artificial respiration, and it recovered completely. On the other hand, many hamsters did not freeze when their body temperature fell below zero. These "supercooled" animals remained quite limp at internal temperatures as low as 6 degrees below zero C. Rewarming from this state resulted in complete recovery of the animal.

A remarkable finding of these experiments was the rarity of cold injury to the surface parts of the animal that had been frozen stiff. Even extremities, such as the ears, very rarely showed signs of frostbite unless manipulated while frozen—a relationship discovered when a group of distinguished visitors to whom the experiments were being demonstrated yielded to the temptation to try to bend the frozen extremities.

At zero temperature the hamster has every appearance of being dead. All of its obvious vital functions have ceased. No doubt less obvious functions, such as enzyme activity, also are arrested or slowed down. However, the animal's many and various processes are certainly not slowed to a uniform degree, and it must be in a biochemically unbalanced and therefore unstable condition. Consequently no prolonged survival can be expected at this temperature. For long-term preservation in the frozen state much lower temperatures will be necessary, presumably not less than those already known to be required for individual cells and tissues. The induction of suspended animation at minus 80 degrees C. will no doubt necessitate the pretreatment of the mammal with some such substance as glycerol at a concentration which according to present knowledge is highly toxic to the whole animal. In other words, the biologist is not yet in sight of achieving suspended animation of a warm-blooded animal at a temperature likely to result in a stable state.

When biologists arrive at this achievement, they will find, of course, that novelists have been there before them and that the icy Valhalla already has several charter members. Among them is Edmond About's *L'homme à l'oreille cassée*. "The Man with the Broken Ear" was one of Napoleon's soldiers, who was frozen stiff during the retreat from Moscow. His body was retrieved and dried from the frozen state by an enterprising scientist, and years later was reconstituted and reanimated by another genius—at no greater cost than the accidental breaking off of one dried ear.

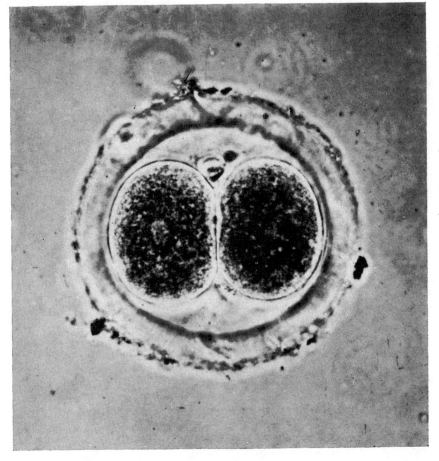

NEWLY FERTILIZED RABBIT EGG divides in laboratory culture after having been treated with glycerol, cooled to –190 degrees C. and kept frozen for 20 hours before thawing.

20

Heat Transfer in Plants

by David M. Gates
December 1965

Plants, like animals, must regulate their temperature in order to function at optimum physiological efficiency. This is accomplished through three mechanisms: radiation, transpiration and convection

There is an old Iowa saying that on a hot summer night one can hear the corn grow. This is no tall tale. Many physiological processes in plants depend primarily on temperature rather than on light, and among them is the enlargement and elongation of cells; if the weather is warm enough, corn will grow at night. A plant's temperature can also affect its photosynthetic activity, even though photosynthesis is primarily dependent on light. The photosynthetic process will slacken or halt if the plant's temperature gets too high or too low; it will reach an optimum at some intermediate temperature level. Evidently the temperature of a plant is a major factor in determining how efficiently the plant functions. But what determines the temperature of a plant?

The answer is that a plant's temperature is determined by its total environment. If the plant gives up more energy than it receives, it will become cooler; if it receives more energy than it gives up, it will become warmer. If the inward and outward flows of energy are equal, the plant is in thermal balance with its environment. Three separate phenomena govern the transfer of energy: radiation, transpiration and convection. In this article I shall first describe each of these processes and then consider some of the means by which they work together to regulate plant temperatures.

Radiation is quantitatively the most important of the three processes. In this connection it has two distinct forms. The first is solar radiation, the primary external source of energy for all physical and biological processes on the earth; this radiation is broadcast over a wide range of wavelengths [*see top illustration on page 180*]. The second is thermal radiation; this is the energy emitted by any object that is warmer than absolute zero. Thermal radiation is broadcast within a much narrower band than solar radiation; objects with temperatures between zero and 50 degrees centigrade, for example, radiate only at infrared wavelengths, and their peak output is at a wavelength of about 10 microns.

Solar energy falls on the earth's upper atmosphere at an average rate of two calories per square centimeter per minute; this energy flow is known as the solar constant. Minor constituents of the atmosphere absorb portions of the sun's radiation at each end of the visible spectrum. The ozone in the upper atmosphere screens out ultraviolet radiation at wavelengths shorter than 2,900 angstrom units (.29 micron); water vapor and carbon dioxide do the same for infrared radiation at wavelengths longer than 22 microns. Although the resulting "window" in the earth's atmosphere is only a narrow band of the sun's entire radiant spectrum, the sun emits most of its energy at these visible wavelengths; consequently a large fraction of the solar constant reaches the earth's surface.

In summer the surface of the earth in the Temperate Zones receives solar energy at a rate of between 1.2 and 1.4 calories per square centimeter per minute; mountains and deserts, which are often free of cloud cover, may receive as much as 1.6 calories. The solar energy that impinges on a tree or a stalk of corn includes not only direct sunlight but also sunlight that has been scattered by the atmosphere and sunlight that is reflected up from the earth's surface or down from clouds. On high mountains, on which intense direct sunlight and sunlight reflected from clouds occasionally fall at the same time, radiation values go as high as 2.2 calories—higher than the solar constant.

Solar radiation affects the earth's surface only between sunrise and sunset; thermal radiation, on the other hand, is always present. Every part of the environment steadily broadcasts energy at infrared wavelengths. If human eyes were sensitive to infrared radiation, the daytime sky would not be blue. Instead broad shafts of different hues (the "colors" of different infrared wavelengths) would be seen extending from the sky to the ground and from the ground to the sky. Even during the day the total amount of thermal energy radiating from the ground and from the atmosphere can equal or exceed the sun's radiation.

If a plant continuously absorbed energy without dissipating any, its temperature would increase steadily until it had suffered heat death. In actuality plants dispose of considerably more than half of the radiant energy they absorb by reradiating it. Plants also transpire; this second process of energy transfer further helps to rid a plant of surplus heat.

Transpiration converts the water in the leaves of a plant from a liquid to a gas; the water vapor then passes from the leaf into the surrounding atmosphere. The process of evaporation consumes energy and the transpiring leaf grows cooler. The water vapor is emitted through the pores in the leaf called stomates; there can be as many as 20,000 of these openings in one square centimeter of leaf surface. Specialized sausage-shaped cells of the leaf epidermis, which are known as guard cells, control the opening and closing of the stomates. Normally the stomates are closed during the night and open during the day.

The effectiveness of transpiration in transferring energy can be judged by the fact that a transpiration rate of only five ten-thousandths of a gram of water per square centimeter per minute gives rise to an energy loss of approximately three-tenths of a calorie. This is enough to lower the temperature of a transpiring leaf by as much as 15 degrees C. The process of transpiration can nonetheless be halted under circumstances that seem to threaten the plant with imminent heat death.

As the morning passes on a warm summer day a plant's sunlit leaves steadily rise in temperature and their photosynthetic activity approaches the optimum rate. At midday, when the heat load is at its peak, each leaf's transpiration rate is also at its highest and produces the maximum cooling effect. In spite of this energy transfer, the temperature of the leaf now rises above the optimum level and the photosynthetic process slows down. The concentration of carbon dioxide in the leaf's guard cells increases; the cells react and close the leaf's stomates. This puts an end to further transpiration; the leaf's temperature rises even higher and photosynthesis stops. At this point the leaf wilts and the plant appears to be on the verge of thermal catastrophe.

Fortunately wilting changes the leaf's orientation with respect to direct sunlight, so that the heat load is reduced. A cooler leaf temperature may allow photosynthesis to resume, thus reducing the concentration of carbon dioxide in the guard cells and causing the stomates to reopen. Normally, however, once the leaf temperature has gone too high the leaf will remain wilted until the sun has fallen low in the sky and the air has cooled.

I witnessed a remarkable example of temperature control by transpiration in August, 1963, when I accompanied W. M. Hiesey, Harold W. Milner and Malcolm A. Nobs of the Carnegie Institution of Washington on a visit to the institution's botanical study area near Yosemite National Park in California's Sierra Nevada range. Our purpose was to record the leaf temperatures of native species of the monkey flower, *Mimulus*, at various altitudes. The day was dry and cloudless. As we climbed the slopes above the tree line, where the alpine species *Mimulus lewisii* grew at an altitude of 10,600 feet, the sunshine was brilliant and the sky had a clarity that is seldom seen at lower altitudes. There was no wind; the air temperature was 19 degrees C. We found that the temperature of the alpine species' sunlit leaves was warmer than the air; it

WARM PINE BRANCH is cooled by rising convection currents. Each needle yields some of its heat conductively to the cooler air in contact with the needle surface. As the air is heated it rises (*shadows above the branch in this schlieren photograph*). Cool air replaces it and repeats the process. The rising current is somewhat turbulent because of a slight wind.

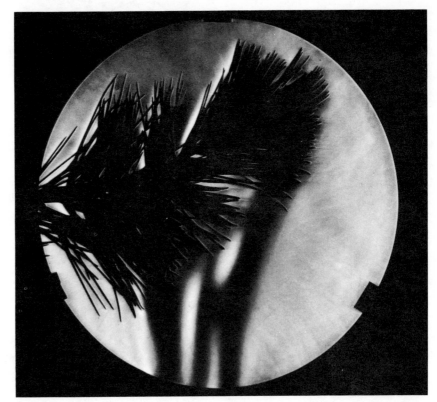

COOL PINE BRANCH that has lost heat by radiation to the night sky is warmed by falling convection currents. Each needle now absorbs heat from the warmer air in contact with its surface. As the air is cooled it sinks (*shadows below the branch in this schlieren photograph*). In both examples the heat transfer helps to stabilize the plant's temperature.

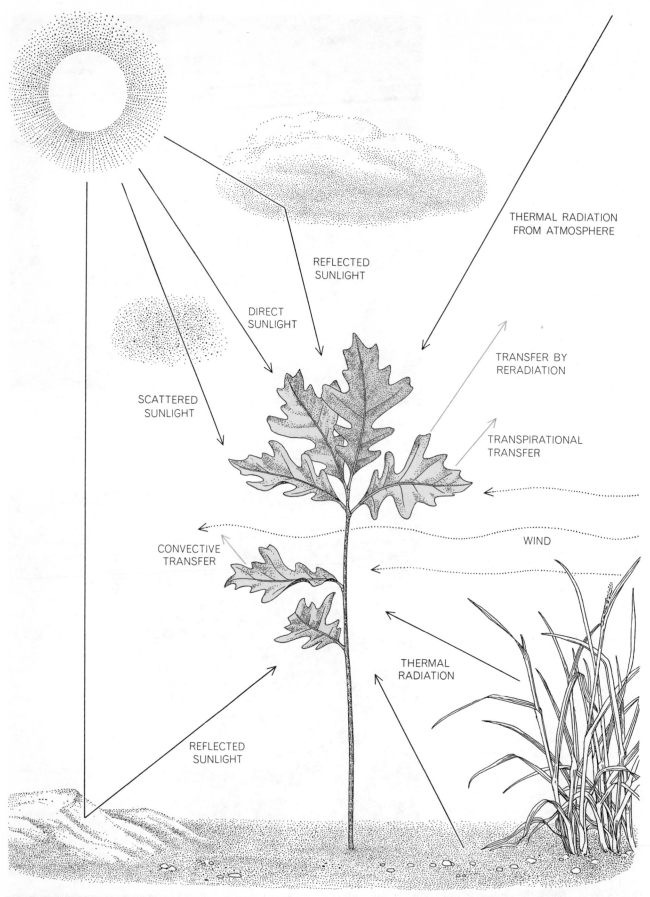

EXCHANGE OF ENERGY between a plant and its environment is shown schematically. Of the two inputs, solar radiation reaches the plant as direct, scattered and reflected sunlight. The second input, the thermal radiation, is emitted at infrared frequencies by the at-mosphere, the ground and other plants and animals. A plant would die of heat if most of this energy were not dissipated (*color*). The bulk of the heat load is reradiated; evaporative cooling by transpiration and heat transfer by convection removes the rest.

ranged between 25 and 28 degrees C. We returned to our jeep and quickly set off to check the temperatures of other monkey flowers growing at lower altitudes. On the edge of the San Joaquin valley, at an altitude of 1,300 feet, the sky was just as clear as it had been above the tree line and the air temperature was 37 degrees C. We examined some flowers of another species, *Mimulus cardinalis*, growing at this comparatively low altitude. I was quite prepared to predict that these sunlit leaves, which had a higher heat load than the alpine species, would yield temperature readings between 42 and 47 degrees C.

Instead the lowland leaves were cooler than the air surrounding them; their temperature ranged from 30 to 35 degrees C., not much warmer than the leaves above the tree line. This finding, contrary to all my previous experience, made me suspect that transpiration was involved. Monkey flowers are water-loving plants. Even the alpine species favor damp soil, and the plants we were examining in the valley grew along the border of a drainage ditch. When we measured the leaf temperatures of other species of plants growing in the same well-watered zone, they proved to be equally low. Then we tested the leaves of a live oak that was growing a few yards away in drier soil; their temperature was above the air temperature, ranging from 40 to 43 degrees C. It seemed likely that an abundant water supply, allowing liberal transpiration, accounted for the other plants' cooler temperatures.

This raised the question of whether the observations could be duplicated with monkey flowers raised in a growth chamber in which the temperature could be varied to match the range of natural conditions from valley to tree line. We placed monkey flower plants in a growth chamber under even and constant illumination and raised the air temperature successively from a low of 10 degrees C. to a high of nearly 60 degrees. The plants' leaf temperatures almost exactly duplicated those measured at various altitudes in the Sierras. It soon appeared that some kind of threshold effect was operating. When the air temperature was about 30 degrees C. or lower, the monkey flower leaves were warmer than the surrounding air; when the air temperature was raised above 30 degrees, the leaves remained cooler than the air.

As the leaf temperature rose, the plants' transpiration rate slowly increased until another kind of threshold

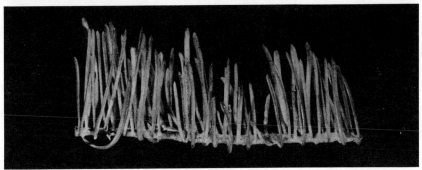

SILVER REPLICAS of conifer twigs were among the casts of various plants that the author made to determine which leaf shapes were the most efficient for heat transfer by convection. The conifers' cylindrical needles proved to be far superior to the flat leaf surfaces.

was reached at 41.5 degrees C. Just below that temperature the transpiration rate was about .0005 gram of water per square centimeter per minute. At the threshold temperature the rate jumped to .0022 gram, an increase of more than 400 percent. It was evident that a dramatic increase in leaf permeability took place at this temperature; the additional cooling effect was enough to prevent the monkey flower leaves from becoming warmer than 42 degrees C. even when the air temperature in the growth chamber was raised nearly 20 degrees higher than that.

A number of other plants—including species of cactus, agave, rhododendron and oleander—were included in these experiments. Their leaves also remained cooler than the air when the temperature was 30 degrees C. or higher, as long as the plants were watered abundantly. None of them, however, performed as dramatically as the monkey flowers.

These experiments clearly demonstrate that transpiration is an efficient mechanism for heat transfer in plants. Independent confirmation of this conclusion came recently from an experiment conducted at Purdue University. There G. D. Cook, J. R. Dixon and A. Carl Leopold painted the leaves of

tomato plants with a substance that prevented the stomates from opening. This treatment of course suppressed transpiration. When the experimenters then measured the temperature of treated and untreated leaves, they found that the transpiring leaves were cooler than the nontranspiring ones by 5 degrees C.

Convection, the third mechanism that allows energy transfer between a plant and its environment, has an important feature not shared by the other two. Radiation raises a plant's temperature and transpiration lowers it; convection will warm a cool plant or cool a warm one with equal facility, depending on whether the air is warmer or cooler than the plant. Convection acts across a thin atmospheric zone, known as the boundary layer, that surrounds all surfaces in still air. The rate at which energy is transferred across the boundary layer depends on the thickness of the layer and on the difference in temperature between the object and the atmosphere.

In still air the boundary layer is often about one centimeter thick. By means of schlieren photography, sometimes called shadow photography, the boundary layer and other regions of small variations in air density can be made

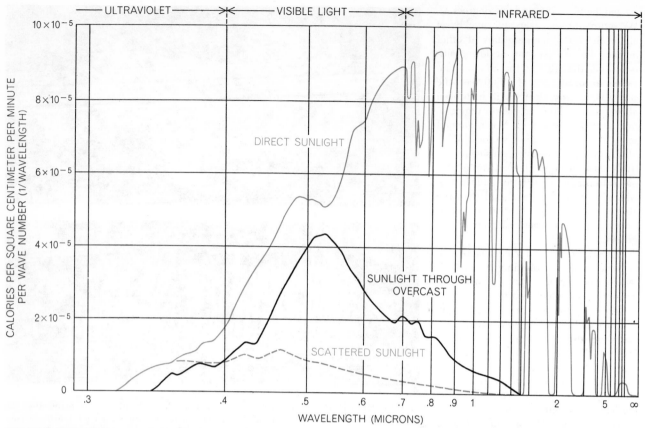

SOLAR RADIATION is not of equal intensity in all parts of the spectrum. Three curves show the variations from ultraviolet to infrared in the case of direct sunlight, of sunlight diffused through an overcast, and of sunlight scattered in the atmosphere. Divisions of the horizontal scale are proportional to frequency of radiation because the energy per quantum is similarly proportional. On this basis direct sunlight reaches peak intensity in the near infrared; how plants escape this heat load is shown in the illustration below.

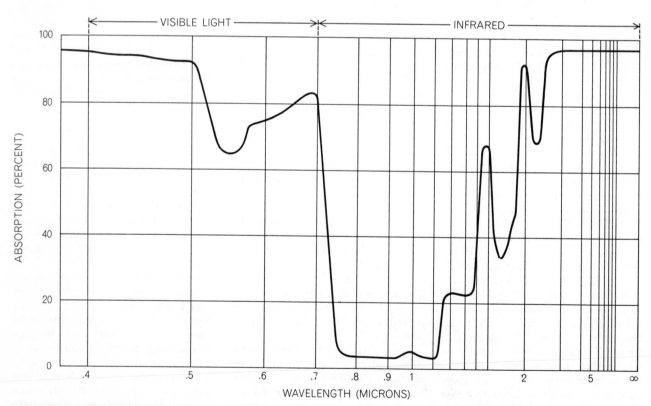

ABSORPTION OF ENERGY by the leaves of a plant averages nearly 90 percent of solar radiation in the spectral range from ultraviolet through visible frequencies. At near-infrared frequencies, however, absorption falls off sharply and stays at a low level throughout the range at which solar radiation is the most intense (*see illustration at top of page*). The curve shows absorption by a leaf of the poplar *Populus deltoides*; as at top, the divisions of the horizontal scale are proportional to the frequency of the radiation.

visible. Schlieren photography soon taught me that the air near the surface of a leaf is never really still. Consider a sunlit leaf becoming warm from solar radiation. If the air is cooler than the leaf, the molecules of gas in physical contact with the leaf's surface will be warmed by conduction and gain energy. The increase in the molecules' energy causes the mass of air to expand and rise like smoke going up a chimney [*see top illustration on page 177*]. As the warm air rises, cooler air replaces it along the leaf's surface; this air in turn is warmed, rises and is replaced, and thus the process of draining heat from the warm leaf continues.

At night, or whenever a leaf is cooler than the air, the situation is reversed. In darkness a leaf may emit more energy than it receives and become cooler than the surrounding air. Convection now forces the warmer air to give up energy to the cooler leaf. As the gas molecules lose energy the air sinks and streams downward to form a cool mass around the base of the plant [*see bottom illustration on page 177*]. This reverse transfer warms the leaf and tends to stabilize its temperature at a point near the temperature of the surrounding air.

A decrease in the efficiency of this energy transfer, which may occur on clear and still nights when the air temperature falls to near the freezing point, can bring disaster to a citrus grove. The leaves and fruit, radiating to the cold sky, can fall several degrees below air temperature and freeze. One way to combat this result is to use wind machines that create strong air currents among the trees. The airflow increases the rate of energy transfer from the warmer atmosphere to the cold trees and keeps them above the freezing temperature.

As the example of the citrus grove suggests, even a small amount of air movement rapidly destroys the boundary layer and thereby increases the rate of conductive heat transfer. As long as the flow of air over the surface of a leaf remains laminar, or unturbulent, the rate of heat transfer remains proportional to the square root of the wind velocity. The flow of air through vegetation, however, quickly becomes turbulent, and when this occurs the transfer rate becomes almost proportional to the actual wind velocity.

In a study of comparative convection efficiencies Frank Kreith, Cam Tibbals and I made precise silver casts of leaves and branches at the University of Colorado and measured the proper-

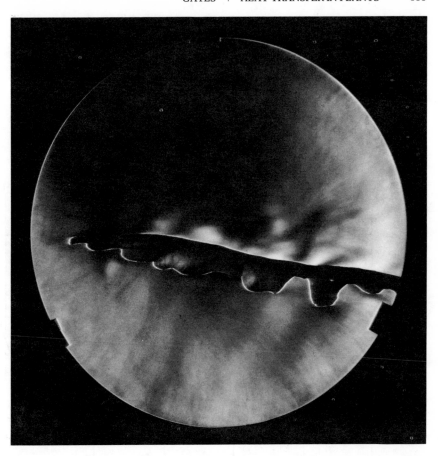

FLOW OF AIR over the leaf of a bur oak (*Quercus macrocarpa*) is revealed in schlieren photograph. Laminar near the leaf tip (*left*), the flow becomes quite turbulent farther along the leaf surface. This accelerates convection and also carries away transpired water vapor.

ties of the casts in a wind tunnel. It soon became clear that, as leaf size increased, the amount of heat transferred to the air by convection per unit of surface area became smaller. The air flowing across the surface of a leaf cast quickly approached the temperature of the surface; as the gradient between the temperature of the air and the temperature of the surface decreased so did the air's capacity for energy transfer over the downstream remainder of the surface.

The wind tunnel studies of silver casts showed that, per unit of surface area, the rate of energy transfer by convection would be many times greater for a pine needle than for a poplar leaf. Moreover, when the pine needle cast and poplar leaf cast were exposed to equal amounts of radiation, the temperature of the pine needle cast remained much closer to the air temperature than did that of the poplar leaf cast. These studies obviously could not account for cooling effects that would have been operated if our silver casts had been capable of transpiration.

The roles that radiation, transpiration and convection play in the life of a plant during the daylight hours are quite different from the roles they play at night. In daylight, for example, the energy budget for a horizontally oriented leaf exposed to the noonday sun in still air has the following breakdown. Direct sunlight and sunlight scattered in the sky irradiate the leaf's upper surface at the rate of 1.4 calories per square centimeter per minute. In addition, perhaps 20 percent of the sunlight that strikes the ground nearby is reflected upward and irradiates the lower surface of the leaf; this adds another .28 calorie per square centimeter per minute to the leaf's potential absorption of solar energy. Actually the leaf absorbs only 60 percent of the solar radiation impinging on it, so that the total load of absorbed heat attributable to solar radiation is 1.01 calories per square centimeter per minute. In addition to this the leaf is exposed to thermal radiation. The flow of energy at infrared wavelengths down from the atmosphere and the flow up from the ground respectively add to the leaf's heat load at the rate of .5 and .6 calorie per square centimeter per minute. The leaf absorbs 100 percent of this radiation, bringing the total input of radiant energy up to a rate of 2.11 calories per square centimeter per minute.

Assuming a leaf temperature of 36 degrees C. and a transpiration rate of .00054 gram per square centimeter per minute, the leaf will start to balance its energy budget by reradiating 68 percent (1.44 calories) of the radiant energy it is absorbing. Transpiration will dissipate an additional 29 percent (.62 calorie) of the load. Only 3 percent of the total (.05 calorie) remains to be disposed of by convection cooling. This is quite a small figure, but we have been assuming that the leaf is in still air; given a wind, convection would account for a much larger part of the energy transfer and might even play a larger role than transpiration.

At night a different series of transfers is required to keep the energy budget in balance. First, with the leaf stomates closed, transpiration is negligible. Second, as the leaf temperature drops below the temperature of the surround-

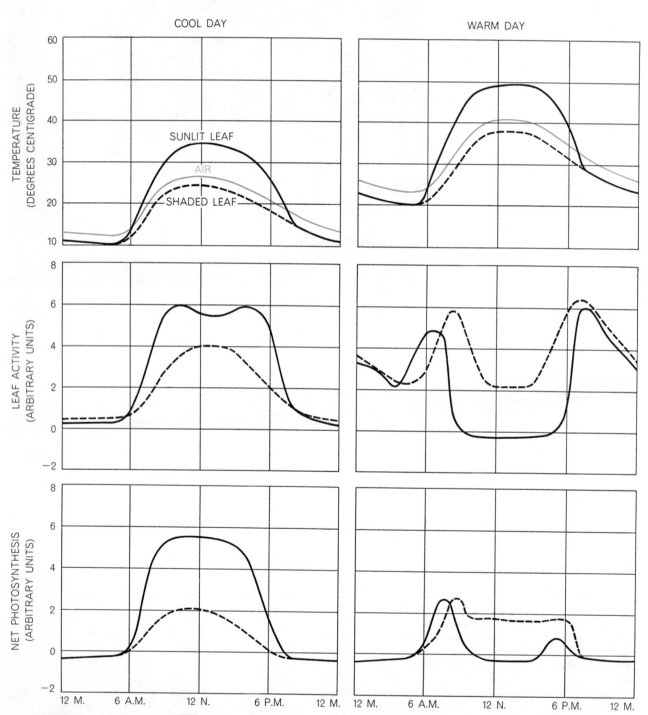

COOL DAY   WARM DAY

COOL SUMMER DAY produces sharply contrasting performances by a sunlit and a shaded leaf. Between sunrise and sunset the sunlit leaf is much warmer than the air (*top*). Its warmth favors physiological activity (*center*); net photosynthesis during the hours of daylight (*bottom*) attains a high level. The shaded leaf never gets enough light to engage in strong photosynthetic activity.

WARM SUMMER DAY produces the opposite effect by bringing the sunlit leaf to the brink of thermal catastrophe. As the leaf temperature rises during the forenoon it soon reaches the level at which biochemical activity is inhibited; photosynthesis stops (*bottom*) and cannot resume until late afternoon. In contrast, the shaded leaf is near optimum temperatures during most of the day.

ing air the convection process ceases to extract energy from the leaf and adds energy instead. Third, between sunset and sunrise the leaf has no solar radiation load to cope with. Although thermal radiation continues throughout the night, even the output of this energy source is somewhat reduced. The night budget has the following breakdown.

Thermal radiation from the atmosphere reaches the leaf at a rate of .4 calorie per square centimeter per minute, compared with the daylight rate of .5 calorie. Thermal radiation from the ground adds .55 calorie per minute, compared with the daylight rate of .6 calorie. The total night heat load from radiation is thus only .95 calorie per square centimeter per minute. Assuming still air, an air temperature of 15 degrees C. and a leaf temperature of 10 degrees, an additional .04 calorie per minute (4 percent of the total heat load) will be supplied to the leaf by convection. The energy input from radiation and convection combined comes to less than one calorie per square centimeter per minute. Reradiation suffices to dispose of this small input and balance the energy budget.

In nature, of course, matters are by no means as simple. There are cool days and warm nights; there are winds, clouds and overcast skies. The variations in energy transfer and fluctuations in leaf temperature under such changing circumstances are evident from a few examples of leaf-temperature readings I made during the summer in the yard of my former home in Boulder, Colo. I found that on overcast days all leaf temperatures remained below the temperature of the air. On cloudy days the leaf temperatures shifted rapidly from 6 to 8 degrees C. above the air temperature to 2 to 4 degrees below it as the clouds alternately exposed and obscured the sun. In the middle of a sunny summer day leaves that were exposed to the sun were warmer than the air by 5 degrees or more; the highest leaf temperature I recorded at Boulder—51 degrees C.—was 22 degrees warmer than the air. Leaves that were in the shade on the same kind of day, however, ranged from close to air temperature to as much as 5 degrees below it.

Each such variation naturally affects the plant's physiological activities. A sunlit leaf's approach to thermal catastrophe was described earlier; the midday heat load broke the leaf's photosynthetic activity into morning and afternoon intervals. A leaf that had been in shade throughout such a day would easily outdo the sunlit leaf in total photosynthetic

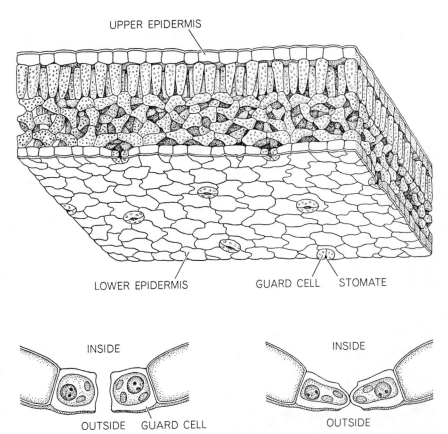

LEAF CROSS SECTION shows the paired sausage-shaped guard cells that control the opening and closing of the stomates, through which water vapor is emitted during transpiration. Stomates are normally closed at night and open during the day. Details show how guard cells bend to form an opening when turgid with moisture (*lower left*) but straighten to close when less turgid (*lower right*). An excess of carbon dioxide also closes the cells.

production. By the same token the pattern of energy transfers during a cool summer day would favor the sunlit leaf and prevent the shaded leaf from reaching the optimum temperature range for photosynthesis.

In spite of such variations there appears to be a balance of factors working in favor of overall plant efficiency. First, the energy transfers we have noted—which tend to keep leaves warmer than cool air but cooler than warm air—serve to maintain the leaves reasonably near the temperature range for optimum physiological activities. Second, taking photosynthesis as an example, the optimum temperature range for the process varies a great deal. For many plants in temperate regions the optimum ranges from 25 degrees C. to 30 degrees; the optimum for arctic and alpine plants can be as low as 15 degrees. One indication of how such optimums may have evolved has been discovered by H. A. Mooney of the University of California at Los Angeles. Mooney and his associates have found that the optimum temperature for photosynthesis depends

at least in part on a plant's own life history. When they were raised under conditions of moderately low temperature, Mooney's experimental plants achieved optimum photosynthesis at one temperature level. When they were acclimatized to warmer temperatures, optimum photosynthesis occurred at temperatures as much as 5 degrees higher than before. This kind of adaptability would be vitally important to plant populations that were extending their range into extreme environments.

In summary, it has been shown that one of the key interactions between a plant and its environment is transfer of energy between the two; the transfers determine the plant's temperature and the plant's temperature affects its physiological efficiency. A question that remains unanswered is why it is that different plants show optimum efficiency at different temperatures. The eventual answer to this question will probably involve the catalytic activity of plant enzymes; when it is found, it should open the door to the understanding of a fundamental mechanism in the ecological system of our planet.

Thermal Pollution and Aquatic Life

by John R. Clark
March 1969

*The increasing use of river and lake waters for industrial
cooling presents a real threat to fish and other organisms.
To avoid an ecological crisis new ways must be found to get
rid of waste heat*

Ecologists consider temperature the primary control of life on earth, and fish, which as cold-blooded animals are unable to regulate their body temperature, are particularly sensitive to changes in the thermal environment. Each aquatic species becomes adapted to the seasonal variations in temperature of the water in which it lives, but it cannot adjust to the shock of abnormally abrupt change. For this reason there is growing concern among ecologists about the heating of aquatic habitats by man's activities. In the U.S. it appears that the use of river, lake and estuarine waters for industrial cooling purposes may become so extensive in future decades as to pose a considerable threat to fish and to aquatic life in general. Because of the potential hazard to life and to the balance of nature, the discharge of waste heat into the natural waters is coming to be called thermal pollution.

The principal contributor of this heat is the electric-power industry. In 1968 the cooling of steam condensers in generating plants accounted for about three-quarters of the total of 60,000 billion gallons of water used in the U.S. for industrial cooling. The present rate of heat discharge is not yet of great consequence except in some local situations; what has aroused ecologists is the ninefold expansion of electric-power production that is in prospect for the coming years with the increasing construction of large generating plants fueled by nuclear energy. They waste 60 percent more energy than fossil-fuel plants, and this energy is released as heat in condenser-cooling water. It is estimated that within 30 years the electric-power industry will be producing nearly two million megawatts of electricity, which will require the disposal of about 20 million billion B.T.U.'s of waste heat per day. To carry off that heat by way of natural waters would call for a flow through power plants amounting to about a third of the average daily freshwater runoff in the U.S.

The Federal Water Pollution Control

HEATED EFFLUENT from a power plant on the Connecticut River is shown in color thermograms (*opposite page*), in which different temperatures are represented by different hues. At the site (*above*) three large pipes discharge heated water that spreads across the river at slack tide (*top thermogram*) and tends to flow downriver at ebb tide (*bottom*). An infrared camera made by the Barnes Engineering Company scans the scene and measures the infrared radiation associated with the temperature at each point in the scene (350 points on each of 180 horizontal lines). Output from an infrared radiometer drives a color modulator, thus changing the color of a beam of light that is scanned across a color film in synchrony with the scanning of the scene. Here the coolest water (*black*) is at 59 degrees Fahrenheit; increasingly warm areas are shown, in three-degree steps, in blue, light blue, green, light green, yellow, orange, red and magenta. The effluent temperature was 87 degrees. A tree (*dark object*) appears in lower thermogram because the camera was moved.

NUCLEAR POWER PLANT at Haddam on the Connecticut River empties up to 370,000 gallons of coolant water a minute through a discharge canal (*bottom*) into the river. In this aerial thermogram, made by HRB-Singer, Inc., for the U.S. Geological Survey's

Administration has declared that waters above 93 degrees Fahrenheit are essentially uninhabitable for all fishes in the U.S. except certain southern species. Many U.S. rivers already reach a temperature of 90 degrees F. or more in summer through natural heating alone. Since the waste heat from a single power plant of the size planned for the future (some 1,000 megawatts) is expected to raise the temperature of a river carrying a flow of 3,000 cubic feet per second by 10 degrees, and since a number of industrial and power plants are likely to be constructed on the banks of a single river, it is obvious that many U.S. waters would become uninhabitable.

A great deal of detailed information is available on how temperature affects the life processes of animals that live in the water. Most of the effects stem from the impact of temperature on the rate of metabolism, which is speeded up by heat in accordance with the van't Hoff principle that the rate of chemical reaction increases with rising temperature. The acceleration varies considerably for particular biochemical reactions and in different temperature ranges, but gener-

ally speaking the metabolic rate doubles with each increase of 10 degrees Celsius (18 degrees F.).

Since a speedup of metabolism increases the animal's need for oxygen, the rate of respiration must rise. F. E. J. Fry of the University of Toronto, experimenting with fishes of the salmon family, found that active fish increased their oxygen consumption as much as fourfold as the temperature of the water was raised to the maximum at which they could survive. In the brown trout the rate of oxygen consumption rose steadily until the lethal temperature of 79 degrees F. was reached; in a species of lake trout, on the other hand, the rate rose to a maximum at about 60 degrees and then fell off as the lethal temperature of 77 degrees was approached. In both cases the fishes showed a marked rise in the basal rate of metabolism up to the lethal point.

The heart rate often serves as an index of metabolic or respiratory stress on the organism. Experiments with the crayfish (*Astacus*) showed that its heart rate increased from 30 beats per minute at a water temperature of 39 degrees F. to 125 beats per minute at 72 degrees and then slowed to a final 65 beats per min-

ute as the water approached 95 degrees, the lethal temperature for this crustacean. The final decrease in heartbeat is evidence of the animal's weakening under the thermal stress.

At elevated temperatures a fish's respiratory difficulties are compounded by the fact that the hemoglobin of its blood has a reduced affinity for oxygen and therefore becomes less efficient in delivering oxygen to the tissues. The combination of increased need for oxygen and reduced efficiency in obtaining it at rising temperatures can put severe stress even on fishes that ordinarily are capable of living on a meager supply of oxygen. For example, the hardy carp, which at a water temperature of 33 degrees F. can survive on an oxygen concentration as low as half a milligram per liter of water, needs a minimum of 1.5 milligrams per liter when the temperature is raised to 95 degrees. Other fishes can exist on one to two milligrams at 39 degrees but need three to four milligrams merely to survive at 65 degrees and five milligrams for normal activity.

The temperature of the water has pronounced effects on appetite, digestion and growth in fish. Tracer experiments

Water Resources Division, temperature is represented by shades of gray. The hot effluent (*white*) is at about 93 degrees F.; ambient river temperature (*dark gray*) is 77 degrees. The line across the thermogram is a time marker for a series of absolute measurements.

with young carp, in which food was labeled with color, established that they digest food four times as rapidly at 79 degrees F. as they do at 50 degrees; whereas at 50 degrees the food took 18 hours to pass through the alimentary canal, at 79 degrees it took only four and a half hours.

The effects of temperature in regulating appetite and the conversion of the food into body weight can be used by hatcheries to maximize fish production in terms of weight. The food consumption of the brown trout, for example, is highest in the temperature range between 50 and 66 degrees. Within that range, however, the fish is so active that a comparatively large proportion of its food intake goes into merely maintaining its body functions. Maximal conversion of the food into a gain in weight occurs just below and just above that temperature range. A hatchery can therefore produce the greatest poundage of trout per pound of food by keeping the water temperature at just under 50 degrees or just over 66 degrees.

It is not surprising to find that the activity, or movement, of fish depends considerably on the water temperature.

By and large aquatic animals tend to raise their swimming speed and to show more spontaneous movement as the temperature rises. In many fishes the temperature-dependent pattern of activity is rather complex. For instance, the sockeye salmon cruises twice as fast in water at 60 degrees as it does in water at 35 degrees, but above 60 degrees its speed declines. The brook trout shows somewhat more complicated behavior: it increases its spontaneous activity as the temperature rises from 40 to 48 degrees, becomes less active between 49 and 66 degrees and above 66 degrees again goes into a rising tempo of spontaneous movements up to the lethal temperature of 77 degrees. Laboratory tests show that a decrease in the trout's swimming speed potential at high temperatures affects its ability to feed. By 63 degrees trout have slowed down in pursuing minnows, and at 70 degrees they are almost incapable of catching the minnows.

That temperature plays a critical role in the reproduction of aquatic animals is well known. Some species of fish spawn during the fall, as temperatures drop; many more species, however,

spawn in the spring. The rising temperature induces a seasonal development of their gonads and then, at a critical point, triggers the female's deposit of her eggs in the water. The triggering is particularly dramatic in estuarine shellfish (oysters and clams), which spawn within a few hours after the water temperature reaches the critical level. Temperature also exerts a precise control over the time it takes a fish's eggs to hatch. For example, fertilized eggs of the Atlantic salmon will hatch in 114 days in water at 36 degrees F. but the period is shortened to 90 days at 45 degrees; herring eggs hatch in 47 days at 32 degrees and in eight days at 58 degrees; trout eggs hatch in 165 days at 37 degrees and in 32 days at 54 degrees. Excessive temperatures, however, can prevent normal development of eggs. The Oregon Fish Commission has declared that a rise of 5.4 degrees in the Columbia River could be disastrous for the eggs of the Chinook salmon.

Grace E. Pickford of Yale University has observed that "there are critical temperatures above or below which fish will not reproduce." For instance, at a temperature of 72 degrees or higher the banded sunfish fails to develop eggs. In the case of the carp, temperatures in the range of 68 to 75 degrees prevent cell division in the eggs. The possum shrimp *Neomysis*, an inhabitant of estuaries, is blocked from laying eggs if the temperature rises above 45 degrees. There is also the curious case of the tiny crustacean *Gammarus*, which at temperatures above 46 degrees produces only female offspring.

Temperature affects the longevity of fish as well as their reproduction. D'Arcy Wentworth Thompson succinctly stated this general life principle in his classic *On Growth and Form:* "As the several stages of development are accelerated by warmth, so is the duration of each and all, and of life itself, proportionately curtailed. High temperature may lead to a short but exhausting spell of rapid growth, while the slower rate manifested at a lower temperature may be the best in the end." Thompson's principle has been verified in rather precise detail by experiments with aquatic crustaceans. These have shown, for example, that *Daphnia* can live for 108 days at 46 degrees F. but its lifetime at 82 degrees is 29 days; the water flea *Moina* has a lifetime of 14 days at 55 degrees, its optimal temperature for longevity, but only five days at 91 degrees.

Other effects of temperature on life processes are known. For example, a century ago the German zoologist Karl Möbius noted that mollusks living in cold waters grew more slowly but attained larger size than their cousins living in warmer waters. This has since been found to be true of many fishes and other water animals in their natural habitats.

Fortunately fish are not entirely at the mercy of variations in the water temperature. By some process not yet understood they are able to acclimate themselves to a temperature shift if it is moderate and not too sudden. It has been found, for instance, that when the eggs of largemouth bass are suddenly transferred from water at 65 or 70 degrees to water at 85 degrees, 95 percent of the eggs perish, but if the eggs are acclimated by gradual raising of the temperature to 85 degrees over a period of 30 to 40 hours, 80 percent of the eggs will survive. Experiments with the possum shrimp have shown that the lethal temperature for this crustacean can be raised by as much as 24 degrees (to a high of 93 degrees) by acclimating it through a series of successively higher temperatures. As a general rule aquatic animals can acclimate to elevated temperatures faster than they can to lowered temperatures.

Allowing for maximum acclimation (which usually requires spreading the gradual rise of temperature over 20 days), the highest temperatures that most fishes of North America can tolerate range from about 77 to 97 degrees F. The direct cause of thermal death is not known in detail; various investigators have suggested that the final blow may be some effect of heat on the nervous system or the respiratory system, the coagulation of the cell protoplasm or the inactivation of enzymes.

Be that as it may, we need to be concerned not so much about the lethal temperature as about the temperatures that may be *unfavorable* to the fish. In the long run temperature levels that adversely affect the animals' metabolism, feeding, growth, reproduction and other vital functions may be as harmful to a fish population as outright heat death.

Studies of the preferred, or optimal, temperature ranges for various fishes have been made in natural waters and in the laboratory [*see illustration, page 190*]. For adult fish observed in nature the preferred level is about 13 degrees F. below the lethal temperature on the average; in the laboratory, where the experimental subjects used (for convenience) were very young fish, the preferred level was 9.5 degrees below the lethal temperature. Evidently young fish need warmer waters than those that have reached maturity do.

The optimal temperature for any water habitat depends not only on the preferences of individual species but also on the well-being of the system as a whole. An ecological system in dynamic balance is like a finely tuned automobile engine, and damage to any component can disable or impair the efficiency of the entire mechanism. This means that if we are to expect a good harvest of fish, the temperature conditions in the water medium must strike a favorable balance for all the components (algae and other plants, small crustaceans, bait fishes and so on) that constitute the food chain producing the harvested fish. For example, above 68 degrees estuarine eelgrass does not reproduce. Above 90 degrees there is extensive loss of bottom life in rivers.

So far there have been few recorded instances of direct kills of fish by thermal pollution in U.S. waters. One recorded kill occurred in the summer of 1968 when a large number of menhaden acclimated to temperatures in the 80's became trapped in effluent water at 93 to 95 degrees during the testing of a new power plant on the Cape Cod Canal. A very large kill of striped bass occurred

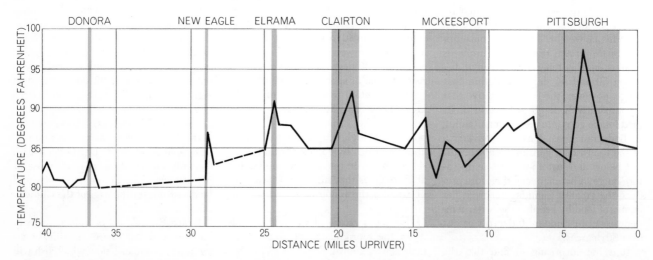

WATER TEMPERATURES can become very high, particularly in summer, along rivers with concentrated industry. The chart shows the temperature of the Monongahela River, measured in August, along a 40-mile stretch upriver from its confluence with the Ohio.

in the winter and early spring of 1963 at the nuclear power plant at Indian Point on the Hudson River. In that instance the heat discharge from the plant was only a contributing factor. The wintering, dormant fish, attracted to the warm water issuing from the plant, became trapped in its structure for water intake, and they died by the thousands from fatigue and other stresses. (Under the right conditions, of course, thermal discharges benefit fishermen by attracting fish to discharge points, where they can be caught with a hook and line.)

Although direct kills attributable to thermal pollution apparently have been rare, there are many known instances of deleterious effects on fish arising from natural summer heating in various U.S. waters. Pollution by sewage is often a contributing factor. At peak summer temperatures such waters frequently generate a great bloom of plankton that depletes the water of oxygen (by respiration while it lives and by decay after it dies). In estuaries algae proliferating in the warm water can clog the filtering apparatus of shellfish and cause their death. Jellyfish exploding into abundant growth make some estuarine waters unusable for bathing or other water sports, and the growth of bottom plants in warm waters commonly chokes shallow bays and lakes. The formation of hydrogen sulfide and other odorous substances is enhanced by summer temperatures. Along some of our coasts in summer "red tides" of dinoflagellates occasionally bloom in such profusion that they not only bother bathers but also may poison fish. And where both temperature and sewage concentrations are high a heavy toll of fish may be taken by proliferating microbes.

This wealth of evidence on the sensitivity of fish and the susceptibility of aquatic ecosystems to disruption by high temperatures explains the present concern of biologists about the impending large increase in thermal pollution. Already last fall 14 nuclear power plants, with a total capacity of 2,782 megawatts, were in operation in the U.S.; 39 more plants were under construction, and 47 others were in advanced planning stages. By the year 2000 nuclear plants are expected to be producing about 1.2 million megawatts, and the nation's total electricity output will be in the neighborhood of 1.8 million megawatts. As I have noted, the use of natural waters to cool the condensers would entail the heating of an amount of water equivalent to a third of the yearly freshwater runoff in the U.S.; during low-flow periods in summer the requirement would be 100 percent of the runoff. Obviously thermal pollution of the waters on such a scale is neither reasonable nor feasible. We must therefore look for more efficient and safer methods of dissipating the heat from power plants.

One might hope to use the heated water for some commercial purpose. Unfortunately, although dozens of schemes have been advanced, no practicable use has yet been found. Discharge water is not hot enough to heat buildings. The cost of transmission rules out piping it to farms for irrigation even if the remaining heat were enough to improve crop production. More promising is the idea of using waste heat in desalination plants to aid in the evaporation process, but this is still only an idea. There has also been talk of improving the efficiency of sewage treatment with waste heat from power plants. Sea farm-

ing may offer the best hope of someday providing a needed outlet for discharges from coastal power plants; pilot studies now in progress in Britain and the U.S. are showing better growth of fish and shellfish in heated waters than in normal waters, but no economically feasible scheme has yet emerged. It appears, therefore, that for many years ahead we shall have to dispose of waste heat to the environment.

The dissipation of heat can be facilitated in various ways by controlling the passage of the cooling water through the condensers. The prevailing practice is to pump the water (from a river, a lake or an estuary) once through the steam-condensing unit, which in a 500-megawatt plant may consist of 400 miles of one-inch copper tubing. The water emerging from the unit has been raised in temperature by an amount that varies from 10 to 30 degrees F., depending on the choice of manageable factors such as the rate of flow. This heated effluent is then discharged through a channel into the body of water from which it was taken. There the effluent, since it is warmer and consequently lighter than the receiving water, spreads in a plume over the surface and is carried off in the direction of the prevailing surface currents.

The ensuing dispersal of heat through the receiving water and into the atmosphere depends on a number of natural factors: the speed of the currents, the turbulence of the receiving water (which affects the rate of mixing of the effluent with it), the temperature difference between the water and the air, the humidity of the air and the speed and direction of the wind. The most variable and most important factor is wind: other things being equal, heat will be dissipated from

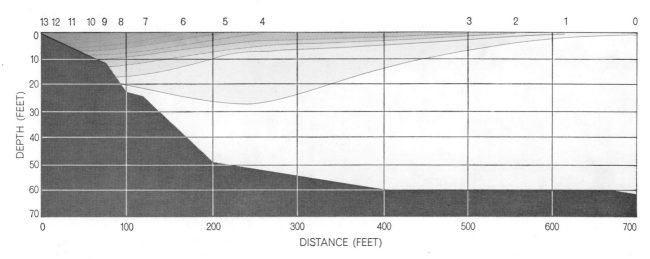

HOT-WATER "PLUME" that would result from an Indian Point nuclear power plant mixes with cooler Hudson River as shown by the one-degree contour lines. This section across the river shows temperature structure that was predicted by engineering studies.

the water to the air by convection three times faster at a wind speed of 20 miles per hour than at a wind speed of five miles.

In regulating the rate of water flow through the condenser one has a choice between opposite strategies. By using a rapid rate of flow one can spread the heat through a comparatively large volume of cooling water and thus keep down the temperature of the effluent; conversely, with a slow rate of flow one can concentrate the heat in a smaller volume of coolant. If it is advantageous to obtain good mixing of the effluent with the receiving water, the effluent can be discharged at some depth in the water rather than at the surface. The physical and ecological nature of the body of water will determine which of these strategies is best in a given situation. Where the receiving body is a swift-flowing river, rapid flow through the condenser and dispersal of the low-temperature effluent in a narrow plume over the water surface may be the most effective way to dissipate the heat into the atmosphere. In the case of a still lake it may be best to use a slow flow through the condenser so that the comparatively small volume of effluent at a high tem-

perature will be confined to a small area in the lake and still transfer its heat to the atmosphere rapidly because of the high temperature differential. And at a coastal site the best strategy may be to discharge an effluent of moderate temperature well offshore, below the ocean-water surface.

There are many waters, however, where no strategy of discharge will avail to make the water safe for aquatic life (and where manipulation of the discharge will also be insufficient to avoid dangerous thermal pollution), particularly where a number of industrial and power plants use the same body of water for cooling purposes. It therefore appears that we shall have to turn to extensive development of devices such as artificial lakes and cooling towers.

Designs for such lakes have already been drawn up and implemented for plants of moderate size. For the 1,000-megawatt power plant of the future a lake with a surface area of 1,000 to 2,000 acres would be required. (A 2,000-acre lake would be a mile wide and three miles long.) The recommended design calls for a lake only a few feet deep at one end and sloping to a depth of 50

feet at the other end. The water for cooling is drawn from about 30 feet below the surface at the deep end and is pumped through the plant at the rate of 500,000 gallons per minute, and the effluent, 20 degrees higher in temperature than the inflow, is discharged at the lake's shallow end. The size of the lake is based on a pumping rate through the plant of 2,000 acre-feet a day, so that all the water of the lake (averaging 15 feet in depth) is turned over every 15 days. Such a lake would dissipate heat to the air at a sufficient rate even in prolonged spells of unfavorable weather, such as high temperature and humidity with little wind.

Artificial cooling lakes need a steady inflow of water to replace evaporation and to prevent an excessive accumulation of dissolved material. This replenishment can be supplied by a small stream flowing into the lake. The lake itself can be built by damming a natural land basin. A lake complex constructed to serve not only for cooling but also for fishing might consist of two sections: the smaller one, in which the effluent is discharged, would be stocked with fishes tolerant to heat, and the water from this basin, having been cooled by exposure to

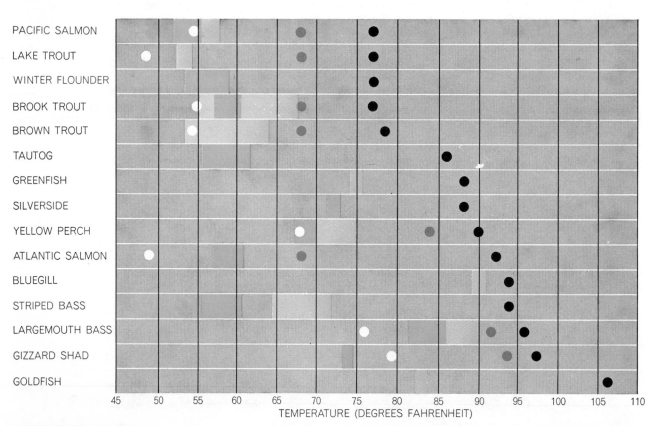

FISHES VARY WIDELY in temperature preference. Preferred temperature ranges are shown here for some species for which they have been determined in the field (*dark colored bands*) and, generally for younger fish, in the laboratory (*light color*). The chart also indicates the upper lethal limit (*black dot*) and upper limits recommended by the Federal Water Pollution Control Administration for satisfactory growth (*colored dot*) and spawning (*white dot*). Temperatures well below lethal limit can be in stress range.

the air, would then flow into a second and larger lake where other fishes could thrive.

In Britain, where streams are small, water is scarce and appreciation of aquatic life is high, the favored artificial device for getting rid of waste heat from power plants has been the use of cooling towers. One class of these towers employs the principle of removing heat by evaporation. The heated effluent is discharged into the lower part of a high tower (300 to 450 feet) with sloping sides; as the water falls in a thin film over a series of baffles it is exposed to the air rising through the tower. Or the water can be sprayed into the tower as a mist that evaporates easily and cools quickly. In either case some of the water is lost to the atmosphere; most of it collects in a basin and is pumped into the waterway or recirculated to the condensers. The removal of heat through evaporation can cool the water by about 20 degrees F.

The main drawback of the evaporation scheme is the large amount of water vapor discharged into the atmosphere. The towers for a 1,000-megawatt power plant would eject some 20,000 to 25,000 gallons of evaporated water per minute—an amount that would be the equivalent of a daily rainfall of an inch on an area of two square miles. On cold days such a discharge could condense into a thick fog and ice over the area in the vicinity of the plant. The "wet" type of cooling tower may therefore be inappropriate in cold climates. It is also ruled out where salt water is used as the coolant: the salt spray ejected from a single large power plant could destroy vegetation and otherwise foul the environment over an area of 160 square miles.

A variation of the cooling-tower system avoids these problems. In this refinement, called the "dry tower" method, the heat is transferred from the cooling water, through a heat exchanger something like an enormous automobile radiator, directly to the air without evaporation. The "dry" system, however, is two and a half to three times as expensive to build as a "wet" system. In a proposed nuclear power plant to be built in Vernon, Vt., it is estimated that the costs of operation and amortization would be $2.1 million per year for a dry system and $800,000 per year for a wet system. For the consumer the relative costs would amount respectively to 2.6 and 1 percent of the bill for electricity.

The public-utility industry, like other industries, is understandably reluctant to incur large extra expenses that add sub-

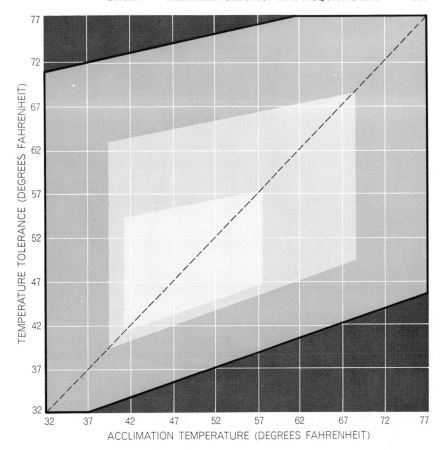

ACCLIMATION extends the temperature range in which a fish can thrive, but not indefinitely. For a young sockeye salmon acclimated as shown by the horizontal scale, spawning is inhibited outside the central (*light color*) zone and growth is poor outside the second zone (*medium color*); beyond the outer zone (*dark color*) lie the lethal temperatures.

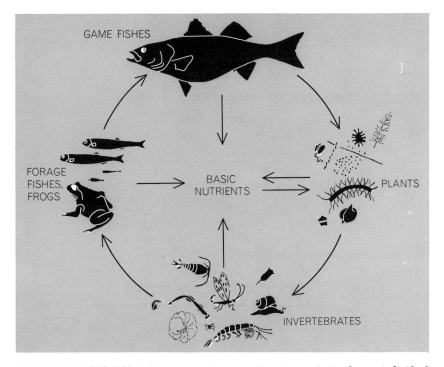

AQUATIC ECOSYSTEM is even more sensitive to temperature variation than an individual fish. A single game-fish species, for example, depends on a food chain involving smaller fishes, invertebrates, plants and dissolved nutrients. Any change in the environment that seriously affects the proliferation of any link in the chain can affect the harvest of game fish.

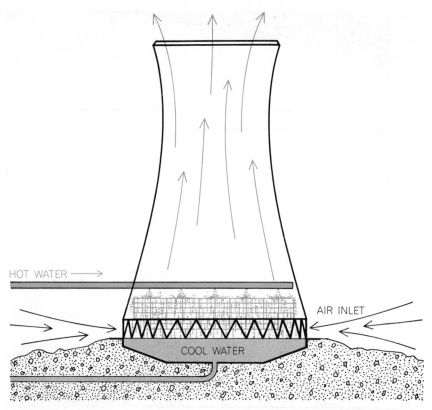

COOLING TOWER is one device that can dissipate industrial heat without dumping it directly into rivers or lakes. This is a "wet," natural-draft, counterflow tower. Hot water from the plant is exposed to air moving up through the chimney-like tower. Heat is removed by evaporation. The cooled water is emptied into a waterway or recirculated through the plant. In cold areas water vapor discharged into the atmosphere can create a heavy fog.

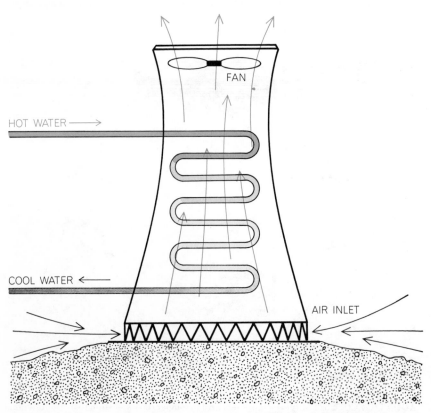

"DRY" COOLING TOWER avoids evaporation. The hot water is channeled through tubing that is exposed to an air flow, and gives up its heat to the air without evaporating. In this mechanical-draft version air is moved through the tower by a fan. Dry towers are costly.

stantially to the cost of its product and services. There is a growing recognition, however, by industry, the public and the Government of the need for protecting the environment from pollution. The Federal Water Pollution Control Administration of the Department of the Interior, with help from a national advisory committee, last year established a provisional set of guidelines for water quality that includes control of thermal pollution. These guidelines specify maximum permissible water temperatures for individual species of fish and recommend limits for the heating of natural waters for cooling purposes. For example, they suggest that discharges that would raise the water temperature should be avoided entirely in the spawning grounds of cold-water fishes and that the limit for heating any stream, in the most favorable season, should be set at five degrees F. outside a permitted mixing zone. The size of such a zone is likely to be a major point of controversy. Biologists would limit it to a few hundred feet, but in one case a power group has advocated 10 miles.

It appears that the problem of thermal pollution will receive considerable attention in the 91st Congress. Senator Edward M. Kennedy has proposed that further licensing of nuclear power-plant construction be suspended until a thorough study of potential hazards, including pollution of the environment, has been made. Senator Edmund S. Muskie's Subcommittee on Air and Water Pollution of the Senate Committee on Public Works last year held hearings on thermal pollution in many parts of the country.

Thermal pollution of course needs to be considered in the context of the many other works of man that threaten the life and richness of our natural waters: the discharges of sewage and chemical wastes, dredging, diking, filling of wetlands and other interventions that are altering the nature, form and extent of the waters. The effects of any one of these factors might be tolerable, but the cumulative and synergistic action of all of them together seems likely to impoverish our environment drastically.

Temperature, Gordon Gunter of the Gulf Coast Research Laboratory has remarked, is "the most important single factor governing the occurrence and behavior of life." Fortunately thermal pollution has not yet reached the level of producing serious general damage; moreover, unlike many other forms of pollution, any excessive heating of the waters could be stopped in short order by appropriate corrective action.

POWER PLANT being completed by the Tennessee Valley Authority on the Green River in Kentucky will be the world's largest coal-fueled electric plant. Its three wet, natural-draft cooling towers, each 437 feet in height and 320 feet in diameter at ground level, will each have a capacity of 282,000 gallons of water a minute, which they will be able to cool through a range of 27.5 degrees.

INDUSTRIAL PLANTS of various kinds can use towers to cool process water. These two five-cell cooling towers were built by the Marley Company for a chemical plant. They are wet, mechanical-draft towers of the cross-flow type: a fan in each stack draws air in through the louvers, across films of falling water and then up. The towers cool 120,000 gallons a minute through a 20-degree range.

V

PRESSURE, GRAVITY, AND LIFE

PRESSURE, GRAVITY, AND LIFE

V

INTRODUCTION

When we think of the interactions between living organisms and their environment, we should distinguish those aspects of the environment that are absolutely essential for life from those that are merely useful to at least some organisms. An adequate supply of matter and radiant energy is obviously essential. A temperature range in which biochemical molecules and their supramolecular structures are stable is also essential.

Seeing, hearing, tasting, smelling, and feeling the environment are essential abilities for many existing organisms but are not necessarily essential for life as such. The viability of an organism in a specific habitat is generally due to the evolutionary adaptations that permit it to exploit an environmental force. Often these adaptations are manifested as the organism's ability to sense environmental stimuli and orient itself with respect to them.

Understanding how living matter responds to light, sound, chemicals, textures, and mechanical stresses will provide us with insight into the potentials of living matter and the life process itself. The property of sensitivity is one of the basic criteria by which we distinguish the living state of matter from the nonliving. The evolutionary significance of an organism's being able to "see" its environment must be very great because this ability appeared early in the development of life forms and has been maintained with magnificent elaboration throughout the three billion years of organic evolution.

Another environmental force that living organisms learned to utilize in sensing their environment is pressure and pressure changes. Our senses of touch and hearing depend on pressure-detecting structures in sensitive cells. Although little is known about how mechanical forces stimulate the pressure-sensitive cells, we suspect that some distortion of an organized, multimolecular structure is the cause of the signal that is generated in response to the force.

The very high pressures that prevail in the depths of the ocean appear to have a direct effect on the state of living protoplasm. Several interesting observations on the protoplasm of amoebas and sea urchin eggs when they are exposed to pressures equal to several-hundred-fold that of normal atmospheric pressure are the subject of the article, "Cells at High Pressure" (page 200), by Douglas Marsland. Since this article was written, further research has revealed many additional details on the changes that occur in protoplasm under high pressures.

Electron microscopic studies have revealed that the ultrastructural elements of the cell, the microtubules, disappear under high pressures, but reappear when the pressure is released. Microtubules are thought to be responsible for maintaining the shape of cells. They are also associated with cellular motility and with the mitotic apparatus. We know that microtubules are constructed by the association of globular protein molecules called tubulins. The reversible polymerization and depolymerization of tubulins is apparently responsive to applied hydrostatic pressure.

Although high pressures can depolymerize vital cell structures, some organisms seem to be immune to such damage. In his article, "Animals of the Abyss" (page 208), Anton F. Bruun describes numerous, quite complex animals that seem to thrive in the depths of the ocean under high hydrostatic pressures. Since these organisms undoubtedly possess a basic cellular design similar to that of all other living cells, it is very interesting to compare their microtubules with those of organisms from the sea surface. We can probably learn much about the process of formation and disappearance of these cellular skeletons by a comparitive study of the animals of the abyss.

Perhaps it is pertinent to suggest that we keep an open mind when we consider the possibilities or impossibility of life in this or that environment. Were we to generalize from the laboratory findings on sea-level organisms (i.e., that microtubules disappear at high pressures), we would conclude that higher forms of life could not exist in the abyss. The versatility and adaptability of living matter often prove to be greater than we suspect. Biologists are attracted to the study of organisms that are adapted to extreme environmental conditions for precisely this reason. We would like to understand, for example, what kind of changes occur in the shape of the tubulin molecules of the animals of the abyss, and how such changes enable them to maintain polymerized microtubules under high pressures.

Pressure is sensed by living organisms not only as a static force per unit area, but also as minute variations in the ambient pressure. What we call the senses of touch and hearing are manifestations of this sensitivity to variations in pressure. At the cellular level the stimulus is probably a mechanical distortion of some delicate structure. Little is known of the nature of the mechanoreceptors and the changes that take place in them when a distorting force is applied. A nerve impulse, however, is the result of the stimulus.

Georg von Békésy, in his article, "The Ear" (page 216), describes one of the most elaborate of biological "instruments," which evolved to receive and respond to waves of air pressure, i.e., sound. The elaborate mechanisms that have evolved for sensing sound are evidence of the great biological advantage with which organisms possessing this sensitivity are endowed. On reading the article I was especially impressed by the section dealing with feedback to the voice. It appears that we subconsciously monitor the sounds that we ourselves generate. For example, we generate a sound; our ears hear it and signal the brain. The brain, in turn, signals our vocal cords and other muscles that participate in sound generation to modulate their activity appropriately. A pianist subconsciously regulates his finger, arm, and foot muscles by signals from the brain, which receives continuous information from the ears on the qualities of the sound being generated.

A rather more elaborate and refined system of sound generation and monitoring has evolved in some animals, especially bats. [See the article, "More About Bat 'Radar' " (Offprint 1121), by Donald R. Griffin.] Bats can navigate in absolute darkness or when blinded and still capture small flying insects efficiently. They do this by emitting a sound beam and echo-locating the objects in their path. At present, serious work is underway to develop similar echo-locating systems for use by blind people. It seems that we have a rudimentary capacity of this type; perhaps with adequate training from early childhood this sense could be made much more useful.

The concluding article in this section, "The Sensitivity of Organisms to Gravity" (page 226), is one that I wrote for this book. You should realize, of course, that the separation of gravity effects from pressure effects is arbitrary. The specific sense receptors used by the organism in orienting itself with respect to the gravity vector apparently respond to changes in pressure that occur when the organism moves in any direction.

22

Cells at High Pressure

by Douglas Marsland
October 1958

Temperature and pressure are basic factors in physical and chemical processes, but it is only recently that biologists have considered the role of pressure in the physical and chemical processes of life

Everyone knows that living things are profoundly influenced by temperature. In winter plants stop growing and cold-blooded creatures such as frogs and snakes are dormant. As the body temperature of any animal falls, its metabolism slows and eventually stops.

It is not generally recognized that pressure also plays a decisive role in metabolism. In every living cell there are many metabolic reactions which are sensitive to pressure. Until recently the pressure factor was largely ignored in physiological research; now pressure studies are beginning to yield evidence which is helping to solve classic problems. Pressure has become a new tool of physiology.

Man and other air-dwelling creatures are not required to tolerate large changes of pressure because our atmosphere is gaseous and relatively light. At sea level the pressure of the atmosphere averages slightly less than 15 pounds per square inch, and its variations never exceed more than about one pound per square inch. Even if an animal were transported from the highest mountain to the deepest cave, the change in the pressure of its environment would be quite small. The indirect effects of this small change would be important, because it would drastically alter the penetration of oxygen and other atmospheric gases into the protoplasm and body fluids. But the direct effects of the change on the vital processes would be almost negligible.

Water-dwelling organisms, on the other hand, are surrounded by a heavy liquid medium, and in the sea the pressure variations may be tremendous. The cumulative weight of water of course increases rapidly with depth—the pressure mounts roughly at the rate of one pound for every two feet of descent. At a depth of 1½ miles, where a wide variety of species are known to live, the pressure is more than 5,000 pounds per square inch, which is equivalent to the

PRESSURE CHAMBER developed in Marsland's laboratory makes it possible to study living cells at pressures up to 15,000 pounds per square inch. In this photograph the two halves of the chamber are unscrewed. The glass window in each half is a quarter-inch thick.

pressure of some 350 atmospheres. Such pressure drastically changes the vital activities of an organism. True deep-sea forms cannot survive very long at surface pressure; conversely, surface forms soon die when subjected to deep-sea conditions. Exactly how does pressure affect the fundamental life processes of various organisms? The question is significant if only because most ancient forms of life were oceanic, and pressure must have played an important part in guiding the evolutionary destiny of many species.

An obvious way to study the effects of pressure on living cells is to subject them to pressure in the laboratory. The development of apparatus to make such studies under controlled conditions began some 75 years ago. Paul Regnard of France became intensely interested in the findings of the French *Talisman* oceanographic expedition, which had succeeded in dredging up quite a variety of organisms from depths of more than seven miles. Here the pressure exceeds 1,000 atmospheres. Regnard undertook an extensive series of pressure studies, on which he reported regularly to the French Academy of Sciences from 1884 to 1888.

The early pressure investigators were handicapped by the fact that the living material could not be examined closely while it was inside the thick-walled pressure chamber. Most of the effects of pressure upon living processes are remarkably transient; the effects rapidly reverse themselves as soon as the pressure is released. Thus many of the early workers reported that pressure had no effect at all. By the time they had removed their cells from the pressure chamber, all evidence of change had disappeared.

The modern epoch of pressure research awaited the development of windowed pressure chambers and other devices which make it possible to study living material continuously while the pressure is maintained. Among the more useful devices have been: (1) a microscope pressure chamber (developed in our laboratory at New York University), which makes it possible to observe pressurized cells at a magnification of 600 diameters; (2) a muscle-nerve pressure chamber (developed by McKeen Cattell, Dayton J. Edwards and Dugald E. S. Brown at the Cornell University Medical College), which permits measurement of the force of contraction in pressurized muscle and of the intensity of electrical discharges in pressurized nerve preparations; and (3) a pressure

CHAMBER IS MOUNTED with a special microscope so that cells within it can be examined at magnifications up to 600 diameters. Pressure is transmitted to chamber through tube at top.

PRESSURE IS APPLIED by means of a modified hydraulic truck jack. Each full stroke of the handle at right raises the pressure in the chamber by 1,000 pounds per square inch.

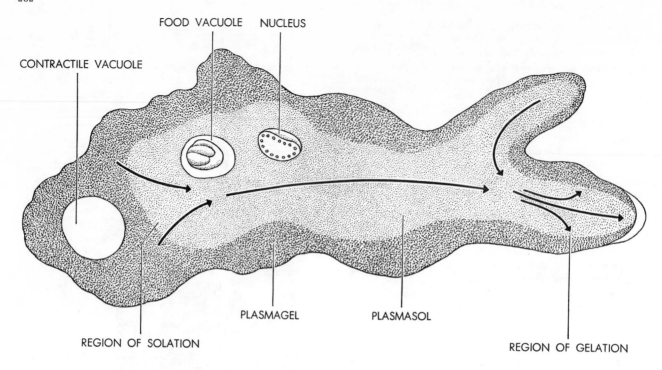

CONTRACTILE VACUOLE

FOOD VACUOLE   NUCLEUS

PLASMAGEL

PLASMASOL

REGION OF SOLATION

REGION OF GELATION

MOTION OF AN AMOEBA is outlined in this simplified drawing. As the amoeba moves toward the right, protoplasm in the sol state (*plasmasol*) streams toward the tip of its pseudopod ("false foot"). At the sides of the pseudopod (*region of gelation*) plasmasol enters the gel state (*plasmagel*). At the other end of the amoeba (*region of solation*), plasmagel enters the sol state.

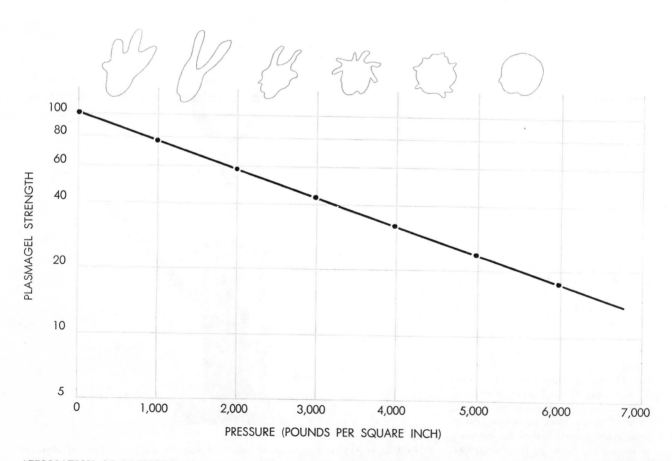

APPLICATION OF PRESSURE decreases the strength of the amoeba's plasmagel layer. The outline drawings at top show how with increasing pressure the shape of the amoeba changes until, at about 6,000 pounds per square inch, it is round and motionless.

centrifuge (devised by Brown at New York University), which enables the investigator to spin cells at high speeds and at pressures up to 1,000 atmospheres.

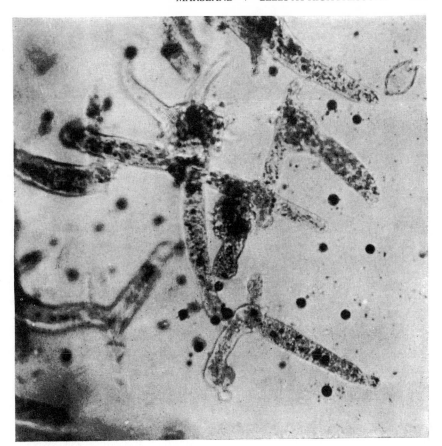

One very simple but informative experiment is to see what happens when a single cell is subjected to high pressure. Take, for example, the fertilized egg cell of the sea urchin. Place the cell in the microscope pressure chamber, which is filled with sea water, and quickly build up the pressure by pumping the handle of a modified hydraulic jack. With each full stroke the pressure goes up some 1,000 pounds per square inch (68 atmospheres). Stop, say, at 6,000 pounds per square inch. When one looks into the microscope, one sees that the cell has not perceptibly changed! To be sure, it has stopped developing, and it will never divide into a two-celled embryo unless the pressure is released in reasonable time. But the cell remains perfectly spherical, and its decrease in size is so small that the change can be detected only with refined measuring instruments.

The experiment illustrates two familiar but basic characteristics of hydrostatic pressure. The first is that in a hydrostatic system—which is of course what we manipulate in such experiments—the pressure is equal in all directions. Thus the pressure does not change the shape of the cell. Moreover, the protoplasm of the cell itself is essentially liquid, so the pressure inside the cell is the same as that outside. If a cell is compressed between two *solid* surfaces, it is easily flattened and can be ruptured by very little pressure (less than one atmosphere).

The second characteristic of hydrostatic pressure is that the compressibility of most liquids is exceedingly small. When pure water is subjected to a pressure of 1,000 atmospheres (about 15,000 pounds per square inch), its volume decreases less than 3 per cent; sea water and protoplasm are even more resistant to pressure. It is no wonder that we can see no difference in the size of a cell when we compress it. But the small decrease in volume has large effects upon the interplay of physical and chemical reactions in the cell.

We know from physics and chemistry that pressure tends to oppose any type of reaction which leads to an increase in the volume of the reacting substances. Conversely, pressure favors reactions which decrease the volume of the system. Take boiling water, for example.

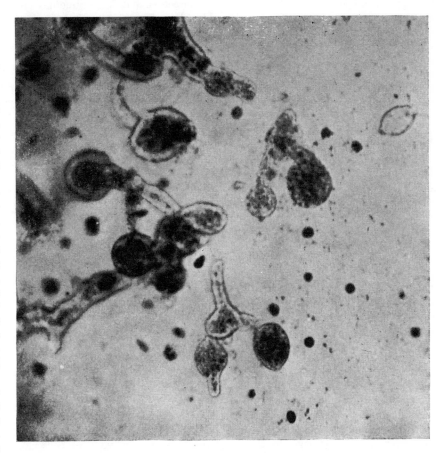

AMOEBAS INSIDE THE PRESSURE CHAMBER are photographed at atmospheric pressure (*top*) and two seconds after 6,000 pounds per square inch had been applied (*bottom*).

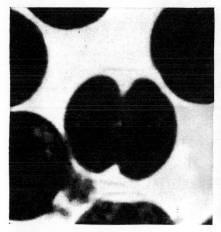 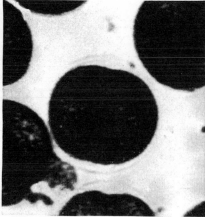 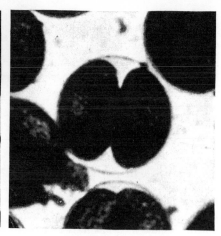

DIVIDING EGG CELL of the sea urchin was photographed in the pressure chamber five seconds after a pressure of 7,000 pounds per square inch had been applied (*left*), four minutes after the pressure had been applied (*middle*) and four minutes after the pres- sure had been released (*right*). In the first picture the cell had begun to pinch in two. In the second the pinching process has been reversed and the cell is almost round again. In the third picture the pinching process has resumed and the cell has nearly divided.

Vaporized water takes up far more space than an equivalent amount of liquid water. The boiling reaction must therefore perform work upon the environment; if the pressure of the environment is increased, the work is also increased. Thus pressure opposes boiling and favors condensation.

The effect of temperature is exactly the opposite. Heat provides the energy by which the reaction can perform work upon the environment. To sum up the effects of pressure and temperature: pressure tends to shift the equilibrium of any reversible reaction in the direction of lesser volume, and temperature tends to shift it in the other direction.

One significant reversible reaction in the cell is the sol-gel reaction, which seesaws back and forth more or less continuously. Parts of the protoplasm may have a sol structure in which their consistency resembles that of water; then these same parts will quickly switch to a gel structure resembling strongly set gelatin. Physiologists have long observed these repetitive protoplasmic changes, but the functional significance of the changes remained rather obscure until high-pressure techniques provided a new approach to this old problem.

Sol-gel reactions within the cell proved to be exceedingly sensitive to both temperature and pressure. It was more surprising to discover that the behavior of the protoplasmic gel was quite different from that of ordinary gelatin. Gelatin is made weaker by an increase of temperature and stronger by an increase of pressure. For a protoplasmic gel the opposite is true: it is made

stronger by an increase of temperature and weaker by an increase of pressure. In fact, if enough pressure is applied, all the protoplasm in the cell has almost the consistency of water.

It has become increasingly clear that gelation is the mechanism by which the cell generates mechanical energy. Protoplasm in the gel state has the power of contraction, and this contractility enables the cell to change shape and to move. In contracting, however, the protoplasm reverts to the sol condition, so that a continuous regeneration of gel structure is necessary if movement is to continue. Moreover, if gelation is prevented, or if existing gels are converted into sols by high pressure or other influences, protoplasmic contractility is lost and the cell becomes immobile.

These general conclusions have emerged from a long series of pressure-temperature experiments on a wide variety of living things. Here, however, we will give the reader just a few examples. These will deal primarily with the mechanical activities of single cells, especially cell division and the movement of the amoeba.

Amoeboid movement (which can also be observed in other animal cells) is perhaps the most primitive form of locomotion. Such movement is not restricted to the locomotion of lowly forms; it is also displayed by many cells in the human body. White blood cells often leave the capillaries and wander in amoeboid fashion through spaces in the tissues in search of infectious bacteria. Many connective-tissue cells also migrate through the tissue spaces. Moreover, in the embryo, before the

tissues assume their adult form, many cells employ amoeboid movement as they migrate to their ultimate stations.

Amoeboid movement has a peculiar fascination when it is observed in an active amoeba under the microscope. As the finger-like pseudopodium (the "false foot") of the amoeba advances, one sees the cytoplasm (the protoplasm between the cell nucleus and the cell wall) flowing steadily through a central channel in the cell. But one also sees that the surface layer of cytoplasm, which lies just inside the outer membrane of the cell, does not participate in the streaming. This layer of cytoplasm maintains a fixed position. It forms a tubular wall around the flowing cytoplasm, and seems to guide the streaming through a definite channel extending toward the tip of the advancing pseudopodium. Thus it is clear from direct observation that the cytoplasm of the amoeba is differentiated into two parts: the plasmasol, a fluid central part which flows while the amoeba is moving; and the plasmagel, a firmly gelated surface layer which surrounds the plasmasol.

Even before 1900 it was suspected that cyclic sol-gel changes might be involved in amoeboid movement. It seemed clear from direct observation that new plasmagel keeps forming and adding itself to the wall of an advancing pseudopodium. It was also apparent that old gel, at the dragging hind end of the amoeba, keeps transforming itself into new sol, which joins the forward-moving stream. Indeed, it was postulated quite early that the motive force which causes streaming originates with a forceful contraction of the plasmagel layer. Before the pressure studies, however, this "contractile hypothesis" was not supported by experimental evidence.

If we observe an actively moving amoeba in the pressure chamber and suddenly raise the pressure to 7,000 pounds per square inch, the effect is quick and dramatic. The pseudopodia collapse and the cell becomes a motionless sphere. The pressure has apparently liquefied the plasmagel system; the protoplasm now behaves like any small droplet of liquid.

So long as the pressure is maintained, no sign of movement or streaming can be detected in the rounded amoeba. Gradually the nucleus, food vacuoles and other relatively heavy bodies in the protoplasm begin to sink into the lower half of the cell. As soon as the pressure is released, however, activity resumes. Within about two minutes a new plasmagel layer forms and begins to contract violently. This causes a vigorous bubbling at the cell surface. Three or four minutes later normal pseudopodia begin to appear, and soon the amoeba resumes its locomotion.

The effect of various pressures on the plasmagel system can be precisely measured with the pressure centrifuge. One compartment of the centrifuge contains cells at high pressure; a second compartment contains a control group of cells at atmospheric pressure. Thus both pressurized and nonpressurized cells can be subjected to the same centrifugal force (up to 50,000 times the force of gravity). When the cells are spun, the relatively heavy bodies such as the food vacuoles tend to move outward through the cytoplasm. The rate at which these structures move in the pressurized cells can then be compared with their rate of movement in the nonpressurized cells. Obviously the more liquid the cytoplasm of the pressurized cells, the faster the heavy structures will move. In this way it was determined that the increase of pressure progressively reduces the viscosity of the cytoplasm of the amoeba. At 6,000 pounds per square inch all signs of pseudopodia and movement have vanished. By similar measurements it was shown that, up to a point, the increase of temperature strengthens the plasmagel system. As the temperature is raised, the pressure necessary to liquefy the plasmagel must be increased.

Where does the cell obtain the energy to build up its plasmagel system? The evidence strongly suggests that the cell obtains this energy from its high-energy phosphate reserves, partic-

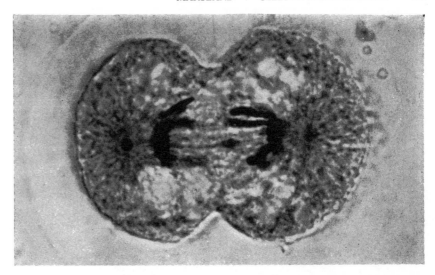

DIVISION OF AN EGG CELL is shown in greater detail. The chromosomes (*stained finger-like bodies*) have been pulled apart by the spindle and the cell has begun to pinch in two.

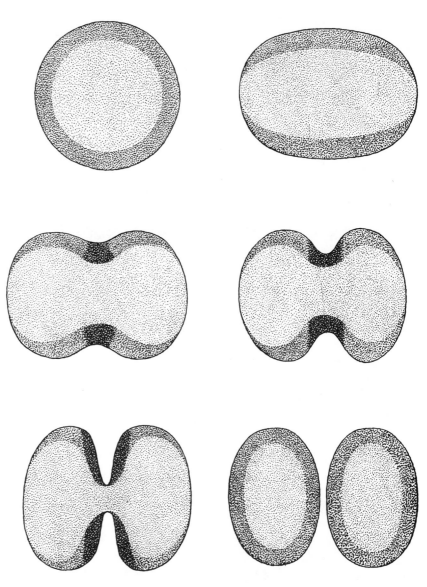

PLASMAGEL CONTRACTION THEORY is illustrated in sequential diagrams. The lighter stippling is the plasmasol; the darker, the plasmagel. In the third, fourth and fifth diagrams the contracting part of the plasmagel (*darkest stippling*) is localized in "cleavage furrow."

ularly adenosine triphosphate (ATP). It has been found, at any rate, that various cells have a stronger gel structure and are better able to perform mechanical work when they are provided with additional quantities of this energy-rich compound. Conversely, the gel structure and work performance are definitely weakened when the phosphate metabolism is inhibited by certain drugs. To this evidence must be added an extremely interesting finding of four workers at the California Institute of Technology: Paul O. P. T'so, Luther Eggman, Jerome R. Vinograd and James Bonner. This group recently suc-

ceeded in extracting from the slime mold Pelomyxa (a giant amoeboid form) a protein called myxomyosin which contracts in the presence of ATP. It seems quite certain that myxomyosin is the chief component of the contractile gel system, at least in this large amoeboid organism.

Precisely how protoplasmic gels contract and perform work is still unknown. One of the simplest suggestions is that protein molecules, which are the most essential constituents of the gel structure, can forcefully fold and unfold. Individual molecules acting in this way could not perform any useful work for the cell as a whole. But if the extended molecules are linked in a larger structure, they can perform work by contracting in concert.

The unfolding of molecules and the formation of linkages between them apparently involve the absorption of energy and a slight increase in molecular volume. It is for this reason that the formation of a protoplasmic gel structure is opposed by higher pressure and favored by higher temperature. The energy which goes into the building of a gel structure can be returned in the form of mechanical work when the gel contracts and reverts to a sol.

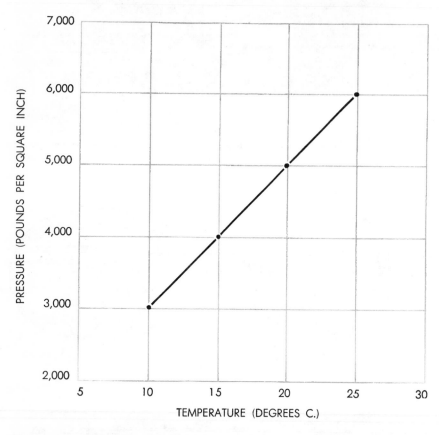

INCREASE OF TEMPERATURE counteracts the effect of pressure on the amoeba. This graph shows that at 10 degrees it takes 3,000 pounds per square inch to stop the motion of the amoeba, and that at 25 degrees it takes 6,000 pounds per square inch to stop the motion.

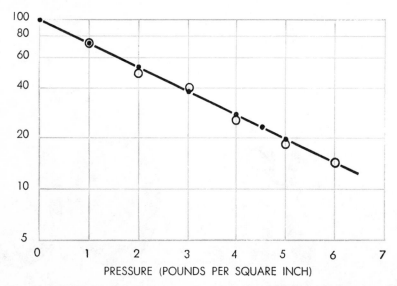

INCREASE OF PRESSURE slows and eventually stops the pinching process in the sea-urchin egg. The dots in this graph show the rate of progress of the cleavage furrow; the circles, the strength of the plasmasol. The vertical coordinate is marked off in arbitrary units.

What do the pressure experiments reveal about cell division, the process by which one cell gives rise to two daughter cells? The cell physiologist customarily studies cell division with egg cells, which are relatively large and can be obtained in abundance from lower animals such as the sea urchin. When these cells are placed in a vessel, and sperm cells are added to them, they begin a long series of divisions. Usually the first division occurs within an hour, and the succeeding divisions follow more rapidly. But given the exact temperature and time of insemination, the experimenter knows precisely when each of the successive divisions is destined to occur.

The division of the egg begins with the division of its nucleus. This involves the appearance and splitting of the chromosomes and the formation of a "spindle" which separates the two sets of daughter chromosomes. The division of the cytoplasm, which follows the division of the nucleus, is more direct. The cell merely pinches itself into two cells by means of a constriction, or "cleavage furrow," which cuts deeper and deeper into the egg toward the center of the spindle.

Pressure studies strongly indicate that

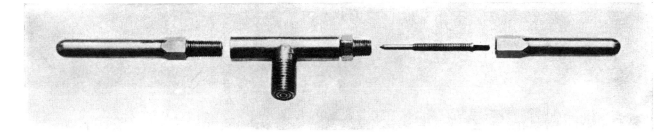

SPECIAL CENTRIFUGE HEAD is used to compare the effects of spinning cells at atmospheric pressure and at high pressure. The cells which are to be subjected to high pressure are placed in the chamber at left in this "exploded" view. The pressure is applied through the vertical fitting, which also joins the head to the centrifuge. After the pressure has been applied, the pressure chamber is sealed off by the needle valve next to the right. The cells at atmospheric pressure are placed in the chamber at far right.

the work performed by the dividing cell —in separating the daughter sets of chromosomes and cutting itself in two— is achieved by means of contractile gel structures. Daniel C. Pease, then working at Princeton University, showed that the spindle is a gel structure which is dissolved at a pressure of 6,000 pounds per square inch, and that slightly lower pressures are adequate to keep the chromosomes from moving toward the ends of the spindle. Apparently the fibers which extend from the ends of the spindle to the chromosomes are contractile gel structures which serve to drag the daughter chromosomes apart. But here we will concentrate on our own experiments dealing with the furrow which cuts through the cytoplasm.

If we place a batch of fertilized eggs in the pressure chamber, we can watch them as their first division proceeds. Then, very late in the process, when the chromosomes are widely separated and when the cleavage furrow has cut almost through the cell, let us suddenly apply a fairly high pressure (6,000 to 7,000 pounds per square inch). The furrowing stops immediately and begins to retreat. Gradually the cell loses its hourglass shape, even though its two daughter cells had almost separated. Within about three minutes the cell, which now has two nuclei, has become a single sphere again.

If the pressure is released as soon as the cell is a sphere, the furrowing will start again and the cytoplasm will divide normally. But it is possible to stop the furrowing a second time by re-applying the pressure. In fact, one can reverse the cytoplasmic division three or four times in succession.

If furrowing is held back for 13 to 15 minutes, the cell displays no further tendency to cleave until it is time for the second division to occur. By now each of the two nuclei within the single mass of cytoplasm has begun to divide

again. When the cytoplasm divides, a double furrow appears and the cell cuts itself directly into four daughter cells.

Experiments with the pressure centrifuge have shown that the egg cell, like the amoeba, has near its surface a layer of gel which undergoes large and repetitive changes. This plasmagel layer is very weak in the unfertilized egg, but within 10 minutes after insemination it becomes more than 20 times stronger.

The mechanism by which the animal cell divides its cytoplasm is still the subject of debate. There is nonetheless strong evidence for the relatively simple view that the mechanism involves a contraction of the plasmagel layer. According to this hypothesis, cleavage results from a progressively developing contraction of the plasmagel in the furrow region [see bottom illustration, page 205]. The process seems to involve a series of sol-gel reactions: new gel steadily mobilizes along the sides of the deepening furrow; old gel at the bottom of the furrow, having spent its contractile force, reverts to a sol. This clears the way for the furrow to finish the work of splitting the cell.

The evidence that tends to substantiate the hypothesis can be outlined as follows. First, furrowing stops whenever the plasmagel strength, as measured in the pressure centrifuge, drops to a certain critical level. Second, the strength of the furrowing reaction, as judged by how quickly the furrow cuts through the cell, is determined by the strength of the plasmagel. Third, increasing the temperature proportionately strengthens the plasmagel and the furrowing reaction, so that higher and higher pressures are required to convert the plasmagel into a sol and block the cleavage. Fourth, egg cells provided with an augmented supply of ATP have a proportionately stronger plasmagel structure and fur-

rowing reaction. Fifth, drugs that inhibit the liberation of energy from ATP reduce the strength of the gel and of the furrowing reaction. Thus both the dividing animal cell and the moving amoeba lead us to the same conclusion: The sol-gel system provides a mechanism which transforms the energy generated by metabolism into mechanical work.

As I have indicated, cell division and amoeboid movement are only examples of the biological processes which have been examined by means of high pressure. High-pressure studies have become a vigorous school of research, initiated in the modern period by Dayton Edwards and McKeen Cattell of the U. S., M. Fontaine of France and U. Ebbecke of Germany. Among the basic processes investigated at high pressure are: bioluminescence, or the production of "cold light" by living things (Dugald Brown, Frank H. Johnson and the author); muscle metabolism and muscle contraction (Brown, Edwards and Cattell); the transmission of nerve impulses (Harry Grundfest); the streaming of protoplasm in plant cells (the author); the mechanism of color change in fish (the author); the effects of the cell nucleus on cell metabolism (Henry I. Hirshfield, Arthur M. Zimmerman and the author); the metabolic control of rhythmic processes such as heartbeat (John A. Kitching and Joseph V. Landau); and the mechanism of enzyme activity (Johnson, Henry Eyring and Milton J. Polissar).

Indeed, pressure has begun to take its place with temperature as a fundamental factor governing physiological processes. Physicists and chemists have always recognized the importance of pressure, as indicated by the basic thermodynamic equation: $PV = nRT$ (pressure times volume equals the Avogadro number times the gas constant times temperature). Physiologists are now following suit.

23 Animals of the Abyss

by Anton F. Bruun
November 1957

When the Danish oceanographic vessel "Galathea" sailed around the world, its investigators sampled the life of the greatest depths with a cable seven and a half miles long

The oceanic abyss of our planet is an immense region. Below 4,000 meters (two and a half miles) of water lies an area as great as all the continents put together. The deepest deeps—under 6,000 meters—amount to an area half as large as the U. S. This vast world is in many ways as strange and remote as another planet. It is a world of total darkness and eternal cold, never more than a few degrees above freezing. The pressure is enormous—up to 1,000 atmospheres and more. As Charles Wyville Thomson remarked when he first sounded the great ocean depths in the famous *Challenger* expedition, it is almost as hard to imagine life existing in these conditions as in fire or in a vacuum. But we know today that there are forms of life—strange forms, to be sure—which thrive in the very deepest

"GALATHEA" is photographed at anchor off Kondul in the Nicobar Islands of the Indian Ocean. Originally the New Zealand vessel H.M.S. *Leith,* she was purchased after World War II by the Danish Navy. Her length is 266 feet; her displacement, 1,600 tons.

trenches of the ocean bottom. From these "hadal" regions (to borrow a word from the Greek Hades) scientific expeditions of the last few years have dredged up a considerable array of living creatures.

It is less than a century since scientists began to explore the deep ocean bottoms. In the 1870s several British and U. S. ships set forth with long ropes to fish at depths of thousands of feet along the sea floors. The round-the-world expedition of the British *Challenger* brought to light an amazing variety and abundance of life in the deep oceans [see "The Voyage of the 'Challenger,'" by Herbert S. Bailey, Jr.; SCIENTIFIC AMERICAN, May, 1953]. Since World War II, elaborately equipped ships have plumbed depths never reached before, and we have now sampled the deepest trenches in the oceans. In 1948 the Swedish *Albatross* expedition fished down to 7,900 meters in the Puerto Rican Trench; in 1949 the U.S.S.R. ship *Vitiaz* reached 8,100 meters in the Kurile-Kamchatka Trench. And on August 21, 1951, the Danish *Galathea*, on its round-the-world deep-sea expedition, brought animals to the surface from the bottom of the Philippine Trench. This is the deepest haul so far: it came from the great depth of 10,190 meters (about 33,600 feet, or more than six miles below the sea surface).

It is not a completely accidental coincidence that the deepest abyss in the oceans and the highest mountain on the earth—Mount Everest—were conquered at the same time. Both achievements required highly developed techniques and equipment which were brought forth by the war. To appreciate what exploring the great ocean deeps involves, you must keep in mind that if Mount Everest were dropped into the Philippine Trench its peak would be a mile and a half below sea level. From the water's surface atop this fantastically deep chasm the ship drops a long, thin line to grope blindly for organisms in the ooze at the bottom of the trench. It is comparable to flying in an airplane six miles up and trying to snag animals on the ground with a grappling hook hung from a miles-long drag line. Naturally, fishing at this distance, we cannot be sure that what our line happens to drag up is a really fair sample of what lies on the bottom. An airplane explorer who pulled up a mouse on his line might conclude that the earth's surface was largely inhabited by tiny rodents. But fortunately the en-

THREE FISH are among the many brought up from great depths by the *Galathea*. At the top is *Typhlonus*, a blind fish with a mouth that can be protruded like a shovel; it was caught at 5,090 meters in the Celebes Sea. In the middle is *Galatheathauma axeli*, a fish with a large luminous organ in its mouth; it was caught at 3,590 meters in the Pacific off Central America. At the bottom is *Bassogigas*, the fish caught at the greatest depth: 7,130 meters.

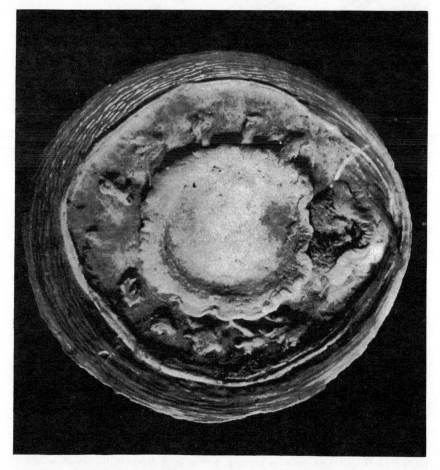

MOLLUSK DISCOVERED BY THE "GALATHEA" was *Neopilina galatheae*. It is a relative of the ancient mollusk *Pilina*, which has been extinct for 350 million years. At the top the mollusk, which has only one shell, is shown from above; at the bottom, from below.

vironment in the deep ocean is so uniform that we can be fairly confident our samples are representative.

I should say a few words about the equipment that makes this fishing possible. First of all, echo-sounding apparatus allows us to locate a deep-sea trench and explore its topography. Without such a survey we would have little chance of fishing successfully along its bottom, for the trenches are narrow furrows with steep, rugged sides. If you imagine trying to drag a net through Yosemite Valley from an airplane six miles above without tearing the net on its rocky sides, you will understand the difficulty. With the echo-sounder we locate the bottom of the trench and check the depth continually during the towing. For dragging the bottom the *Galathea* used either a dredge or a trawl net. We had the benefit of powerful winches and high-quality steel wire for the line, in contrast to the hemp ropes and primitive winches used by the pioneers. Our line was a single steel wire about seven and a half miles long, tapering from a thickness of nearly two thirds of an inch at the upper end to a little over a quarter of an inch at the bottom end. The wire could bear a load up to four and a half tons, and the weight and drag of our gear was only one and a half tons. Very complicated calculations are required to find out how much wire must be paid out to reach the bottom, what the drag of the gear and the friction between the long wire and the water will be at a given towing speed, and so on. Here we had the invaluable help of B. Kullenberg, who had taken part in the round-the-world expedition of the Swedish *Albatross* in 1947 and 1948.

Before I describe the animals we found, let us consider a few facts about their deep-sea environment. In the sea the pressure increases by about one atmosphere (roughly 14.7 pounds per square inch) for every 10 meters of depth. Thus at 4,000 meters the pressure is about 400 atmospheres; at 10,000 meters, 1,000 atmospheres. It is almost impossible for us to conceive what living under such pressures must entail. The human body gets into serious difficulties at pressures of only three or four atmospheres (corresponding to a depth of 100 to 140 feet in the water). The deep-dwelling sea animals—worms, mollusks, crustaceans and so forth—have no air bladders, but the great pressure imposes other serious physiological problems. We know very little about what these problems really are; all we know is that a number of animals have somehow managed to evolve mechanisms which enable them to live comfortably under the enormous pressure of 1,000 atmospheres or more.

Next, these animals have to live in complete and eternal darkness. Sunlight does not penetrate more than a few hundred meters below the ocean surface. In the dark regions lower down there are some luminous fishes and other animals, but even this chemical light largely disappears below 1,500 meters.

At great depths it is not only uniformly dark but also uniformly cold—at the Equator as well as near the poles. The salinity of the water also is uniform. Its content of oxygen seems to be quite sufficient to support life for breathing animals. As for food, the deep-sea animals must depend on organic matter falling from the surface levels of the ocean, because the lack of light excludes any manufacture of organic substances by photosynthetic plants. Contrary to a common impression, there is no rain of animal carcasses from above: these are devoured long before they reach the bottom, except for an occasional dead whale or giant shark. What does come down is excrement, the cast skins of crustaceans, the horny beaks of squids, water-logged driftwood, fragments of plants and the like. All this organic matter serves as food for bacteria, which in turn become nutritious eating for bottom animals.

So the darkness is uniform; the temperature is more or less uniform; similarly the salinity, the food (though it varies in quantity) and the habitat (mainly a soft, clayey ooze). Moreover, at the level around 2,000 meters the animals have unhindered mobility from ocean to ocean, because of the lack of physically different zones or topographical barriers. It is not surprising, therefore, to find the animals more or less alike in a general way. Whereas the surface animals of the sea may vary all the way from coral organisms to arctic seals, the deeper bottom dwellers are mostly small burrowing forms: bristle worms, bivalves, various other crustaceans, sea cucumbers, brittle stars and so on. Many of these middle-level animals are not very different from their surface cousins, except that they are blind and rather colorless—like the pale, blind inhabitants of caves, another dark habitat.

From the depths between 2,000 and 6,000 meters the *Galathea* fished up thousands of specimens of animals, many of which had previously been considered rare. Except for the fact that they were more abundant than had been thought, most of these species had little novelty. But occasionally we caught a new species that had never been brought up before. The most peculiar fish we netted was a luminous angler fish, taken at a depth of 3,590 meters in the Pacific off Central America. Like certain other deep-sea anglers, this fish had a luminous organ to attract its prey, but its organ was inside the mouth [*see middle drawing on page 209*]!

Far more interesting was another species of animal caught for the first time by the *Galathea* expedition. It was a small, limpet-like mollusk which turned out to be a living relative of an ancient mollusk, *Pilina*, that has been extinct for some 350 million years. The discovery was no less surprising than if we had brought up a living trilobite—the ancient and extinct arthropod that is believed to have been one of the earliest animal inhabitants of the earth. Our mollusk (named *Neopilina galatheae*) is so different from the present-day mollusks that it must be placed in a special class of its own. As a living fossil it may be as important a clue to the evolution of mollusks as the egg-laying duck-billed platypus is for studying the origin of mammals.

Below about 6,000 meters, what we usually think of as fishes (fishes with backbones) practically disappear. Very few such fishes have been found at greater depths—only six by the *Galathea* and two by the *Vitiaz*. The deeper bottom-dwellers therefore are secure from

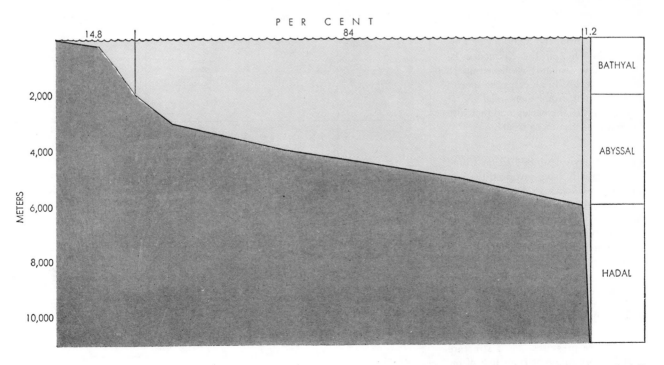

AREA OF THE OCEANS less than 2,000 meters deep (bathyal) is 14.8 per cent of their total area; area between 2,000 and 6,000 meters deep (abyssal), 84 per cent; area deeper than 6,000 meters (hadal), 1.2 per cent. The abyss is thus the world's largest ecological unit.

predaceous fish. Indeed, the great pressure shelters them from invaders of any sort. Since only animals adapted to the extreme pressure can live there, they enjoy a protected niche undisturbed by competition from newcomers.

In the very deep trenches the isolation of the inhabitants is practically perfect. Once an animal has become adapted to the trench-bottom pressure (from 700 to more than 1,000 atmospheres), it cannot rise to a higher level and therefore cannot migrate out of the trench. It lives, in effect, on an island as remote as a mountain peak from contact with the rest of the world. And just as unique species of plants and animals evolve on islands and on mountain peaks, we should expect each trench to have its own peculiar population. That is exactly what we find.

After the *Galathea* had surveyed the Philippine Trench and defined the bottom, it lowered its 12,000-meter line and made a pass along the trench at a speed of two knots. When the trawl was finally pulled up, the explorers were delighted to see clay on the framework and stones in the net, sure evidence that the trawl had reached the bottom. The first animal noticed in this haul was a sea anemone—a whitish growth on a stone. At first glance it looked like a typical sea anemone, but on closer analysis it proved to be so unusual in some properties that it represents an entirely new family. Altogether we found eight species of animals in the Philippine Trench: sea anemones, sea cucumbers, bivalve mollusks, amphipod crustaceans and bristle worms. And from the 8,200-meter Kermadec Trench (off the Kermadec Islands of New Zealand) we fished up 18 different species. In all cases the species were different from their relatives at higher levels of the ocean, and they also differed from trench to trench.

What I have described so far would not greatly have surprised the pioneers on the *Challenger*. Superficially the animal inhabitants of the deep trenches do not look very different from the types living somewhat higher up. Nor do the bacteria that we found on the floor of the Philippine Trench. But their make-up certainly must be unusual. The bacteria are truly pressure-loving organisms, unable to exist at all at pressures of less than 1,000 atmospheres. Somehow they are constructed in a particular way to live at these pressures. Their structure must be very different from that of organisms at sea level. Claude E. ZoBell and Richard Y. Morita of the Scripps

Institution of Oceanography, who took part in our expedition, have been cultivating these bacteria in high-pressure chambers, and we can expect very important results from their studies.

I should mention two other important facts that have been learned from the

explorations of the deep sea. The *Galathea* expedition investigated the production of organic food in the oceans. For this a new measure devised by E. Steemann Nielsen of Copenhagen was used. It estimates productivity by the amount of carbon dioxide assimilated by sea

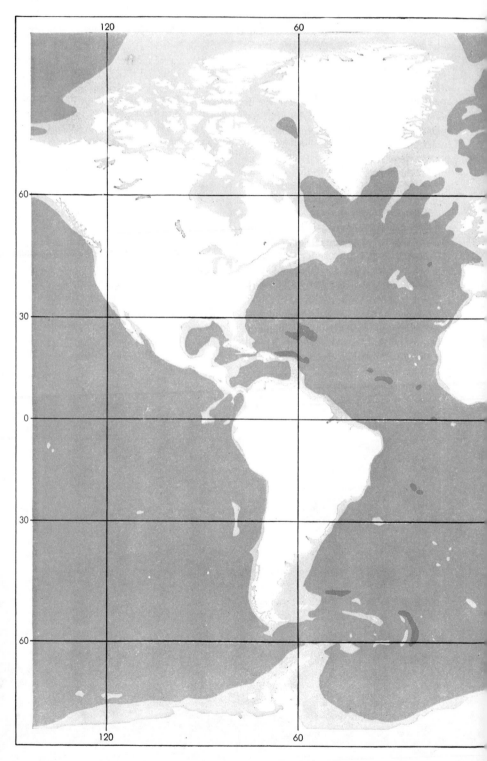

DEPTH OF THE OCEANS is roughly indicated by this map of the world. The waters less than 2,000 meters deep (bathyal) are indicated by the lightest colored area. The waters be-

plants. Using radioactive carbon 14 as a tracer, Steemann Nielsen measured the plant assimilation of carbon dioxide in samples of water drawn from several depths all along the route of the expedition. He found that organic production varies greatly from region to region but that no area is completely barren: there are no "deserts" in the sea. Steemann Nielsen estimated from his measurements that all the seas together produce about 40 billion tons of organic matter per year—about the same amount as the total annual production by plants on land. This finding is an important step toward answering the question about how much food we may eventually get from the sea [see "Food from the Sea," by Gordon A. Riley; SCIENTIFIC AMERICAN, October, 1949].

The other interesting finding is that

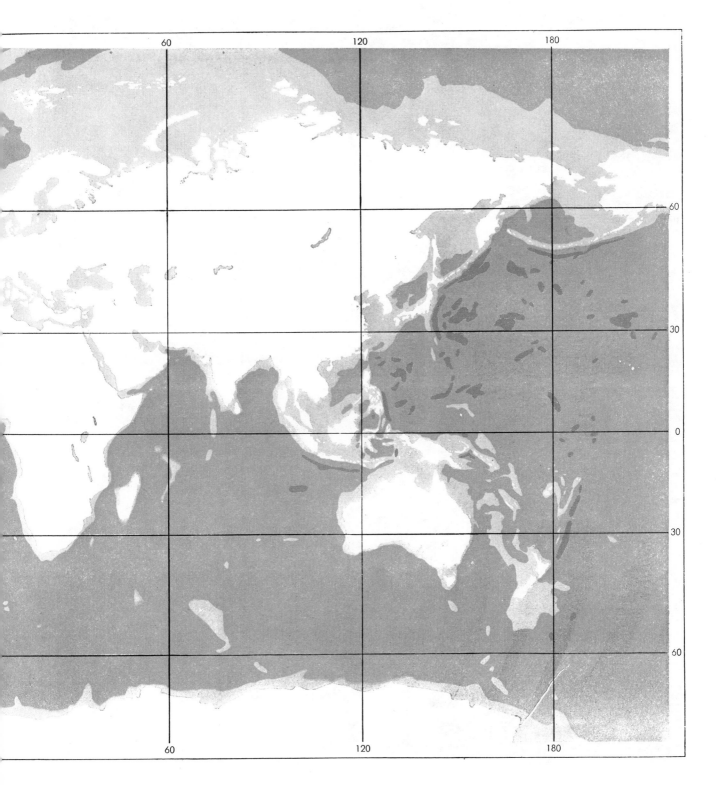

tween 2,000 and 6,000 meters deep (abyssal) are indicated by the darker colored area. The waters deeper than 6,000 meters (hadal) are indicated by the darkest colored area. The deepest haul made by the _Galathea_ was at depth of 10,190 meters near the Philippines.

with the coming of the ice ages on the earth, the temperature in the deep seas dropped sharply. The average temperature of the water at great depths seems to have fallen from about 10 degrees centigrade to two degrees. In other words, life in the deep sea must have been subjected to what amounted to a cold-shock treatment. Undoubtedly this had catastrophic effects, destroying all deep-sea life except the species that could resist the cold or escape to warmer levels of the ocean.

Those of us who are interested in studying the undersea world have some concern just now about the possibility that man may soon start dumping radioactive wastes from atomic energy plants in the deep sea. Such deposits on the ocean bottom might change its living population as drastically as the cold of

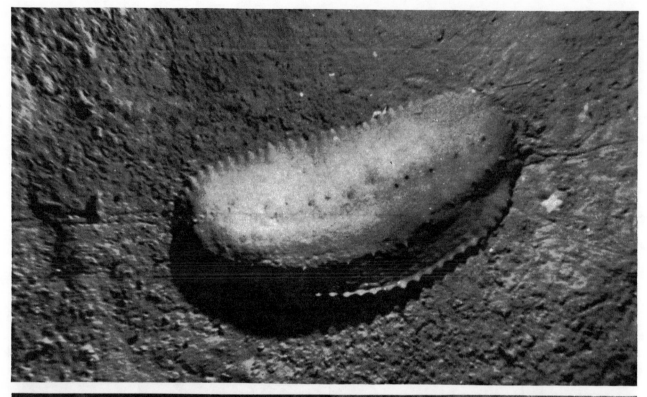

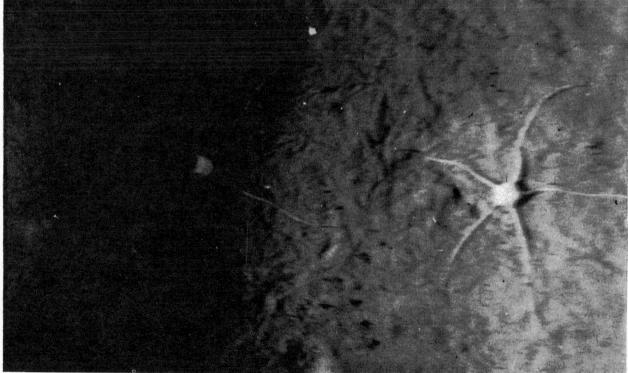

ABYSSAL ANIMALS were photographed, not from the *Galathea*, but by investigators of the Lamont Geological Observatory of Columbia University. At upper left is a sea cucumber, photographed at a depth of 2,000 meters. At upper right is another sea cucumber

the ice ages. We should very much like to get a good look at the animals before that happens.

We have literally only scratched the surface of the world of fabulous life and pressure at the bottom of the deep sea trenches. To go further in that costly exploration we shall need international cooperation. The oceans are international, and so is oceanology. The nations of the world should collaborate in this science, through an agency set up on the same plan, say, as the new European Organization for Nuclear Research in Switzerland. People everywhere will benefit from studies of the sea, for we shall learn much about the oceans' food resources, about our long-term climate and about the remarkable ability of organisms to live under fantastic pressure.

and a buried starfish, photographed at a depth of 3,904 meters. At lower left are two brittle stars, photographed at 2,100 meters. At lower right is a fish swimming above the bottom, photographed at 1,600 meters. All the photographs were made in the North Atlantic.

24

The Ear

by Georg von Békésy
August 1957

It translates sound between some 16 cycles per second and 20,000 into nerve impulses. Together with the auditory centers of the brain, it is an instrument that is not only remarkably sensitive but also selective

Even in our era of technological wonders, the performances of our most amazing machines are still put in the shade by the sense organs of the human body. Consider the accomplishments of the ear. It is so sensitive that it can almost hear the random rain of air molecules bouncing against the eardrum. Yet in spite of its extraordinary sensitivity the ear can withstand the pounding of sound waves strong enough to set the body vibrating. The ear is equipped, moreover, with a truly impressive selectivity. In a room crowded with people talking, it can suppress most of the noise and concentrate on one speaker. From the blended sounds of a symphony orchestra the ear of the conductor can single out the one instrument that is not performing to his satisfaction.

In structure and in operation the ear is extraordinarily delicate. One measure of its fineness is the tiny vibrations to which it will respond. At some sound frequencies the vibrations of the eardrum are as small as one billionth of a centimeter—about one tenth the diameter of the hydrogen atom! And the vibrations of the very fine membrane in the inner ear which transmits this stimulation to the auditory nerve are nearly 100 times smaller in amplitude. This fact alone is enough to explain why hearing has so long been one of the mysteries of physiology. Even today we do not know how these minute vibrations stimulate the nerve endings. But thanks to refined electro-acoustical instruments we do know quite a bit now about how the ear functions.

What are the ear's abilities? We can get a quick picture of the working condition of an ear by taking an audiogram, which is a measure of the threshold of hearing at the various sound frequencies. The hearing is tested with pure tones at various frequencies, and the audiogram tells how much sound pressure on the eardrum (*i.e.*, what intensity of sound) is necessary for the sound at each frequency to be just barely audible. Curiously, the audiogram curve often is very much the same for the various members of a family; possibly this is connected in some way with the similarity in the shape of the face.

The ear is least sensitive at the low frequencies: for instance, its sensitivity for a tone of 100 cycles per second is 1,000 times lower than for one at 1,000 cycles per second. This comparative insensitivity to the slower vibrations is an obvious physical necessity, because otherwise we would hear all the vibrations of our own bodies. If you stick a finger in each ear, closing it to air-borne sounds, you hear a very low, irregular tone, produced by the contractions of the muscles of the arm and finger. It is interesting that the ear is just insensitive enough to low frequencies to avoid the disturbing effect of the noises produced by muscles, bodily movements, etc. If it were any more sensitive to these frequencies than it is, we would even hear the vibrations of the head that are produced by the shock of every step we take when walking.

On the high-frequency side the range that the ear covers is remarkable. In childhood some of us can hear well at frequencies as high as 40,000 cycles per second. But with age our acuteness of hearing in the high-frequency range steadily fails. Normally the drop is almost as regular as clockwork: testing several persons in their 40s with tones at a fixed level of intensity, we found that over a period of five years their upper limit dropped about 80 cycles per second every six months. (The experiment was quite depressing to most of the partici-

pants.) The aging of the ear is not difficult to understand if we assume that the elasticity of the tissues in the inner ear declines in the same way as that of the skin: it is well known that the skin becomes less resilient as we grow old—a

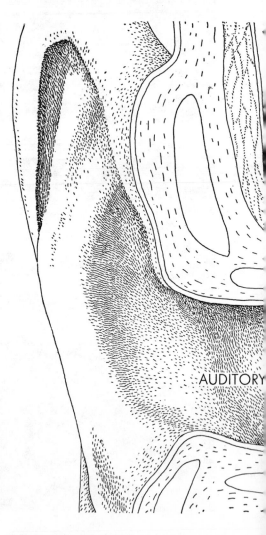

AUDITORY

PARTS OF THE EAR are illustrated in somewhat simplified cross section. Be-

phenomenon anyone can test by lifting the skin on the back of his hand and measuring the time it takes to fall back.

However, the loss of hearing sensitivity with age may also be due to nerve deterioration. Damage to the auditory nervous system by extremely loud noises, by drugs or by inflammation of the inner ear can impair hearing. Sometimes after such damage the hearing improves with time; sometimes (*e.g.*, when the damaging agent is streptomycin) the loss is permanent. Unfortunately a physician cannot predict the prospects for recovery of hearing loss, because they vary from person to person.

Psychological factors seem to be involved. Occasionally, especially after an ear operation, a patient appears to improve in hearing only to relapse after a short time. Some reports have even suggested that operating on one ear has improved the unoperated ear as well. Since such an interaction between the two ears would be of considerable neuro-logical interest, I have investigated the matter, but I have never found an improvement in the untreated ear that could be validated by an objective test.

Structure of the Ear

To understand how the ear achieves its sensitivity, we must take a look at the anatomy of the middle and the inner ear. When sound waves start the eardrum (tympanic membrane) vibrating, the vibrations are transmitted via certain small bones (ossicles) to the fluid of the inner ear. One of the ossicles, the tiny stirrup (weighing only about 1.2 milligrams), acts on the fluid like a piston, driving it back and forth in the rhythm of the sound pressure. These movements of the fluid force into vibration a thin membrane, called the basilar membrane. The latter in turn finally transmits the stimulus to the organ of Corti, a complex structure which contains the endings of the auditory nerves. The question im-mediately comes up: Why is this long and complicated chain of transmission necessary?

The reason is that we have a formidable mechanical problem if we are to extract the utmost energy from the sound waves striking the eardrum. Usually when a sound hits a solid surface, most of its energy is reflected away. The problem the ear has to solve is to absorb this energy. To do so it has to act as a kind of mechanical transformer, converting the large amplitude of the sound pressure waves in the air into more forceful vibrations of smaller amplitude. A hydraulic press is such a transformer: it multiplies the pressure acting on the surface of a piston by concentrating the force of the pressure upon a second piston of smaller area. The middle ear acts exactly like a hydraulic press: the tiny footplate of the stirrup transforms the small pressure on the surface of the eardrum into a 22-fold greater pressure on the fluid of the inner ear. In this way the

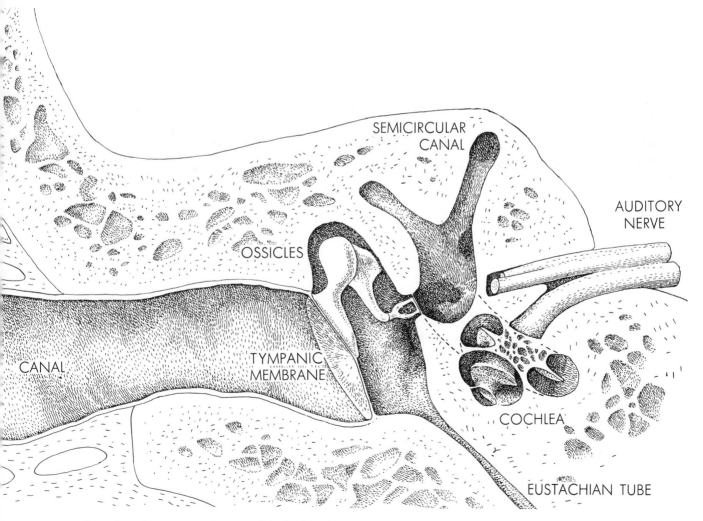

tween the eardrum (tympanic membrane) and the fluid-filled inner ear are the three small bones (ossicles) of the middle ear. The audi-tory nerve endings are in an organ (*not shown*) between the plate of bone which spirals up the cochlea and the outer wall of the cochlea.

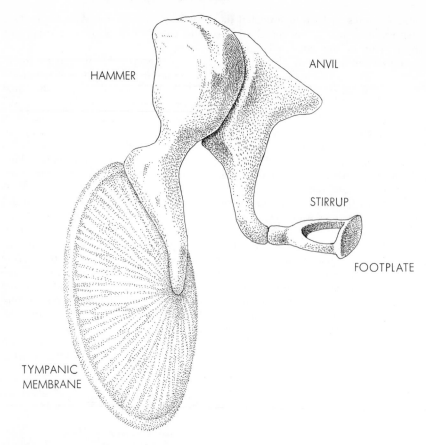

HAMMER

ANVIL

STIRRUP

FOOTPLATE

TYMPANIC
MEMBRANE

THREE OSSICLES transmit the vibrations of the tympanic membrane to the inner ear. The footplate of stirrup, surrounded by a narrow membrane, presses against inner-ear fluid.

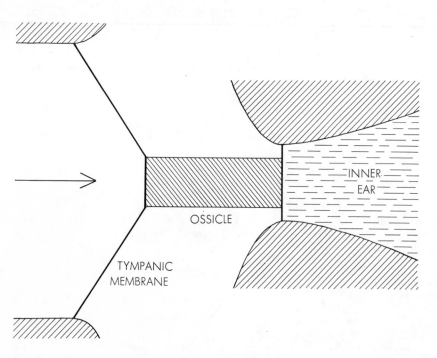

INNER
EAR

OSSICLE

TYMPANIC
MEMBRANE

HOW OSSICLES ACT as a piston pressing against the fluid of the inner ear is indicated by this drawing. Pressure of the vibrations of tympanic membrane are amplified 22 times.

ear absorbs the greater part of the sound energy and transmits it to the inner ear without much loss.

But it needs another transformer to amplify the pressure of the fluid into a still larger force upon the tissues to which the nerves are attached. I think the ear's mechanism for this purpose is very ingenious indeed. It is based on the fact that a flat membrane, stretched to cover the opening of a tube, has a lateral tension along its surface. This tension can be increased tremendously if pressure is applied to one side of the membrane. And that is the function of the organ of Corti. It is constructed in such a way that pressure on the basilar membrane is transformed into shearing forces many times larger on the other side of the organ [*see diagram at bottom of opposite page*]. The enhanced shearing forces rub upon extremely sensitive cells attached to the nerve endings.

The eardrum is not by any means the only avenue through which we hear. We also hear through our skull, which is to say, by bone conduction. When we click our teeth or chew a cracker, the sounds come mainly by way of vibrations of the skull. Some of the vibrations are transmitted directly to the inner ear, by-passing the middle ear. This fact helps in the diagnosis of hearing difficulties. If a person can hear bone-conducted sounds but is comparatively deaf to air-borne sounds, we know that the trouble lies in the middle ear. But if he hears no sound by bone conduction, then his auditory nerves are gone, and there is no cure for his deafness. This is an old test, long used by deaf musicians. If a violin player cannot hear his violin even when he touches his teeth to the vibrating instrument, then he knows he suffers from nerve deafness, and there is no cure.

Speaking and Hearing

Hearing by bone conduction plays an important role in the process of speaking. The vibrations of our vocal cords not only produce sounds which go to our ears via the air but also cause the body to vibrate, and the vibration of the jawbone is transmitted to the ear canal. When you hum with closed lips, the sounds you hear are to a large degree heard by bone conduction. (If you stop your ears with your fingers, the hum sounds much louder.) During speaking and singing, therefore, you hear two different sounds—one by bone conduction and the other by air conduction. Of course another listener hears only the air-conducted sounds. In these sounds

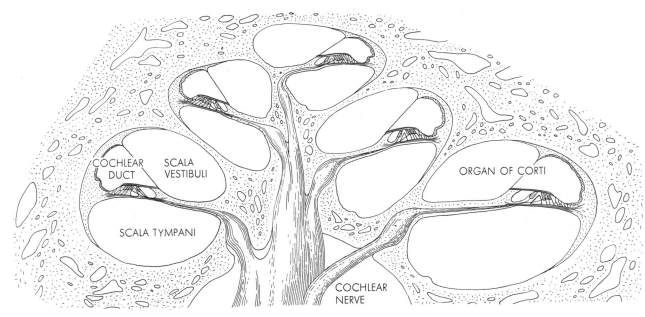

TUBE OF THE COCHLEA, coiled like the shell of a snail, is depicted in cross section. The plate of bone which appears in the cross section on pages 216 and 217 juts from the inside of the tube. Between it and the outside of the tube is the sensitive organ of Corti.

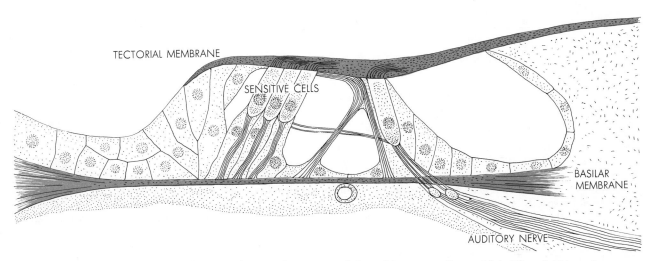

ORGAN OF CORTI lies between the basilar and tectorial membranes. Within it are sensitive cells which are attached to a branch of the auditory nerve (*lower right*). When fluid in scala tympani (*see drawing at top of page*) vibrates, these cells are stimulated.

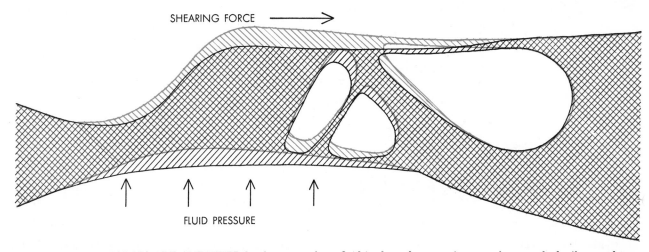

HOW VIBRATION FORCES ARE AMPLIFIED by the organ of Corti is indicated by this drawing. When the vibration of the fluid in the scala tympani exerts a force on the basilar membrane, a larger shearing force is brought to bear on tectorial membrane.

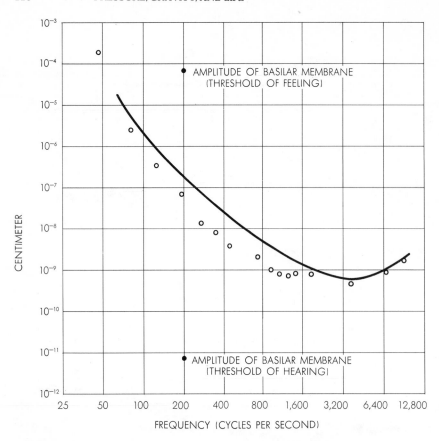

SENSITIVITY OF THE EAR is indicated by this curve, in which the amplitude of the vibrations of the tympanic membrane in fractions of a centimeter is plotted against the frequency of sound impinging on the membrane. Diameter of hydrogen atom is 10^{-8} centimeter.

cording of our voice may strike us as very thin and disappointing. From this point of view we have to admire the astonishing performance of an opera singer. The singer and the audience hear rather different sounds, and it is a miracle to me that they understand each other so well. Perhaps young singers would progress faster if during their training they spent more time studying recordings of their voices.

Feedback to the Voice

The control of speaking and singing involves a complicated feedback system. Just as feedback between the eyes and the muscles guides the hand when it moves to pick up an object, so feedback continually adjusts and corrects the voice as we speak or sing. When we start to sing, the beginning of the sound tells us the pitch, and we immediately adjust the tension of the vocal cords if the pitch is wrong. This feedback requires an exceedingly elaborate and rapid mechanism. How it works is not yet entirely understood. But it is small wonder that it takes a child years to learn to speak, or that it is almost impossible for an adult to learn to speak a foreign language with the native accents.

Any disturbance in the feedback immediately disturbs the speech. For instance, if, while a person is speaking, his speech is fed back to him with a time delay by means of a microphone and receivers at his ears, his pronunciation and accent will change, and if the delay interval is made long enough, he will find it impossible to speak at all.

some of the low-frequency components of the vocal cords' vibrations are lost. This explains why one can hardly recognize his own voice when he listens to a recording of his speech. As we normally hear ourselves, the low-frequency vibrations of our vocal cords, conducted to our own ears by the bones, make our speech sound much more powerful and dynamic than the pure sound waves heard by a second person or through a recording system. Consequently the re-

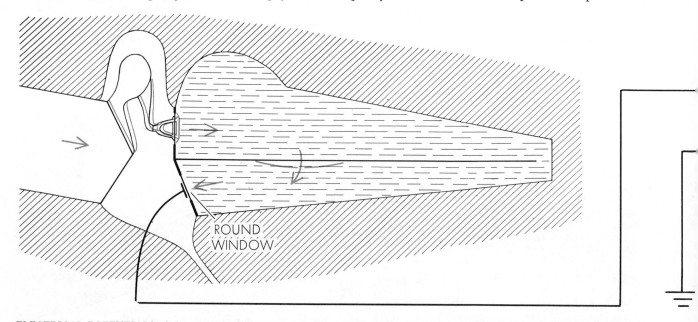

ELECTRICAL POTENTIALS of the microphonic type generated by the inner ear of an experimental animal can be detected by this arrangement. At left is a highly schematic diagram of the ear; the cochlea is represented in cross section by the fluid-filled chamber and the organ of Corti by the horizontal line in this chamber. When the vibrations of the eardrum are transmitted to the organ of Corti,

This phenomenon affords an easy test for exposing pretended deafness. If the subject can continue speaking normally in the face of a delayed feedback through the machine to his ears, we can be sure that he is really deaf.

The same technique can be used to assess the skill of a pianist. A piano player generally adjusts his touch to the acoustics of the room: if the room is very reverberant, so that the music sounds too loud, he uses a lighter touch; if the sound is damped by the walls, he strengthens his touch. We had a number of pianists play in a room where the damping could be varied, and recorded the amplitude of the vibrations of the piano's sounding board while the musicians played various pieces. When they played an easy piece, their adjustment to the acoustics was very clear: as the sound absorption of the room was increased, the pianist played more loudly, and when the damping on the walls was taken away, the pianist's touch became lighter. But when the piece was difficult, many of the pianists concentrated so hard on the problems of the music that they failed to adjust to the feedback of the room. A master musician, however, was not lost to the sound effects. Taking the technical difficulties of the music in stride, he was able to adjust the sound level to the damping of the room with the same accuracy as for an easy piece. Our rating of the pianists by this test closely matched their reputation among musical experts.

In connection with room acoustics, I should like to mention one of the ear's most amazing performances. How is it

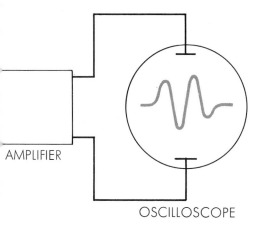

AMPLIFIER

OSCILLOSCOPE

INDIFFERENT ELECTRODE IN THE MUSCLE

its microphonic potentials can be picked up at the round window of the cochlea and displayed on the face of an oscilloscope (right).

that we can locate a speaker, even without seeing him, in a bare-walled room where reflections of his voice come at us from every side? This is an almost unbelievable performance by the ear. It is as if, looking into a room completely lined with mirrors, we saw only the real figure and none of the hundreds of reflected images. The eye cannot suppress the reflections, but the ear can. The ear is able to ignore all the sounds except the first that strikes it. It has a built-in inhibitory mechanism.

Suppressed Sounds

One of the most important factors that subordinate the reflected sounds is the delay in their arrival; necessarily they come to the ear only after the sound that has traveled directly from the speaker to the listener. The reflected sounds reinforce the loudness and tone volume of the direct sound, and perhaps even modify its localization, but by and large, they are not distinguishable from it. Only when the delay is appreciable does a reflected sound appear as a separate unit—an echo. Echoes often are heard in a large church, where reflections may lag more than half a second behind the direct sound. They are apt to be a problem in a concert hall. Dead walls are not desirable, because the music would sound weak. For every size of concert room there is an optimal compromise on wall reflectivity which will give amplification to the music but prevent disturbing echoes.

In addition to time delay, there are other factors that act to inhibit some sounds and favor others. Strong sounds generally suppress weaker ones. Sounds in which we are interested take precedence over those that concern us less, as I pointed out in the examples of the speaker in a noisy room and the orchestra conductor detecting an errant instrument. This brings us to the intimate collaboration between the ear and the nervous system.

Any stimulation of the ear (e.g., any change in pressure) is translated into electrical messages to the brain via the nerves. We can therefore draw information about the ear from an analysis of these electrical impulses, now made possible by electronic instruments. There are two principal types of electric potential that carry the messages. One is a continuous, wavelike potential which has been given the name microphonic. In experimental animals such as guinea pigs and cats the microphonics are large enough to be easily measured (they range up to about half a millivolt). It

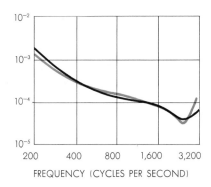

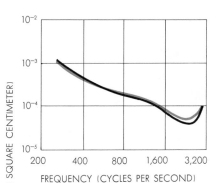

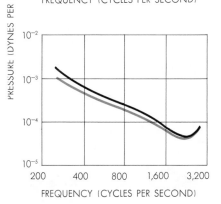

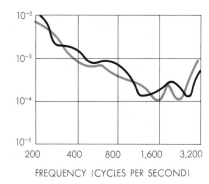

AUDIOGRAMS plot the threshold of hearing (in terms of pressure on the tympanic membrane) against the frequency of sound. The first three audiograms show the threshold for three members of the same family; the fourth, the threshold for an unrelated person. The black curves represent the threshold for one ear of the subject; the colored curves, for the other ear of the same subject. The audiogram curves indicate that in normal hearing the threshold in both ears, and the threshold in members of the same family, are remarkably similar.

7

222

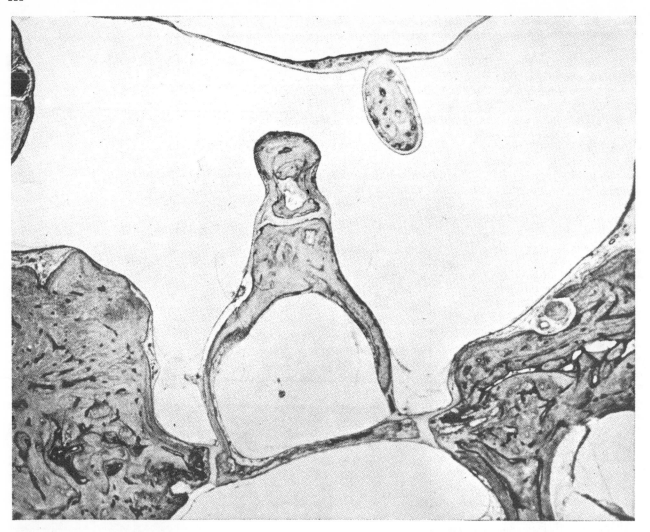

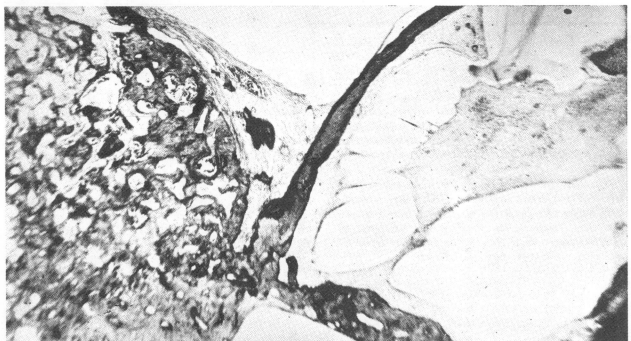

STIRRUP of the normal human ear is enlarged 19 times in the photograph at the top of this page. The thin line at the top of the photograph is the tympanic membrane seen in cross section. The hammer and anvil do not appear. The narrow membrane around the footplate of the stirrup may be seen as a translucent area between the footplate and the surrounding bone. The photograph at the bottom shows the immobilized footplate of an otosclerotic ear. In this photograph only the left side of the stirrup appears; the footplate is the dark area at the bottom center. The membrane around the footplate has been converted into a rigid bony growth.

has turned out that the magnitude of the microphonics produced in the inner ear is directly proportional to the displacements of the stirrup footplate that set the fluid in the inner ear in motion. The microphonics therefore permit us to determine directly to what extent the sound pressure applied to the eardrum is transmitted to the inner ear, and they have become one of the most useful tools for exploring sound transmission in the middle ear. For instance, there used to be endless discussion of the simple question: Just how much does perforation of the eardrum affect hearing? The question has now been answered with mathematical precision by experiments on animals. A hole of precisely measured size is drilled in the eardrum, and the amount of hearing loss is determined by the change in the microphonics. This type of observation on cats has shown that a perforation about one millimeter in diameter destroys hearing at the frequencies below 100 cycles per second but causes almost no impairment of hearing in the range of frequencies above 1,000 cycles per second. From studies of the physical properties of the human ear we can judge that the findings on animals apply fairly closely to man also.

The second type of electric potential takes the form of sharp pulses, which appear as spikes in the recording instrument. The sound of a sharp click produces a series of brief spikes; a pure tone generates volleys of spikes, generally in the rhythm of the period of the tone. We can follow the spikes along the nerve pathways all the way from the inner ear up to the cortex of the brain. And when we do, we find that stimulation of specific spots on the membrane of the inner ear seems to be projected to corresponding spots in the auditory area of the cortex. This is reminiscent of the projection of images on the retina of the eye to the visual area of the brain. But in the case of the ear the situation must be more complex, because there are nerve branches leading to the opposite ear and there seem to be several auditory projection areas on the surface of the brain. At the moment research is going on to find out how the secondary areas function and what their purpose is.

Detecting Pitch

The orderly projection of the sensitive area of the inner ear onto the higher brain levels is probably connected with the resolution of pitch. The ear itself can analyze sounds and separate one tone from another. There are limits to this

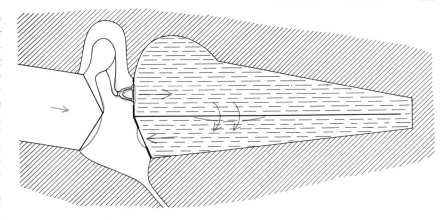

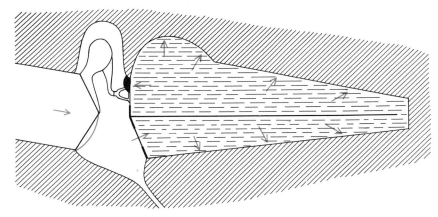

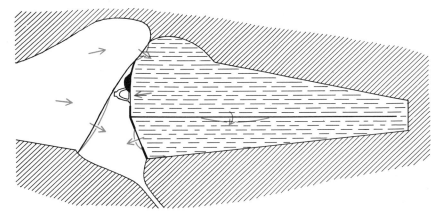

FENESTRATION OPERATION can alleviate the effects of otosclerosis. The drawing at the top schematically depicts the normal human ear as described in the caption for the illustration on page 220 and 221. The pressure on the components of the ear is indicated by the colored arrows. The drawing in the middle shows an otosclerotic ear; the otosclerotic growth is represented as a black protuberance. Because the stirrup cannot move, the pressure on the tympanic membrane is transmitted to the organ of Corti only through the round window of the cochlea; and because the fluid in the cochlea is incompressible, the organ of Corti cannot vibrate. The drawing at the bottom shows how the fenestration operation makes a new window into the cochlea to permit the organ of Corti to vibrate freely.

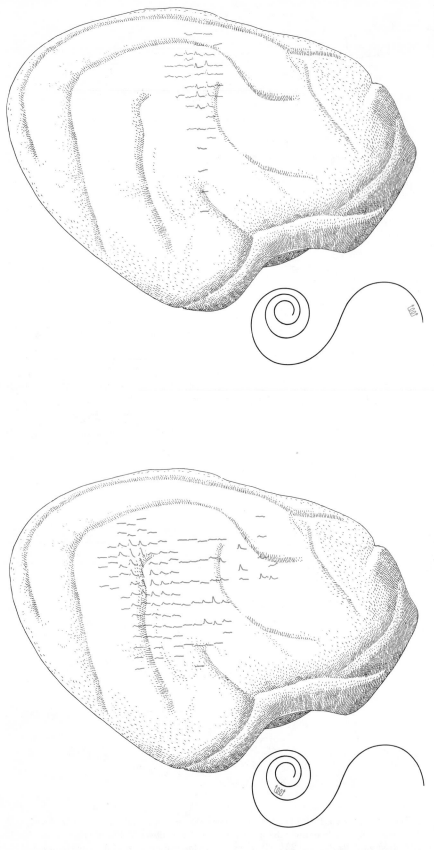

ability, but if the frequencies of the tones presented are not too close together, they are discriminated pretty well. Long ago this raised the question: How is the ear able to discriminate the pitch of a tone? Many theories have been argued, but only within the last decade has it been possible to plan pertinent experiments.

In the low-frequency range up to 60 cycles per second the vibration of the basilar membrane produces in the auditory nerve volleys of electric spikes synchronous with the rhythm of the sound. As the sound pressure increases, the number of spikes packed into each period increases. Thus two variables are transmitted to the cortex: (1) the number of spikes and (2) their rhythm. These two variables alone convey the loudness and the pitch of the sound.

Above 60 cycles per second a new phenomenon comes in. The basilar membrane now begins to vibrate unequally over its area: each tone produces a maximal vibration in a different area of the membrane. Gradually this selectivity takes over the determination of pitch, for the rhythm of the spikes, which indicates the pitch at low frequencies, becomes irregular at the higher ones. Above 4,000 cycles per second pitch is determined entirely by the location of the maximal vibration amplitude along the basilar membrane. Apparently there is an inhibitory mechanism which suppresses the weaker stimuli and thus sharpens considerably the sensation around the maximum. This type of inhibition can also operate in sense organs such as the skin and the eye. In order to see sharply we need not only a sharp image of the object on the retina but also an inhibitory system to suppress stray light entering the eye. Otherwise we would see the object surrounded by a halo. The ear is much the same. Without inhibitory effects a tone would sound like a noise of a certain pitch but not like a pure tone.

We can sum up by saying that the basilar membrane makes a rough, mechanical frequency analysis, and the auditory nervous system sharpens the analysis in some manner not yet understood. It is a part of the general functioning of the higher nerve centers, and it will be understood only when we know more about the functioning of these centers. If the answer is found for the ear, it will probably apply to the other sense organs as well.

Deafness

Now let us run briefly over some of

NERVE IMPULSES due to the electrical stimulation of the organ of Corti were localized on the surface of the brain of a cat. The spirals below each of these drawings of a cat's brain represent the full length of the organ of Corti. The pairs of colored arrows on each spiral indicate the point at which the organ was stimulated. The colored peaks superimposed on the brains represent the electrical potentials detected by an electrode placed at that point.

the types of hearing disorders, which have become much more understandable as a result of recent experimental researches.

Infections of the ear used to be responsible for the overwhelming majority of the cases of deafness. Ten years ago in a large city hospital there was a death almost every day from such infections. Thanks to antibiotics, they can now be arrested, and, if treated in time, an ear infection is seldom either fatal or destructive of hearing, though occasionally an operation is necessary to scoop out the diseased part of the mastoid bone.

The two other principal types of deafness are those caused by destruction of the auditory nerves and by otosclerosis (a tumorous bone growth). Nerve deafness cannot be cured: no drug or mechanical manipulation or operation can restore the victim's hearing. But the impairment of hearing caused by otosclerosis can usually be repaired, at least in part.

Otosclerosis is an abnormal but painless growth in a temporal bone (*i.e.*, at the side of the skull, near the middle ear). If it does not invade a part of the ear that participates in the transmission of sound, no harm is done to the hearing. But if the growth happens to involve the stirrup footplate, it will reduce or even completely freeze the footplate's ability to make its piston-like movements; the vibrations of the eardrum then can no longer be transmitted to the inner ear. An otosclerotic growth can occur at any age, may slow down for many years, and may suddenly start up again. It is found more often in women than in men and seems to be accelerated by pregnancy.

Immobilization of the stirrup blocks the hearing of air-borne sound but leaves hearing by bone conduction unimpaired. This fact is used for diagnosis. A patient who has lost part of his hearing ability because of otosclerosis does not find noise disturbing to his understanding of speech; in fact, noise may even improve his discrimination of speech. There is an old story about a somewhat deaf English earl (in France it is a count) who trained his servant to beat a drum whenever someone else spoke, so that he could understand the speaker better. The noise of the drum made the speaker

raise his voice to the earl's hearing range. For the hard-of-hearing earl the noise of the drum was tolerable, but for other listeners it masked what the speaker was saying, so that the earl enjoyed exclusive rights to his conversation.

Difficulty in hearing air-borne sound can be corrected by a hearing aid. Theoretically it should be possible to compensate almost any amount of such hearing loss, because techniques for amplifying sound are highly developed, particularly now with the help of the transistor. But there is a physiological limit to the amount of pressure amplification that the ear will stand. Heightening of the pressure eventually produces an unpleasant tickling sensation through its effect on skin tissue in the middle ear. The sensation can be avoided by using a bone-conduction earphone, pressed firmly against the surface of the skull, but this constant pressure is unpleasant to many people.

Operations

As is widely known, there are now operations (*e.g.*, "fenestration") which can cure otosclerotic deafness. In the 19th century physicians realized that if they could somehow dislodge or loosen the immobilized stirrup footplate, they might restore hearing. Experimenters in France found that they could sometimes free the footplate sufficiently merely by pressing a blunt needle against the right spot on the stirrup. Although it works only occasionally, the procedure seems so simple that it has recently had a revival of popularity in the U. S. If the maneuver is successful (and I am told that 30 per cent of these operations are) the hearing improves immediately. But unfortunately the surgeon cannot get a clear look at the scene of the operation and must apply the pushing force at random. This makes the operation something of a gamble, and the patient's hearing may not only fail to be improved but may even be reduced. Moreover, the operation is bound to be ineffectual when a large portion of the footplate is fixed. There are other important objections to the operation. After all, it involves the breaking of bone, to free the adhering part of the stirrup. I do not think that

bone-breaking can be improved to a standard procedure. In any case, precision cutting seems to me always superior to breaking, in surgery as in mechanics. This brings us to the operation called fenestration.

For many decades it has been known that drilling a small opening, even the size of a pinhead, in the bony wall of the inner ear on the footplate side can produce a remarkable improvement in hearing. The reason, now well understood, is quite simple. If a hole is made in the bone and then covered again with a flexible membrane, movements of the fluid in, for instance, the lateral canal of the vestibular organ can be transmitted to the fluid of the inner ear, and so vibrations are once again communicable from the middle to the inner ear. In the typical present fenestration operation the surgeon bores a small hole in the canal wall with a dental drill and then covers the hole with a flap of skin. The operation today is a straightforward surgical procedure, and all its steps are under accurate control.

Hazards to Hearing

I want to conclude by mentioning the problem of nerve deafness. Many cases of nerve deafness are produced by intense noise, especially noise with high-frequency components. Since there is no cure, it behooves us to look out for such exposures. Nerve deafness creeps up on us slowly, and we are not as careful as we should be to avoid exposure to intense noise. We should also be more vigilant about other hazards capable of producing nerve deafness, notably certain drugs and certain diseases.

We could do much to ameliorate the tragedy of deafness if we changed some of our attitudes toward it. Blindness evokes our instant sympathy, and we go out of our way to help the blind person. But deafness often goes unrecognized. If a deaf person misunderstands what we say, we are apt to attribute it to lack of intelligence instead of to faulty hearing. Very few people have the patience to help the deafened. To a deaf man the outside world appears unfriendly. He tries to hide his deafness, and this only brings on more problems.

25

The Sensitivity of Organisms to Gravity

Aharon Gibor

Gravity is a physical force in our environment that always existed as it does today. Therefore the evolution of all living organisms has taken place under the influence of this force. Consider the emergence of living organisms from the sea to inhabit the land and air. One of the major obstacles such organisms had to overcome was the greatly increased effective weight of their bodies outside the buoyant watery environment. The evolution of rigid skeletons and strengthened muscles are good examples of biological adaptation to the force of gravity.

Almost all environmental forces are utilized biologically either as a source of energy or for such purposes as motion, orientation, and information-gathering. The vectorial force of gravity is sensed and used for orientation by most living organisms.

Interactions of organisms with environmental forces are logically subdivided into three processes: (1) sensing of the force by some component of the organism; (2) transduction of the stimulus to other components of the organism; and (3) response or counteraction of the organism to the stimulus. The sensing, transduction, and response of many diverse organisms to gravity have been studied for many years. In recent years we have gained much knowledge on the details of these processes in some simple organisms.

Complex organisms have special gravity-sensing organs. In mammals they are the semicircular canals of the inner ear (see Fig. 1). When an animal moves its head with respect to the gravity vector, the hydro-dynamic pressure of flowing fluid in these canals is sensed by specialized cells. The sensing cells transmit nerve impulses to the brain, where appropriate responses to the stimulus are initiated by sending nerve impulses to the relevant muscles.

Instead of the hydraulic type of sensing, many invertebrates use a solid particle called a *statolith*. The statolith is suspended on the cilia of numerous cells that line the cavity of a special organ called a *statocyst* (see Fig. 2). The differential pressure, or stretch, exerted on the different cilia when the animal changes its orientation with respect to the gravity vector causes the stimulated cells to generate nerve impulses. These are transmitted to the central nervous system, which then commands the appropriate body reactions.

For gravity sensors the segmented insects utilize special hairy plates in their joints; these plates respond to changes in orientation of the segments in question. For example, the position of the head with respect to the thorax is sensed by hair plates that surround the neck region of the insect (see Fig. 3). Some aquatic insects utilize small air bubbles in their joints—like the bubble levels used for leveling instruments—to stimulate the sensitive hairs.

Complex, multicellular animals are not readily amenable to physiological analysis at the cellular, subcellular, and molecular levels. Plants, particularly the lower, simple plants, permit a more direct approach to cellular and subcellular analysis of the sensing of and response to the force of gravity. In this article I will deal primarily with some recent findings from studies of gravity effects on the growth of plant cells. Such investigations, like all good model studies, should permit us to generalize (with caution) to other biological systems.

Let us first examine the responses of unicellular

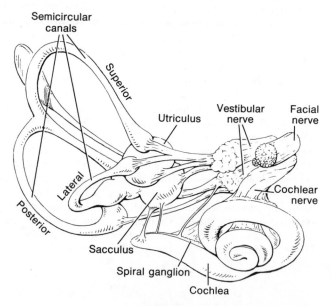

Figure 1
The labyrinth of the human inner ear. Note the orientations of the three semicircular canals. The motion of fluid within these canals stimulates sensitive cells. The sensations of movement, acceleration, and orientation are due largely to this sense organ.

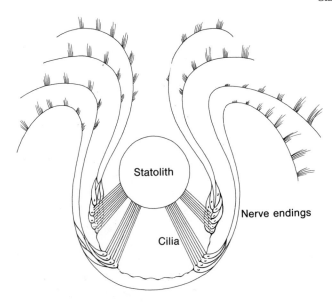

Figure 2

A simple statocyst of a Coelenterate animal. The solid statolith rests on the cilia of the sense cells. Different cells are stimulated when the animal changes its orientation.

organisms to gravity. The protoplasm of all living cells is composed of a large variety of particles and aggregates of particles suspended in a heterogeneous matrix. Particles tend to float or sink with respect to other cell components, depending on their relative densities. Brownian motion on the one hand, and the binding of particles to cellular skeletons such as the endoplasmic reticulum or microtubules on the other, tend to oppose the stratification of particles according to their densities. The spatial relations among cellular organelles and structures, although not determined by gravity, are undoubtedly affected by this ever-present force. Experimental cell biologists take advantage of the differences in density among cellular organelles by exposing the cells to high gravitational forces in high-speed centrifuges; this stratifies the organelles.

Experimentally, a living cell, suspended in water or air, can be reoriented with respect to the gravity vector. The cell can now sense the gravitational force, either by its differential effect on particles or organelles within

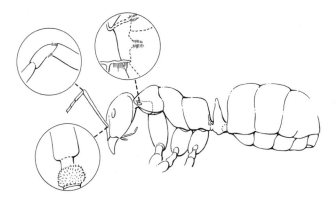

Figure 3

The locations of some of the sensitive hair plates in the joints of an ant. The differential pressure exerted by the segments of the body on these hairs indicates the orientation of the body with respect to the gravity vector.

the cell, or by the stress that it imposes on the cell walls owing to differences in density between the whole cell and the suspending medium. That the first of these alternatives is probably true is indicated by studies on the responses of two simple, lower plants. One study was on the upward vertical growth of the spore-bearing unicellular organ—the *sporangiophore*—of a simple fungus, *Phycomyces*; the other was on the downward vertical growth of the unicellular *rhizoid* (a simple structure that functions like a root) of an alga, *Chara*.

Phycomyces can be grown either in liquid or on plates of gelled liquid. Their vegetative growth forms a mat of randomly oriented, interconnected tubules called a *mycelium*. The sporangiophores grow vertically upward from the mycelium. Such growth, in the direction opposite to the gravity vector, is called negative geotropism; positive geotropism is growth in the direction of the gravity vector. The sporangiophore is a single cell, about 0.1 mm in diameter, and can reach a length of 20 cm. Its upper end is swollen to form a spherical structure, about 0.5 mm in diameter, called the *sporangium*, in which a large number of spores are formed. The sporangiophore grows very rapidly—up to 3 mm per hour; the growth is localized to the region of the cell just below the sporangium. Sporangiophores that are placed horizontally continue to grow and curve upward within hours, despite the asymmetric force exerted on the slender cell by the weight of the massive sporangium at its tip. The question is whether the cell senses gravity by means of intracellular structures or in terms of distortion pressure on the cell wall itself.

In an elegant experiment, David S. Dennison, of Dartmouth College, demonstrated that the geotropic response is not due to the weight of the sporangium on the slender sporangiophore. He immersed the sporangiophore in a biologically inert fluorocarbon liquid with a density of 1.87 g/ml. In this liquid the sporangium, whose density is much less than 1.87, tends to float rather than exert weight on the sporangiophore. If gravity were sensed in terms of distortion of the cell wall, one would expect that, whereas the sporangiophore shows negative geotropism in air, it would show the opposite response when immersed in the dense liquid, and grow downward.

The result of the experiment was an unaltered negative geotropic growth of the sporangiophore. This clearly indicates that the gravity sensor is not the rigid cell wall, but more likely an intracellular structure. (For the results of a related experiment by Dennison, see Fig. 4.)

Which intracellular structure might be the gravity sensor? As in all other living cells, the protoplasm of the sporangiophore contains a large variety of particles and organelles. Detailed microscopic studies failed to show a correlation between the distribution of any of these particles within the growing region of the cell and the angle at which the cell was placed. Dennison suggested that the vacuole of the cell might function as a gravity sensor. As in most differentiated plant cells, most of the volume of the cell is occupied by a vacuole, with the active protoplasm forming a thin layer between the cell wall and the vacuole. Because the vacuolar fluid is lighter than the surrounding protoplasm, the vacuole tends to rise in a horizontally placed cell, thereby causing an

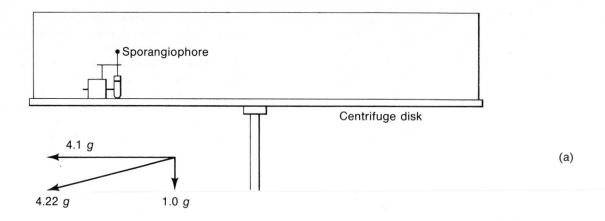

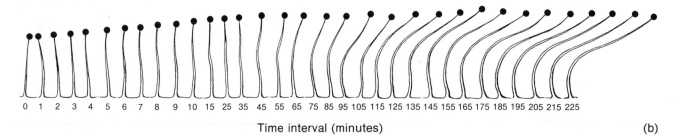

Time interval (minutes)   (b)

Figure 4
An experiment by David S. Dennison to demonstrate the negative geotropic growth of sporangiophores. (A) Sporangiophore mounted on a centrifuge disc. The centrifugal force was equal to 4.1 *g*. The direction and magnitude of the resultant force acting on the sporangiophore are shown in the force diagram. (B) Appearance of the sporangiophore at the indicated times during continuous growth in the centrifugal field. The direction of growth is 180° to the resultant force vector shown above.

uneven distribution of the protoplasm. The layer of protoplasm on the lower side of the cell becomes thicker while the upper layer becomes thinner. As a consequence, the lower side of the cell tends to grow more rapidly than the upper side, causing the cell to curve upward. Though this is an attractive hypothesis, there is not yet any experimental evidence to support it.

More direct correlations between geotropic response and the spatial distribution of specific intracellular particles were obtained in the studies on the rhizoids of the alga *Chara*. The rhizoid cells are cylindrical, about 0.03 millimeter in diameter, and grow to a length of several millimeters. Most of the protoplasm of the cell is concentrated in the apical region, the only region where growth occurs. A large number of refractile granules are usually found in this region.

Andreas Sievers, at the University of Bonn, found that if a growing rhizoid cell is placed in a horizontal position, the cell tip curves downward within half an hour. Careful microscopic studies revealed that, whereas in normally growing cells the refractile particles accumulate at the center of the growing tip, they are displaced downward in cells that are placed horizontally, and accumulate on the now lower side of the tip (see Fig. 5). An electron-microscopic study of the growth zone of such cells revealed that the accumulated refractile bodies on one side of the cell cause a displacement of cytoplasmic organelles toward the opposite side.

The growth of cell walls of plants is known to be correlated with an abundance of Golgi vesicles in the vicinity of the growth region. Sievers was able to show that the accumulation of refractile bodies on one side of the cell deflects the streaming cytoplasm; Golgi vesicles thus accumulate toward the upper side of the cell. This, he argues, causes a more rapid growth of the upper side and the bending downward of the growing tip.

We do not know the entire physiological role of these refractile bodies. In the context of their presumed function as gravity sensors, they are called statoliths, a term that we introduced earlier. In its broadest sense, it denotes any biological structure that functions as a gravity sensor by virtue of its differing density from that of the surrounding medium. Further correlation between the refractile bodies and the positive geotropic response of the rhizoid cells of *Chara* was obtained in experiments in which the heavy granules were removed from the growing tip by centrifugation. Cells that did not contain the refractile bodies did not respond to gravity when they were placed horizontally.

Recently, Sievers reported that the refractile bodies consist of $BaSO_4$ crystals. This is a remarkable feat of biological concentration by this alga, since the amount of barium ions in the fresh waters where *Chara* usually grows is at the lower limit of detection. The accumulation of $BaSO_4$ particles at one side of the growing tip might have effects other than the deflection of the streaming cytoplasm. It has been suggested that the extra pressure exerted on the lower cytoplasmic membrane is the crucial effect on the cell. The essential membrane properties in the pressurized region of the cell will then be sufficiently altered to cause a reduction of growth in that region.

Ulrich Friedrich and Rainer Hartel, of the University of Freiburg, have demonstrated that the degree of curvature of *Chara* rhizoids obtained after the lateral deflection of the BaSO₄ particles by centrifugation is proportional to the applied force. The deflection of the particles required centrifugation for only 10 seconds at 10 *g*. However, the degree of curvature is higher in cells centrifuged for 10 minutes at 10 *g* than in cells centrifuged for 10 minutes at 5 *g*. These results suggest that the pressure effects on the membrane, rather than the displacement of the particles to one side, ultimately cause the differential elongation of the two sides of the cell.

The suggestion that localized pressure on the cell membrane is responsible for its differential response in a gravitational field is not novel. Studies on higher plants also point to effects on the membranes of the cells in their response to gravity. Charles Darwin was one of the early experimenters on geotropism in higher plants. He discovered that intact roots, when placed horizontally, rapidly changed their direction of growth and curved downward. On the other hand, roots that were decapitated and then placed horizontally failed to bend. It took about a century of research to reach a satisfactory explanation for these observations.

With the establishment of the role of the shoot apex and root tip in the production of growth hormones, the *auxins*, it became important to find whether the geotropic responses of roots and shoots can be related to the production and/or distribution of auxins in them. In 1926 N. Cholodney and, independently, Frits Went suggested that the curvature of roots and shoots results from lateral movement of auxins from the upper to the lower part of a horizontally placed plant organ. The positive geotropic response of roots and the opposite response of shoots were explained by the then-established fact that these plant organs respond differently to given concentrations of auxins. In general, concentrations that stimulated the elongation of shoots were found to inhibit the elongation of roots. Thus an increase in concentration of auxins at the lower side of a horizontal shoot causes the accelerated elongation of that side and a curving upward, whereas the opposite occurs in a horizontal root.

Experimental verification of this hypothesis had to await the introduction of highly sensitive tracer techniques. In the early 1960s Barbara Gillespie and Kenneth Thimann, at Harvard University, performed the following experiment. A young, growing shoot of an oat seedling was cut off and its apex was removed. A small block of agar containing radioactive auxin, called the donor block, was substituted for the missing shoot apex. The basal part of the shoot was cut longitudinally a short way up the shoot and a partition was inserted between the two halves. Two blocks of agar, called receiving blocks, were placed at the basal ends of the two half-shoots. The migration of the auxin from the apical to the basal end of the shoot was followed by counting the radioactive auxin that accumulated in receiving blocks. When the experimental shoots stood vertically it was found that the two receiving blocks

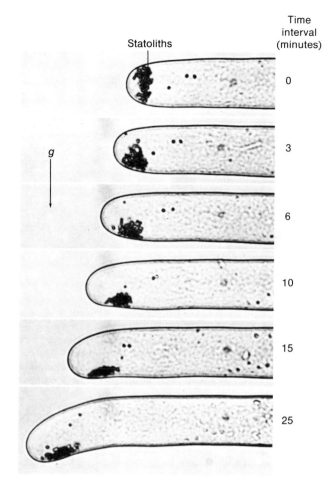

Figure 5
Time-lapse photographs of a growing rhizoid tip of *Chara*, beginning immediately upon placing the cell horizontally. Note the rapid redistribution of the statoliths in the tip and the curving downward of the growing tip.
(Courtesy of Prof. Andreas Sievers.)

accumulated equal amounts of auxin. When the shoots were placed horizontally, however, the lower block was found to have accumulated 1.6 times as much auxin as the upper block (see Fig. 6).

The results of this experiment strongly support the hypothesis that under the influence of gravity more auxin moves to the lower side of the shoot. But, as it was performed, the experiment is open to criticism in that gravity might affect the relative rates of absorption of auxin from the donor block and secretion into the receiving blocks rather than the actual migration from one side of the shoot to the other.

To show that movement did take place, Mary Helen Goldsmith and Malcolm B. Wilkins, of the University of Glasgow, introduced a small but significant variation in the design of the experiment. Instead of having the donor block cover the entire apical surface area of the decapitated shoot, they used a smaller donor block that covered only half that area. In all other respects the experiment was identical to that of Gillespie and Thimann (see Fig. 7).

The results of the experiment with the shoot standing vertically indicated that there was a small amount of lateral migration of the auxin. The receiving block placed directly under the donor block contained 72% of the total translocated auxin, while the other receiving

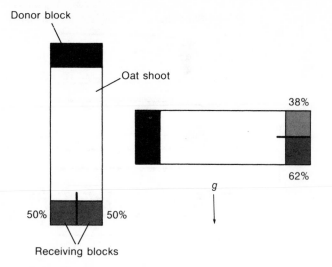

Figure 6
An experiment by Barbara Gillespie and Kenneth Thimann. Agar donor blocks containing radioactive auxin were placed on the apical ends of decapitated oat shoots. Pairs of agar receiving blocks were placed on the split basal ends. The two receiving blocks were counted for accumulated auxin. The distribution of the translocated auxins between these two blocks is shown.

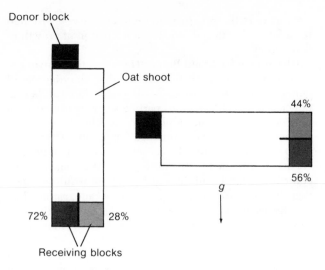

Figure 7
Refinement on the experiment shown in Figure 6, introduced by Mary Helen Goldsmith and Malcolm B. Wilkins. The increased lateral migration of auxin is clearly shown by covering only one-half the apical surface area of the shoot with the donor block and counting, as before, the accumulated auxin in the two receiving blocks.

block contained the remaining 28%. However, when the shoot was placed horizontally with the donor block on the upper side of the shoot, the lower receiving block contained 56% of the translocated auxin. Thus gravity does increase the rate of movement of auxin into the lower side of the shoot.

We are left with the question of the control of lateral migration in such multicellular plant organs. We do not know whether auxin alone or other substances as well are differentially translocated to the lower side of a horizontal plant organ. And we know very little of what causes this unequal movement of molecules among cells. We might think of the lateral migration of auxin as the transduction process in the physiological response to gravity. But what is the gravity sensor that initiates the response?

There is considerable evidence that starch grains, which are found in most plant cells, are the statoliths. The density of starch grains is about 1.5 g/ml, which is considerably greater than that of the cytoplasm, and they vary in diameter from 1 to 100 micrometers. It is a readily observable fact that these rather large, heavy cellular particles are displaced to the lower side of the cell. If we place a root tip in a horizontal position, the starch particles will accumulate on what is now the lower side, sometimes within minutes.

A direct correlation between the presence of starch grains and the sensitivity of plants to gravity has been demonstrated in numerous studies. For example, starch can be depleted from plant cells by keeping the cells for long periods in the dark or at reduced temperature. Similarly, treating plants with chemicals such as sulfur dioxide or potassium alum causes depletion of the starch reserves of the cells. Plants so treated are found to lose their geotropic sensitivity in direct pro-

portion to the extent of the loss of their starch grains. Upon return to normal conditions, the plants resynthesize and accumulate starch. This is accompanied by a restoration of the sensitivity to gravity.

There are exceptions, however. Some plants in which starch grains cannot be demonstrated by the standard iodine staining test are still susceptible to gravity. It is therefore possible that other dense particles can also function as statoliths.

The question remains as to how the displacement of the starch grains in the cells causes the apparent increased lateral migration of auxins. It seems reasonable to assume that here too the increased pressure exerted by the heavy particles on particular regions of the cell membrane causes changes in the membrane's properties. In this case we assume that auxin transport to the adjacent cell is facilitated by the compressed region of the cell membrane. When the shoot is in a horizontal position the lateral walls of the cells become the "floor" and "ceiling." Starch grains accumulate on the new floor and facilitate the movement of auxins through it to the cell below. We are still far from understanding the mechanisms that govern the migration of auxins across cells.

It appears that, like so many other intriguing biological phenomena, the response to gravity is also a membrane phenomenon. Whether and how the slight application of pressure modifies the properties of the cell membrane will have to await our fuller understanding of the functions of the normal, uncompressed membrane. Drugs are now known that specifically abolish the geotropic responses of roots. It is yet to be determined whether these drugs affect the auxin distribution and/or membrane properties of such tissues.

EPILOGUE

EPILOGUE

INTRODUCTION

In selecting the articles for this volume, I attempted to present two main ideas. The first is that living matter is composed of substances that are readily available in the inanimate environment. Not only the chemical elements but also the intermolecular forces that cause the organization of living matter are the same as those that characterize the nonliving world.

The other idea is that of the versatility and adaptability of organisms. Living matter has acquired the ability to sense and respond in a biologically meaningful way to practically all existing conditions on the surface of the earth and to utilize most forms of available energy in the environment. We attribute this "fitness" of living organisms for environmental conditions to the process of evolution. But how do we explain biological evolution?

Life processes lead to the reproduction of living matter. At the same time, biochemical instabilities cause new forms — mutants — to arise. The environment itself is not stable, but changeable. Given a variable environment and genetic variations in the population of living organisms, competition and natural selection of the fittest organisms are inevitable. Evolution is thus a remarkable, creative process of the biosphere as a whole; it results from the continuous interplay between the instabilities of the animate and inanimate portions of the biosphere.

In the concluding article, "The Search for Extraterrestrial Intelligence" (page 234), Carl Sagan and Frank Drake present a well-reasoned argument for the likelihood of intelligent life in the universe, and describe how to search for it. Planets with chemical and physical conditions similar to those on earth are abundant in the universe. It is as reasonable to expect to find life on such planets as it is to expect to find rocks that are identical in mineral composition to rocks found on earth.

Another level of expectation is to find not just living organisms but highly intelligent beings. Here again, our basic assumption is that intelligence is a fundamental evolutionary consequence that becomes manifest at a certain level of biological complexity. Some degree of intelligence appears in many forms of life on earth. Given ample time, the evolution of higher intelligence is as much to be expected as the evolution of complex sense organs such as eyes and ears. The continuing exploration of the planets of our solar system may soon tell us how reasonable these expectations are.

26

The Search for Extraterrestrial Intelligence

by Carl Sagan and Frank Drake
May 1975

There can be little doubt that civilizations more advanced than the earth's exist elsewhere in the universe. The probabilities involved in locating one of them call for a substantial effort

Is mankind alone in the universe? Or are there somewhere other intelligent beings looking up into their night sky from very different worlds and asking the same kind of question? Are there civilizations more advanced than ours, civilizations that have achieved interstellar communication and have established a network of linked societies throughout our galaxy? Such questions, bearing on the deepest problems of the nature and destiny of mankind, were long the exclusive province of theology and speculative fiction. Today for the first time in human history they have entered into the realm of experimental science.

From the movements of a number of nearby stars we have now detected unseen companion bodies in orbit around them that are about as massive as large planets. From our knowledge of the processes by which life arose here on the earth we know that similar processes must be fairly common throughout the universe. Since intelligence and technology have a high survival value it seems likely that primitive life forms on the planets of other stars, evolving over many billions of years, would occasionally develop intelligence, civilization and a high technology. Moreover, we on the earth now possess all the technology necessary for communicating with other civilizations in the depths of space. Indeed, we may now be standing on a threshold about to take the momentous step a planetary society takes but once: first contact with another civilization.

In our present ignorance of how common extraterrestrial life may actually be, any attempt to estimate the number of technical civilizations in our galaxy is necessarily unreliable. We do, however, have some relevant facts. There is reason to believe that solar systems are formed fairly easily and that they are abundant in the vicinity of the sun. In our own solar system, for example, there are three miniature "solar systems": the satellite systems of the planets Jupiter (with 13 moons), Saturn (with 10) and Uranus (with five). It is plain that however such systems are made, four of them formed in our immediate neighborhood.

The only technique we have at present for detecting the planetary systems of nearby stars is the study of the gravitational perturbations such planets induce in the motion of their parent star. Imagine a nearby star that over a period of decades moves measurably with respect to the background of more distant stars. Suppose it has a nonluminous companion that circles it in an orbit whose plane does not coincide with our line of sight to the star. Both the star and the companion revolve around a common center of mass. The center of mass will trace a straight line against the stellar background and thus the luminous star will trace a sinusoidal path. From the existence of the oscillation we can deduce the existence of the companion. Furthermore, from the period and amplitude of the oscillation we can calculate the period and mass of the companion. The technique is only sensitive enough, however, to detect the perturbations of a massive planet around the nearest stars.

The single star closest to the sun is Barnard's star, a rather dim red dwarf about six light-years away. (Although Alpha Centauri is closer, it is a member of a triple-star system.) Observations made by Peter van de Kamp of the Sproul Observatory at Swarthmore College over a period of 40 years suggest that Barnard's star is accompanied by at least two dark companions, each with about the mass of Jupiter.

There is still some controversy over his conclusion, however, because the observations are very difficult to make. Perhaps even more interesting is the fact that of the dozen or so single stars nearest the sun nearly half appear to have dark companions with a mass between one and 10 times the mass of Jupiter. In

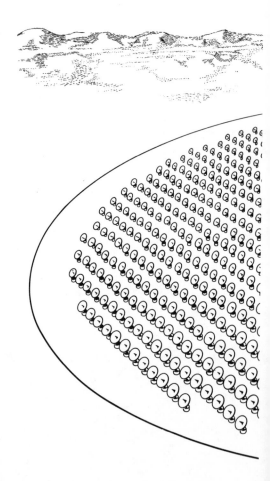

"CYCLOPS," an array of 1,500 radio antennas each 100 meters in diameter, is one system that has been proposed as a tool for detecting signals from extraterrestrial civ-

addition many theoretical studies of the formation of planetary systems out of contracting clouds of interstellar gas and dust imply that the birth of planets frequently if not inevitably accompanies the birth of stars.

We know that the master molecules of living organisms on the earth are the proteins and the nucleic acids. The proteins are built up of amino acids and the nucleic acids are built up of nucleotides. The earth's primordial atmosphere was, like the rest of the universe, rich in hydrogen and in hydrogen compounds. When molecular hydrogen (H_2), methane (CH_4), ammonia (NH_3) and water (H_2O) are mixed together in the presence of virtually any intermittent source of energy capable of breaking chemical bonds, the result is a remarkably high yield of amino acids and the sugars and nitrogenous bases that are the chemical constituents of the nucleotides. For example, from laboratory experiments we can determine the amount of amino acids produced per photon of ultraviolet radiation, and from our knowledge of stellar evolution we can calculate the amount of ultraviolet radiation emitted by the sun over the first billion years of the existence of the earth. Those two rates enable us to compute the total amount of amino acids that were formed on the primitive earth. Amino acids also break down spontaneously at a rate that is dependent on the ambient temperature. Hence we can calculate their steady-state abundance at the time of the origin of life. If amino acids in that abundance were mixed into the oceans of today, the result would be a 1 percent solution of amino acids. That is approximately the concentration of amino acids in the better brands of canned chicken bouillon, a solution that is alleged to be capable of sustaining life.

The origin of life is not the same as the origin of its constituent building blocks, but laboratory studies on the linking of amino acids into molecules resembling proteins and on the linking of nucleotides into molecules resembling nucleic acids are progressing well. Investigations of how short chains of nucleic acids replicate themselves in vitro have even provided clues to primitive genetic codes for translating nucleic acid information into protein information, systems that could have preceded the elaborate machinery of ribosomes and activating enzymes with which cells now manufacture protein.

The laboratory experiments also yield a large amount of a brownish polymer that seems to consist mainly of long hydrocarbon chains. The spectroscopic properties of the polymer are similar to those of the reddish clouds on Jupiter, Saturn and Titan, the largest satellite of Saturn. Since the atmospheres of these objects are rich in hydrogen and are similar to the atmosphere of the primitive earth, the coincidence is not surprising. It is nonetheless remarkable. Jupiter, Saturn and Titan may be vast planetary laboratories engaged in prebiological organic chemistry.

Other evidence on the origin of life comes from the geological record of the earth. Thin sections of sedimentary rocks

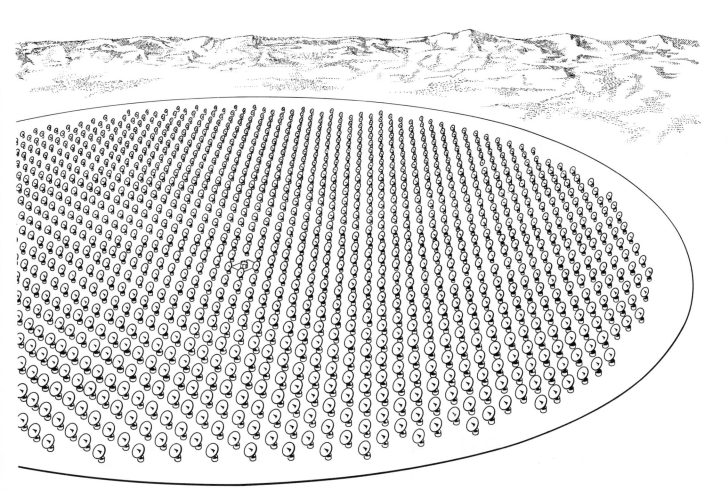

ilizations. The individual antennas would be connected to one another and to a large computer system. The effective signal-collecting area of the system would be hundreds of times greater than that of any existing radio telescope and would be capable of detecting even such relatively weak signals as the internal radio-frequency communications of a civilization as far away as several hundred light-years. Control building shown in center of array includes observatory with telescope operating at visible wavelengths.

between 2.7 and 3.5 billion years old reveal the presence of small inclusions a hundredth of a millimeter in diameter. These inclusions have been identified by Elso S. Barghoorn of Harvard University and J. William Schopf of the University of California at Los Angeles as bacteria and blue-green algae. Bacteria and blue-green algae are evolved organisms and must themselves be the beneficiaries of a long evolutionary history. There are no rocks on the earth or on the moon, however, that are more than four billion years old; before that time the surface of both bodies is believed to have melted in the final stages of their accretion. Thus the time available for the origin of life seems to have been short: a few hundred million years at the most. Since life originated on the earth in a span much shorter than the present age of the earth, we have additional evidence that the origin of life has a high probability, at least on planets with an abundant supply of hydrogen-rich gases, liquid water and

sources of energy. Since those conditions are common throughout the universe, life may also be common.

Until we have discovered at least one example of extraterrestrial life, however, that conclusion cannot be considered secure. Such an investigation is one of the objectives of the Viking mission, which is scheduled to land a vehicle on the surface of Mars in the summer of 1976, a vehicle that will conduct the first rigorous search for life on another planet. The *Viking* lander carries three separate experiments on the metabolism of hypothetical Martian microorganisms, one experiment on the organic chemistry of the Martian surface material and a camera system that might just conceivably detect macroscopic organisms if they exist.

Intelligence and technology have developed on the earth about halfway through the stable period in the lifetime of the sun. There are obvious selective

advantages to intelligence and technology, at least up to the present evolutionary stage when technology also brings the threats of ecological catastrophes, the exhaustion of natural resources and nuclear war. Barring such disasters, the physical environment of the earth will remain stable for many more billions of years. It is possible that the number of individual steps required for the evolution of intelligence and technology is so large and improbable that not all inhabited planets evolve technical civilizations. It is also possible—some would say likely—that civilizations tend to destroy themselves at about our level of technological development. On the other hand, if there are 100 billion suitable planets in our galaxy, if the origin of life is highly probable, if there are billions of years of evolution available on each such planet and if even a small fraction of technical civilizations pass safely through the early stages of technological adolescence, the number of technological civi-

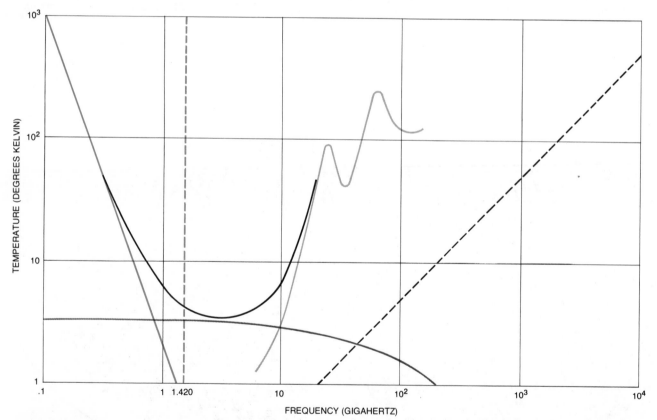

RADIO SPECTRUM of the sky as it is seen from the earth is quite noisy. Any radio telescope picks up the three-degree-Kelvin background radiation (*gray line*), the remnant of the primordial fireball of the "big bang." The background radiation begins to fall off at about 60 gigahertz (billion cycles per second). At that frequency the quantum noise associated with all electromagnetic radiation (*broken black line*) begins to predominate, and the total noise level rises. Noise from within our galaxy (*dark colored line*) is due mainly to synchrotron radiation emitted by particles spiraling in around the lines of force in magnetic fields. Together these three

sources of noise define a broad quiet region in the radio spectrum, between about one gigahertz and 100 gigahertz, that would be nearly the same for observers in the neighborhood of the sun and observers in similar regions of the galaxy. The earth's atmosphere is also a source of noise (*light colored line*) because molecules of water and oxygen absorb and reradiate energy at 22 gigahertz and 60 gigahertz. All sources of noise added together yield the curve in black, representing the total sky noise detected on the earth. The broken vertical line in color is frequency of spin-flip of the electron in un-ionized hydrogen atom at frequency of 1.420 gigahertz.

lizations in the galaxy today might be very large.

It is obviously a highly uncertain exercise to attempt to estimate the number of such civilizations. The opinions of those who have considered the problem differ significantly. Our best guess is that there are a million civilizations in our galaxy at or beyond the earth's present level of technological development. If they are distributed randomly through space, the distance between us and the nearest civilization should be about 300 light-years. Hence any information conveyed between the nearest civilization and our own will take a minimum of 300 years for a one-way trip and 600 years for a question and a response.

Electromagnetic radiation is the fastest and also by far the cheapest method of establishing such contact. In terms of the foreseeable technological developments on the earth, the cost per photon and the amount of absorption of radiation by interstellar gas and dust, radio waves seem to be the most efficient and economical method of interstellar communication. Interstellar space vehicles cannot be excluded a priori, but in all cases they would be a slower, more expensive and more difficult means of communication.

Since we have achieved the capability for interstellar radio communication only in the past few decades, there is virtually no chance that any civilization we come in contact with will be as backward as we are. There also seems to be no possibility of dialogue except between very long-lived and patient civilizations. In view of these circumstances, which should be common to and deducible by all the civilizations in our galaxy, it seems to us quite possible that one-way radio messages are being beamed at the earth at this moment by radio transmitters on planets in orbit around other stars.

To intercept such signals we must guess or deduce the frequency at which the signal is being sent, the width of the frequency band, the type of modulation and the star transmitting the message. Although the correct guesses are not easy to make, they are not as hard as they might seem.

Most of the astronomical radio spectrum is quite noisy [*see illustration on opposite page*]. There are contributions from interstellar matter, from the three-degree-Kelvin background radiation left over from the early history of the universe, from noise that is fundamentally associated with the operation of any detector and from the absorption of radia-

INVESTIGATOR	OBSERVATORY	DATE	FREQUENCY OR WAVELENGTH	TARGETS
DRAKE	N.R.A.O.	1960	1,420 MEGAHERTZ	EPSILON ERIDANI TAU CETI
TROITSKY	GORKY	1968	21 AND 30 CENTIMETERS	12 NEARBY SUNLIKE STARS
VERSCHUUR	N.R.A.O.	1972	1,420 MEGAHERTZ	10 NEARBY STARS
TROITSKY	EURASIAN NET-WORK, GORKY	1972 TO PRESENT	16, 30 AND 50 CENTIMETERS	PULSED SIGNALS FROM ENTIRE SKY
ZUCKERMAN PALMER	N.R.A.O.	1972 TO PRESENT	1,420 MEGAHERTZ	~600 NEARBY SUNLIKE STARS
KARDASHEV	EURASIAN NET-WORK, I.C.R.	1972 TO PRESENT	SEVERAL	PULSED SIGNALS FROM ENTIRE SKY
BRIDLE FELDMAN	A.R.O.	1974 TO PRESENT	22.2 GIGAHERTZ	SEVERAL NEARBY STARS
DRAKE SAGAN	ARECIBO	1975 (IN PROGRESS)	1,420, 1,653 AND 2,380 MEGAHERTZ	SEVERAL NEAR-BY GALAXIES

ATTEMPTS TO DETECT SIGNALS beamed toward the earth by other civilizations have so far been unsuccessful, but the number of stars that have been examined is less than .1 percent of the number that would have to be investigated if there were to be a reasonable statistical chance of discovering one extraterrestrial civilization. "N.R.A.O." is the National Radio Astronomy Observatory in Green Bank, W.Va.; "Gorky" is the 45-foot antenna at Gorky University in the U.S.S.R.; "Eurasian network" is a network of omnidirectional antennas in the U.S.S.R. that is being operated jointly by V. S. Troitsky of Gorky University and N. S. Kardashev of the Institute for Cosmic Research (I.C.R.) of the Academy of Sciences of the U.S.S.R.; "A.R.O." is the Algonquin Radio Observatory at Algonquin Park in Canada; "Arecibo" is 1,000-foot radio-radar antenna at Arecibo Observatory in Puerto Rico.

tion by the earth's atmosphere. This last source of noise can be avoided by placing a radio telescope in space. The other sources we must live with and so must any other civilization.

There is, however, a pronounced minimum in the radio-noise spectrum. Lying at the minimum or near it are several natural frequencies that should be discernible by all scientifically advanced societies. They are the resonant frequencies emitted by the more abundant molecules and free radicals in interstellar space. Perhaps the most obvious of these resonances is the frequency of 1,420 megahertz (millions of cycles per second). That frequency is emitted when the spinning electron in an atom of hydrogen spontaneously flips over so that its direction of spin is opposite to that of the proton comprising the nucleus of the hydrogen atom. The frequency of the spin-flip transition of hydrogen at 1,420 megahertz was first suggested as a channel for interstellar communication

in 1959 by Philip Morrison and Giuseppe Cocconi. Such a channel may be too noisy for communication precisely because hydrogen, the most abundant interstellar gas, absorbs and emits radiation at that frequency. The number of other plausible and available communication channels is not large, so that determining the right one should not be too difficult.

We cannot use a similar logic to guess the bandwidth that might be used in interstellar communication. The narrower the bandwidth is, the farther a signal can be transmitted before it becomes too weak for detection. On the other hand, the narrower the bandwidth is, the less information the signal can carry. A compromise is therefore required between the desire to send a signal the maximum distance and the desire to communicate the maximum amount of information. Perhaps simple signals with narrow bandwidths are sent to enhance the probability of the signals' being received. Perhaps information-rich signals with broad bandwidths are sent in order to achieve rapid and extensive communication. The broad-bandwidth signals would be intended for those enlightened civilizations that have invested major resources in large receiving systems.

When we actually search for signals, it is not necessary to guess the exact bandwidth, only to guess the minimum bandwidth. It is possible to communicate on many adjacent narrow bands at once. Each such channel can be studied individually, and the data from several adjacent channels can be combined to yield the equivalent of a wider channel without any loss of information or sensitivity. The procedure is relatively easy with the aid of a computer; it is in fact routinely employed in studies of pulsars. In any event we should observe the maximum number of channels because of the possibility that the transmitting civilization is not broadcasting on one of the "natural" frequencies such as 1,420 megahertz.

We do not, of course, know now which star we should listen to. The most conservative approach is to turn our receivers to stars that are rather similar to the sun, beginning with the nearest. Two nearby stars, Epsilon Eridani and Tau Ceti, both about 12 light-years away, were the candidates for Project Ozma, the first search with a radio telescope for extraterrestrial intelligence, conducted by one of us (Drake) in 1960. Project Ozma, named after the ruler of Oz in L. Frank Baum's children's stories, was "on the air" for four weeks at 1,420 megahertz. The results were negative. Since then there have been a number of other studies. In spite of some false alarms to the contrary, none has been successful. The lack of success is not unexpected. If there are a million technical civilizations in a galaxy of some 200 billion stars, we must turn our receivers to 200,000 stars before we have a fair statistical chance of detecting a single extraterrestrial message. So far we have listened to only a few more than 200 stars. In other words, we have mounted only .1 percent of the required effort.

Our present technology is entirely adequate for both transmitting and receiving messages across immense interstellar distances. For example, if the 1,000-foot radio telescope at the Arecibo Observatory in Puerto Rico were to transmit information at the rate of one bit (binary digit) per second with a bandwidth of one hertz, the signal could

be received by an identical radio telescope anywhere in the galaxy. By the same token, the Arecibo telescope could detect a similar signal transmitted from a distance hundreds of times greater than our estimate of 300 light-years to the nearest extraterrestrial civilization.

A search of hundreds of thousands of stars in the hope of detecting one message would require remarkable dedication and would probably take several decades. It seems unlikely that any existing major radio telescope would be given over to such an intensive program to the exclusion of its usual work. The construction of one radio telescope or more that would be devoted perhaps half-time to the search seems to be the

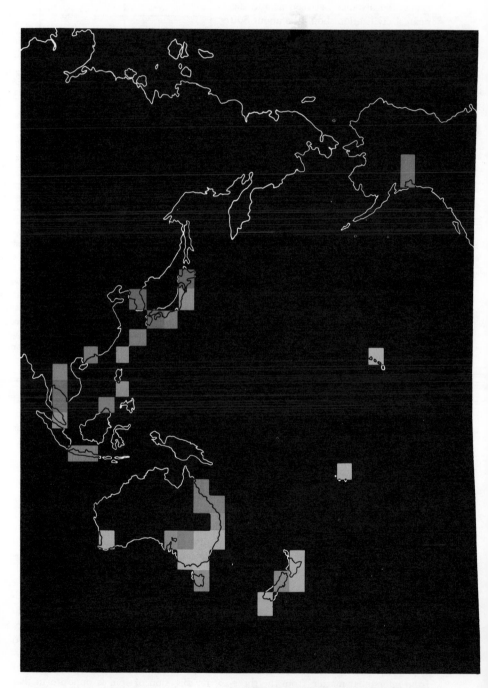

BRIGHTNESS TEMPERATURES
(MILLIONS OF DEGREES K.)

20	300	3,000
90	900	9,000

EARTH IS BRIGHT at the frequencies between 40 and 220 megahertz because of the radiation from FM radio and VHF television broadcasts. The power radiated by the stations is shown averaged over squares five degrees in longitude by five degrees in latitude. The radio brightness is equivalent to

only practical method of seeking out extraterrestrial intelligence in a serious way. The cost would be some tens of millions of dollars.

So far we have been discussing the reception of messages that a civilization would intentionally transmit to the earth. An alternative possibility is that

we might try to "eavesdrop" on the radio traffic an extraterrestrial civilization employs for its own purposes. Such radio traffic could be readily apparent. On the earth, for example, a new radar system employed with the telescope at the Arecibo Observatory for planetary studies emits a narrow-bandwidth signal that, if it were detected from another

star, would be between a million and 10 billion times brighter than the sun at the same frequency. In addition, because of radio and television transmission, the earth is extremely bright at wavelengths of about a meter [*see illustration on these two pages*]. If the planets of other civilizations have a radio brightness comparable to the earth's

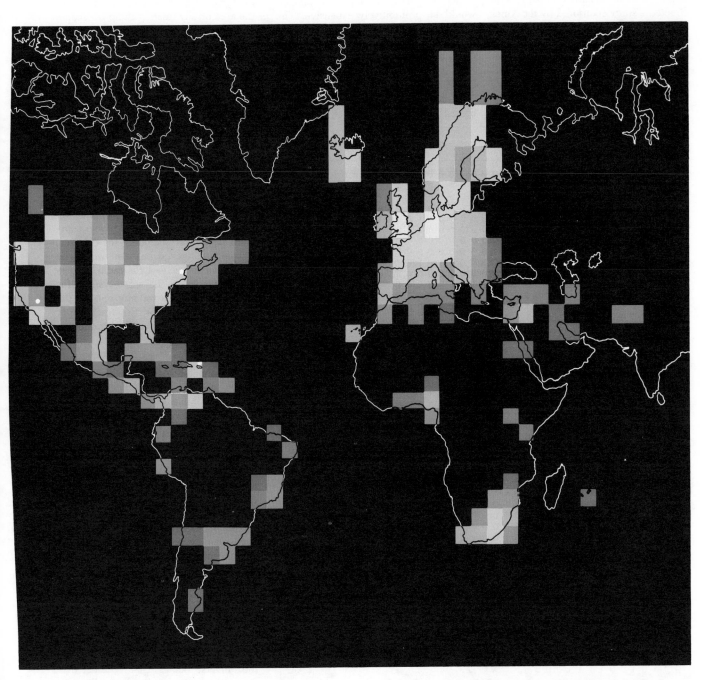

the temperature to which each area on the earth would have to be raised in order to produce the actual radio emission observed. The three brightest areas are the locations of three particularly powerful radar systems: the radio-radar antenna of the Haystack Observatory in Massachusetts, operating at a wavelength of 3.75 centimeters and giving a brightness temperature of 2.3×10^{20} degrees K., the 1,000-foot radio-radar antenna of the Arecibo Observatory, operat-

ing at a wavelength of 12.6 centimeters and giving a brightness temperature of 1.4×10^{21} degrees K., and the 210-foot antenna of the Jet Propulsion Laboratory at Goldstone, Calif., operating at a wavelength of 12.6 centimeters and giving a brightness temperature of 6.2×10^{19} degrees. Systems radiate so much power that at those wavelengths and in the direction of their beam they are brighter than the sun and should be detectable over interstellar distances.

240 EPILOGUE

from television transmission alone, they should be detectable. Because of the complexity of the signals and the fact that they are not beamed specifically at the earth, however, the receiver we would need in order to eavesdrop would have to be much more elaborate and sensitive than any radio-telescope system we now possess.

One such system has been devised in a preliminary way by Bernard M. Oliver of the Hewlett-Packard Company, who directed a study sponsored by the Ames Research Center of the National Aeronautics and Space Administration. The system, known as Cyclops, would consist of an enormous radio telescope connected to a complex computer system. The computer system would be designed particularly to search through the data from the telescope for signals bearing the mark of intelligence, to combine numerous adjacent channels in order to construct signals of various effective bandwidths and to present the results of the automatic analyses for all conceivable forms of interstellar radio communication in a way that would be intelligible to the project scientists.

To construct a radio telescope of enormous aperture as a single antenna would be prohibitively expensive. The Cyclops system would instead capitalize on our ability to connect many individual antennas to act in unison. This concept is already the basis of the Very Large Array now under construction in New Mexico. The Very Large Array consists of 27 antennas, each 82 feet in diameter, arranged in a Y-shaped pattern whose three arms are each 10 miles long. The Cyclops system would be much larger. Its current design calls for 1,500 antennas each 100 meters in diameter, all electronically connected to one another and to the computer system. The array would be as compact as possible but would cover perhaps 25 square miles.

The effective signal-collecting area of the system would be hundreds of times the area of any existing radio telescope, and it would be capable of detecting even relatively weak signals such as television transmissions from civilizations several hundred light-years away. Moreover, it would be the instrument par excellence for receiving signals specifically directed at the earth. One of the greatest virtues of the Cyclops system is that no technological advances would be required in order to build it. The necessary electronic and computer techniques are already well developed. We would need only to build a vast number of items we already build well. The Cyclops system not only would have enormous power for searching for extraterrestrial intelligence but also would be an extraordinary tool for radar studies of the bodies in the solar system, for traditional radio astronomy outside the solar system and for the tracking of space vehicles to distances beyond the reach of present receivers.

The estimated cost of the Cyclops system, ranging up to $10 billion, may make it prohibitively expensive for the time being. Moreover, the argument in favor of eavesdropping is not completely persuasive. Half a century ago, before radio transmissions were commonplace, the earth was quiet at radio wavelengths. Half a century from now, because of the development of cable television and communication satellites that relay signals in a narrow beam, the earth may again be quiet. Thus perhaps for only a century out of billions of years do planets such as the earth appear remarkably bright at radio wavelengths. The odds of our discovering a civilization during that short period in its history may not be good enough to justify the construction of a system such as Cyclops. It may well be that throughout the universe beings usually detect evidence of extraterrestrial intelligence with more traditional radio telescopes. It nonetheless seems clear that our own chances of finding extraterrestrial intelligence will improve if we consciously attempt to find it.

How could we be sure that a particu-

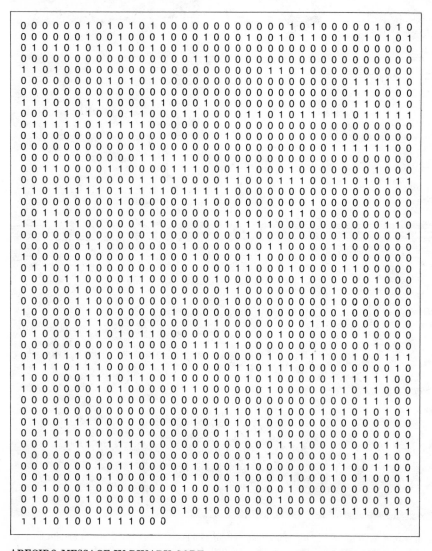

ARECIBO MESSAGE IN BINARY CODE was transmitted in 1974 toward the Great Cluster in Hercules from the 1,000-foot antenna at Arecibo. The message is decoded by breaking up the characters into 73 consecutive groups of 23 characters each and arranging the groups in sequence one under the other, reading right to left and then top to bottom. The result is a visual message (see illustration on opposite page) that can be more easily interpreted by making each 0 of binary code represent a white square and each 1 a black square.

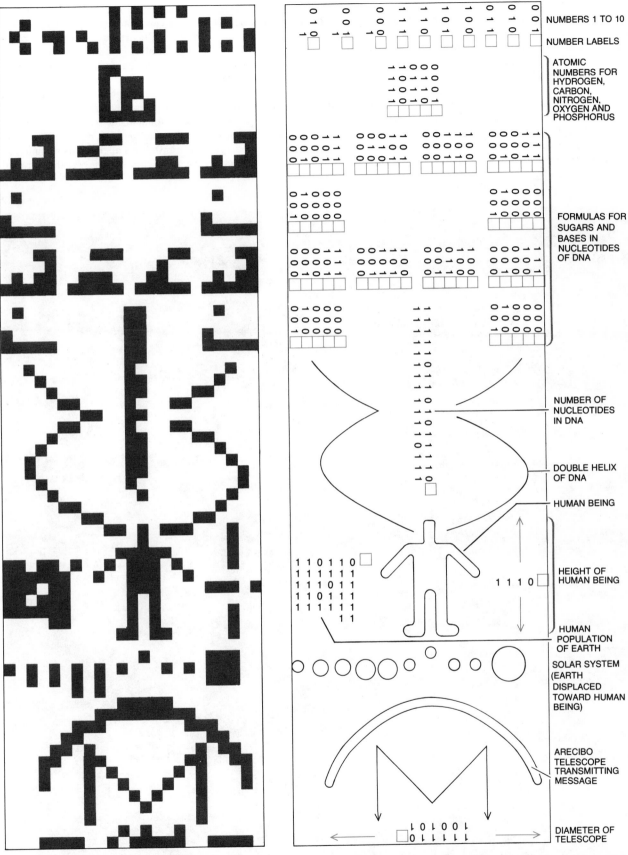

ARECIBO MESSAGE IN PICTURES and accompanying translation shows the binary version of the message decoded. Each number that is used is marked with a label that indicates its start. When all the digits of a number cannot be fitted into one line, the digits for which there is no room are written under the least significant digit. (The message must be oriented in three different ways for all the numbers shown to be read.) The chemical formulas are those for the components of the DNA molecule: the phosphate group, the deoxyribose sugar and the organic bases thymine, adenine, guanine and cytosine. Both the height of the human being and the diameter of the telescope are given in units of the wavelength that is used to transmit the message: 12.6 centimeters.

lar radio signal was deliberately sent by an intelligent being? It is easy to design a message that is unambiguously artificial. The first 30 prime numbers, for example, would be difficult to ascribe to some natural astrophysical phenomenon. A simple message of this kind might be a beacon or announcement signal. A subsequent informative message could have many forms and could consist of an enormous number of bits. One method of transmitting information, beginning simply and progressing to more elaborate concepts, is pictures [see illustration on preceding page].

One final approach in the search for extraterrestrial intelligence deserves mention. If there are indeed civilizations thousands or millions of years more advanced than ours, it is entirely possible that they could beam radio communications over immense distances, perhaps even over the distances of intergalactic space. We do not know how many advanced civilizations there might be compared with the number of more primitive earthlike civilizations, but many of these older civilizations are bound to be in galaxies older than our own. For this reason the most readily detectable radio signals from another civilization may come from outside our galaxy. The relatively small number of such extragalactic transmitters might be more than compensated for by the greater strength

THOUSAND-FOOT ANTENNA of the radio-radar system at the Arecibo Observatory is made of perforated aluminum panels whose spherical shape is accurate to within 1/8 inch over the antenna's entire area of 20 acres. The triangular structure suspended above the antenna holds the receiver and the transmitter for the system. Control rooms and office buildings are to lower right of antenna.

of their signals. At the appropriate frequency they could even be the brightest radio signals in the sky. Therefore an alternative to examining the nearest stars of the same spectral type as the sun is to examine the nearest galaxies. Spiral galaxies such as the Great Nebula in Andromeda are obvious candidates, but the elliptical galaxies are much older and more highly evolved and could conceivably harbor a large number of extremely advanced civilizations.

There might be a kind of biological law decreeing that there are many paths to intelligence and high technology, and that every inhabited planet, if it is given enough time and it does not destroy itself, will arrive at a similar result. The biology on other planets is of course expected to be different from our own because of the statistical nature of the evolutionary process and the adaptability of life. The science and engineering, however, may be quite similar to ours, because any civilization engaged in interstellar radio communication, no matter where it exists, must contend with the same laws of physics, astronomy and radio technology that we do.

Should we be sending messages ourselves? It is obvious that we do not yet know where we might best direct them. One message has already been transmitted to the Great Cluster in Hercules by the Arecibo radio telescope, but only as a kind of symbol of the capabilities of our existing radio technology. Any radio signal we send would be detectable over interstellar distances if it is more than about 1 percent as bright as the sun at the same frequency. Actually something close to 1,000 such signals from our everyday internal communications have left the earth every second for the past two decades. This electromagnetic frontier of mankind is now some 20 light-years away, and it is moving outward at the speed of light. Its spherical wave front, expanding like a ripple from a disturbance in a pool of water and inadvertently carrying the news that human beings have achieved the capacity for interstellar discourse, envelops about 20 new stars each year.

We have also sent another kind of message: two engraved plaques that ride aboard *Pioneer 10* and *Pioneer 11*. These spacecraft, the first artifacts of mankind that will escape from the solar system, will voyage forever through our galaxy at a speed of some 10 miles per second. *Pioneer 10* was accelerated to the velocity of escape from the solar system by the gravitational field of Jupiter

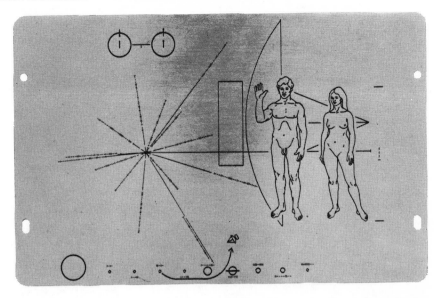

ENGRAVED PLAQUE on the *Pioneer* spacecraft to Jupiter is another message that has been dispatched beyond the solar system. Meaning of symbols is given in text of article.

on December 3, 1973. *Pioneer 11* swung past Jupiter on December 4, 1974, and will travel on to Saturn before it is accelerated on a course to the far side of the galaxy.

Identical plaques for each vehicle were designed by us and Linda Salzman Sagan. Each plaque measures six by nine inches and is made of gold-anodized aluminum. These engraved cosmic greeting cards bear the location of the earth and the time the spacecraft was built and launched. The sun is located with respect to 14 pulsars. The precise periods of the pulsars are specified in binary code to allow them to be identified. Since pulsars are cosmic clocks that are running down at a largely constant rate, the difference in the pulsar periods at the time one of the spacecraft is recovered and the periods indicated on the plaque will enable any technically sophisticated civilization to deduce the year the vehicle was sent on its epic journey. Units of time and distance are specified in terms of the frequency of the hydrogen spin-flip at 1,420 megahertz. In order to identify the exact location of the spacecraft's launch a diagram of the solar system is given. The trajectory of the spacecraft is shown as it leaves the third planet, the earth, and swings by the fifth planet, Jupiter. (The diversion of *Pioneer 11* past Saturn had not been planned when the plaques were prepared.) Last, the plaques show images of a man and a woman of the earth in 1973. An attempt was made to give the images panracial characteristics. Their heights are shown

with respect to the spacecraft and are also given by a binary number stated in terms of the wavelength of the spectral line at 1,420 megahertz (21 centimeters).

These plaques are destined to be the longest-lived works of mankind. They will survive virtually unchanged for hundreds of millions, perhaps billions, of years in space. When plate tectonics has completely rearranged the continents, when all the present landforms on the earth have been ground down, when civilization has been profoundly transformed and when human beings may have evolved into some other kind of organism, these plaques will still exist. They will show that in the year we called 1973 there were organisms, portrayed on the plaques, that cared enough about their place in the hierarchy of all intelligent beings to share knowledge about themselves with others.

How much do we care? Enough to devote an appreciable effort with existing telescopes to search for life elsewhere in the universe? Enough to take a major step such as Project Cyclops that offers a greater chance of carrying us across the threshold, to finally communicate with a variety of extraterrestrial beings who, if they exist, would inevitably enrich mankind beyond imagination? The real question is not how, because we know how; the question is when. If enough of the beings of the earth cared, the threshold might be crossed within the lifetime of most of those alive today.

BIBLIOGRAPHIES

I ELEMENTAL COMPOSITION OF THE BIOMASS

1. The Chemical Elements of Life

TRACE ELEMENTS IN BIOCHEMISTRY. H. J. M. Bowen. Academic Press, 1966.

CONTROL OF ENVIRONMENTAL CONDITIONS IN TRACE ELEMENT RESEARCH: AN EXPERIMENTAL APPROACH TO UNRECOGNIZED TRACE ELEMENT REQUIREMENTS. Klaus Schwarz in Trace Element Metabolism in Animals, edited by C. F. Mills, E. & S. Livingstone, 1970.

THE PROTEINS: METALLOPROTEINS, VOL. V. Edited by Bert L. Vallee and Warren E. C. Wacker. Academic Press, 1970.

CERULOPLASMIN: A LINK BETWEEN COPPER AND IRON METABOLISM. Earl Frieden in Bioinorganic Chemistry, Advances in Chemistry Series 100. American Chemical Society, 1971.

TRACE ELEMENTS IN HUMAN AND ANIMAL NUTRITION. E. J. Underwood. Academic Press, 1971.

BIOGEOCHEMISTRY OF SULFUR ISOTOPES: PROCEEDINGS OF A NATIONAL SCIENCE FOUNDATION SYMPOSIUM, April 12–14, 1962. Edited by Mead LeRoy Jensen. Yale University, Department of Geology, 1962.

THE CALCIUM, MAGNESIUM, POTASSIUM, AND SODIUM BUDGETS FOR A SMALL FORESTED ECOSYSTEM. G. E. Likens, F. H. Bormann, N. M. Johnson and R. S. Pierce in Ecology, Vol. 48, No. 5, pages 772–785; Late Summer, 1967.

THE SULFUR CYCLE IN LAKE WATERS DURING THERMAL STRATIFICATION. M. Stuiver in Geochimica et Cosmochimica Acta, Vol. 31, No. 11, pages 2151–2167; November, 1967.

PRODUCTION AND MINERAL CYCLING IN TERRESTRIAL VEGETATION. L. E. Rodin and N. I. Bazilevich. Edited by G. E. Fogg. Oliver & Boyd, Ltd, 1968.

2. Mineral Cycles

THE YEARLY CIRCULATION OF CHLORIDE AND SULFUR IN NATURE: METEOROLOGICAL, GEOCHEMICAL AND PEDOLOGICAL IMPLICATIONS, PART I. Erik Eriksson in Tellus, Vol. 11, No. 4, pages 375–403; November, 1959.

THE YEARLY CIRCULATION OF CHLORIDE AND SULFUR IN NATURE: METEOROLOGICAL, GEOCHEMICAL AND PEDOLOGICAL IMPLICATIONS, PART II. Erik Eriksson in Tellus, Vol. 12, No. 1, pages 63–109; February, 1960.

3. The Nutrient Cycles of an Ecosystem

NUTRIENT CYCLING F. H. Bormann and G. E. Likens in Science, Vol. 155, No. 3761, pages 424–429; January 27, 1967.

FOREST TRANSPIRATION REDUCTION BY CLEARCUTTING AND CHEMICAL TREATMENT. Robert S. Pierce in Proceedings of the Northeastern Weed Control Conference, Vol. 23; pages 344–349; 1969.

EFFECTS OF FOREST CUTTING AND HERBICIDE TREATMENT ON NUTRIENT BUDGETS IN THE HUBBARD BROOK WATERSHED-ECOSYSTEM.

Gene E. Likens, F. Herbert Bormann, Noye M. Johnson, D. W. Fisher and Robert S. Pierce in *Ecological Monographs*, Vol. 40, No. 1, pages 23–47; Winter, 1970.

4. Trace-Element Deserts

THE DIAGNOSIS OF MINERAL DEFICIENCIES IN PLANTS BY VISUAL SYMPTOMS. T. Wallace. Chemical Publishing Company, 1953.

MECHANISM OF ACTION OF MICRONUTRIENT ELEMENTS IN ENZYME SYSTEMS. W. D. McElroy and A. Nason in *Annual Review of Plant Physiology*, Vol. 5, pages 1–30; 1954.

MICRONUTRIENTS IN CROP VIGOR. Perry R. Stout in *Journal of Agricultural and Food Chem-istry*, Vol. 4, No. 12, pages 1,000–1,006; December, 1956.

TRACE ELEMENTS IN HUMAN AND ANIMAL NUTRITION. E. J. Underwood. Academic Press, Inc., 1956.

5. The Ecology of Desert Plants

ECOLOGY OF DESERT PLANTS. F. W. Went and M. Westergaard in *Ecology*, Vol. 29, No. 3, pages 242–253; July, 1948. Vol. 30, No. 1, pages 1–13 and 26–38; January, 1949.

VEGETATION AND FLORA OF THE SONORAN DESERT. Forrest Shreve and Ira L. Wiggins. Carnegie Institution of Washington, 1951.

II THE STRUCTURE OF LIVING MATTER

6. High-Energy Reactions of Carbon

FREE CARBON ATOM CHEMISTRY. Colin MacKay and Richard Wolfgang in *Science*, Vol. 148[2], No. 3672, pages 899–907; May 14, 1965.

REACTIONS OF ACCELERATED CARBON IONS AND ATOMS. Richard M. Lemmon in *Accounts of Chemical Research*, Vol. 6, No. 2, pages 65–73; February, 1973.

THE NEW WORLD OF GALACTOCHEMISTRY. R. D. Brown in *Chemistry in Britain*, Vol. 9, No. 10, pages 450–455; October, 1973.

7. The Force between Molecules

DIRECT MEASUREMENT OF MOLECULAR ATTRACTION BETWEEN SOLIDS SEPARATED BY A NARROW GAP. B. V. Derjaguin, I. I. Abrikossova and E. M. Lifshitz in *Quarterly Reviews*, Vol. X, No. 3, pages 295–329; 1956.

THE GENERAL THEORY OF MOLECULAR FORCES. F. London in *Transactions of the Faraday Society*, Vol. XXXIII, Part I, pages 8–26; January, 1937.

THE INFLUENCE OF RETARDATION OF THE LONDON-VAN DER WAALS FORCES. H. B. G. Casimir and D. Polder in *The Physical Review*, Vol. 73, No. 4, pages 360–372; February 15, 1948.

VAN DER WAALS FORCES. H. Margenau in *Reviews of Modern Physics*, Vol. 11, No. 1, pages 1–35; January, 1939.

ZUR THEORIE UND SYSTEMATIK MOLEKULARKRÄFTE. F. London in *Zeitschrift für Physik*, Band 63, Heft 3 and 4, pages 245–279; July 1930.

8. Water

WATER ISN'T H_2O. A. M. Buswell in *Journal of the American Water Works Association*, Vol. 30, No. 9, pages 1433–1441; September, 1938.

9. The Structure of Cell Membranes

MEMBRANES OF MITOCHONDRIA AND CHLOROPLASTS. Edited by Efraim Racker. Van Nostrand Reinhold Company, 1969.

STRUCTURE AND FUNCTION OF BIOLOGICAL MEMBRANES. Edited by Lawrence I. Rothfield. Academic Press, 1971.

MEMBRANE MOLECULAR BIOLOGY. Edited by C. F. Fox and A. Keith. Sinauer Associates, Stamford, Conn., 1972.

III RADIANT ENERGY AND LIFE

10. The Chemical Effects of Light

LIGHT-SENSITIVE SYSTEMS: CHEMISTRY AND APPLICATION OF NONSILVER HALIDE PHOTOGRAPHIC PROCESSES. Jaromir Kosar. John Wiley & Sons, 1965.

THE MIDDLE ULTRAVIOLET: ITS SCIENCE AND TECHNOLOGY. Edited by A. E. S. Green. John Wiley & Sons, 1966.

ENERGY TRANSFER FROM HIGH-LYING EXCITED STATES. Gisela K. Oster and H. Kallmann in *Journal de Chimie Physique èt de Physico-Chimie Biologique*, Vol. 64, No. 1, pages 28–32; January, 1967.

PHOTOPOLYMERIZATION OF VINYL MONOMERS. Gerald Oster and Nan-Loh Yang in *Chemical Reviews*, Vol. 68, No. 2, pages 125–151; March 25, 1968.

FLASH PHOTOLYSIS AND SOME OF ITS APPLICATIONS. George Porter in *Science*, Vol. 160, No. 3834, pages 1299–1307; June 21, 1968.

11. Life and Light

PHOTOSYNTHESIS AND RELATED PROCESSES. Eugene I. Rabinowitch. Interscience Publishers, Inc., 1945–1956.

RADIATION BIOLOGY. VOLUME III: VISIBLE AND NEAR-VISIBLE LIGHT. Edited by Alexander Hollaender. McGraw-Hill Book Company, Inc., 1956.

VISION AND THE EYE. M. H. Pirenne. The Pilot Press Ltd., 1948.

THE VISUAL PIGMENTS. H. J. A. Dartnall. Methuen & Co. Ltd., 1957.

12. Ultraviolet Radiation and Nucleic Acid

DISAPPEARANCE OF THYMINE PHOTODIMER IN ULTRAVIOLET IRRADIATED DNA UPON TREATMENT WITH A PHOTOREACTIVATING ENZYME FROM BAKER'S YEAST. Daniel L. Wulff and Claud S. Rupert in *Biochemical and Biophysical Research Communications*, Vol. 7, No. 3, pages 237–240; April, 1962.

THE EFFECTS OF U.V.-IRRADIATION ON NUCLEIC ACIDS AND THEIR COMPONENTS. R. Beukers and W. Berends in *Biochimica et Biophysica Acta*, Vol. 49, No. 1, pages 181–189; April, 1961.

EVIDENCE THAT ULTRAVIOLET-INDUCED THYMINE DIMERS IN DNA CAUSE BIOLOGICAL DAMAGE. Richard B. Setlow and Jane K. Setlow in *Proceedings of the National Academy of Sciences of the U.S.A.*, Vol. 48, No. 7, pages 1250–1257; July, 1962.

PHOTOCHEMISTRY OF NUCLEIC ACIDS AND THEIR CONSTITUENTS. D. Shugar in *The Nucleic Acids*, Vol. III, edited by E. Chargaff and J. N. Davidson. Academic Press, Inc., 1960.

13. The Effects of Light on the Human Body

PREVENTION OF HYPERBILIRUBINEMIA OF PREMATURITY BY PHOTOTHERAPY. J. F. Lucey, M. Ferreiro and J. Hewitt in *Pediatrics*, Vol. 41, pages 1047–1056; 1968.

NATURAL AND SYNTHETIC SOURCES OF CIRCULATING 25-HYDROXYVITAMIN D IN MAN. John G. Haddad and Theodore J. Hahn in *Nature*, Vol. 244, No. 5417, pages 515–516; August 24, 1973.

THE EFFECTS OF LIGHT ON MAN AND OTHER MAMMALS. Richard J. Wurtman in *Annual Review of Physiology*, Vol. 37, pages 467–483; 1975.

DAILY RHYTHM IN HUMAN URINARY MELATONIN. H. J. Lynch, R. J. Wurtman, M. A. Moskowitz, M. C. Archer and M. H. Ho in *Science*, Vol. 187, No. 4172, pages 169–171; January 17, 1975.

14. Light and Plant Development

CONTROL OF GROWTH AND REPRODUCTION BY LIGHT AND DARKNESS. Sterling B. Hendricks in *American Scientist*, Vol. 44, No. 3, pages 229–247; July, 1956.

DETECTION, ASSAY, AND PRELIMINARY PURIFICATION OF THE PIGMENT CONTROLLING PHOTORESPONSIVE DEVELOPMENT OF PLANTS. W. L. Butler, K. H. Norris, H. W. Siegelman and S. B. Hendricks in *Proceedings of the National Academy of Sciences*, Vol. 45, No. 12, pages 1,703–1,708; December, 1959.

INFLUENCE OF LIGHT ON PLANT GROWTH. M. W. Parker and H. A. Borthwick in *Annual Review of Plant Physiology*, Vol. 1, pages 43–58; 1950.

PHYSIOLOGY OF SEED GERMINATION. E. H. Toole, S. B. Hendricks, H. A. Borthwick and Vivian K. Toole in *Annual Review of Plant Physiology*, Vol. 7, pages 299–324; 1956.

15. The Infrared Receptors of Snakes

PROPERTIES OF AN INFRA-RED RECEPTOR. T. H. Bullock and F. P. J. Diecke in *The Journal of Physiology*, Vol. 134, No. 1, pages 47–87; October 29, 1956.

MEN AND SNAKES. Ramona and Desmond Morris. Hutchinson of London, 1965.

RADIANT HEAT RECEPTION IN SNAKES. T. H. Bullock and R. Barrett in *Communication in Behavioral Biology*, Part A, Vol. 1, pages 19–29; January, 1968.

THE PIT ORGANS OF SNAKES. Robert Barrett in *Biology of the Reptilia—Morphology B: Vol. II*. Academic Press, 1970.

SNAKE INFRARED RECEPTORS: THERMAL OR PHOTOCHEMICAL MECHANISM? John F. Harris and R. Igor Gamow in *Science*, Vol. 172, No. 3989, pages 1252–1253; June 18, 1971.

IV   TEMPERATURE AND LIFE

16. Heat and Life

THE KINETIC BASIS OF MOLECULAR BIOLOGY. Frank H. Johnson, Henry Eyring and Milton J. Polissar. John Wiley & Sons, Inc., 1954.

TEMPERATURE AND HUMAN LIFE. C.-E. A. Winslow and L. P. Herrington. Princeton University Press, 1949.

LIFE AND DEATH AT LOW TEMPERATURES. B. J. Luyet and P. M. Gehenio. Biodynamica, 1940.

17. Heat Death

HEAT DEATH, HEAT INJURY, AND TOXIC FACTOR. L. V. Heilbrunn, D. L. Harris, P. G. Le Fevre, W. L. Wilson and A. A. Woodward in *Physiological Zoology*, Vol. 19, No. 4, pages 404–429; October, 1946.

18. Adaptations to Cold

BODY INSULATION OF SOME ARCTIC AND TROPICAL MAMMALS AND BIRDS. P. F. Scholander, Vladimir Walters, Raymond Hock and Laurence Irving in *The Biological Bulletin*, Vol. 99, No. 2, pages 225–236; October, 1950.

BODY TEMPERATURES OF ARCTIC AND SUBARCTIC BIRDS AND MAMMALS. Laurence Irving and John Krog in *Journal of Applied Physiology*, Vol. 6, No. 11, pages 667–680; May, 1954.

EFFECT OF TEMPERATURE ON SENSITIVITY OF THE FINGER. Laurence Irving in *Journal of Applied Physiology*, Vol. 18, No. 6, pages 1201–1205; November, 1963.

METABOLISM AND INSULATION OF SWINE AS BARE-SKINNED MAMMALS. Laurence Irving, Leonard J. Peyton and Mildred Monson in *Journal of Applied Physiology*, Vol. 9, No. 3, pages 421–426; November, 1956.

PHYSIOLOGICAL INSULATION OF SWINE AS BARE-SKINNED MAMMALS. Laurence Irving in *Journal of Applied Physiology*, Vol. 9, No. 3, pages 414–420; November, 1956.

TERRESTRIAL ANIMALS IN COLD: INTRODUCTION. Laurence Irving in *Handbook of Physiology, Section 4: Adaptation to the Environment*. American Physiological Society, 1964.

19. The Freezing of Living Cells

EFFECTS OF LOW TEMPERATURES ON LIVING CELLS AND TISSUES. Audrey U. Smith in *Biological Applications of Freezing and Drying*. Academic Press, Inc., 1954.

LIFE AND DEATH AT LOW TEMPERATURES. B. J. Luyet and P. M. Gehenio. Biodynamica, 1940.

PRESERVATION OF LIVING CELLS AT LOW TEMPERATURES. A. S. Parkes in *Lectures on the Scientific Basis of Medicine*. John de Graff, Inc., 1953.

20. Heat Transfer in Plants

CONVECTION PHENOMENA FROM PLANTS IN STILL AIR. David M. Gates and Charles M. Benedict in *American Journal of Botany*, Vol. 50, No. 6 (Part I), pages 563–573; July, 1963.

ENERGY EXCHANGE IN THE BIOSPHERE. David M. Gates. Harper & Row, Publishers, 1962.

HEAT TRANSFER BETWEEN THE PLANT AND THE ENVIRONMENT. K. Raschke in *Annual Review of Plant Physiology*, Vol. 2, pages 111–126; 1960.

RADIATION AND CONVECTION IN CONIFERS. E. C. Tibbals, Ellen K. Carr, David M. Gates and Frank Kreith in *American Journal of Botany*, Vol. 51, No. 5, pages 529–538; May–June, 1964.

21. Thermal Pollution and Aquatic Life

THE PHYSIOLOGY OF FISHES. Edited by Margaret E. Brown. Academic Press, Inc., 1957.

THE PHYSIOLOGY OF CRUSTACEA, VOL. I: METABOLISM AND GROWTH. Edited by Talbot H. Waterman. Academic Press, Inc., 1960.

FISH AND RIVER POLLUTION. J. R. Erichsen Jones. Butterworths, 1964.

A FIELD AND LABORATORY INVESTIGATION OF THE EFFECT OF HEATED EFFLUENTS ON FISH. J. S. Alabaster in *Fishery Investigations, Ministry of Agriculture, Fisheries, and Food*, Series I, Vol. 6, No. 4; 1966.

THERMAL POLLUTION—1968. Subcommittee on Air and Water Pollution of the Committee on Public Works. U.S. Government Printing Office, 1968.

V PRESSURE, GRAVITY, AND LIFE

22. Cells at High Pressure

THE KINETIC BASIS OF MOLECULAR BIOLOGY. Frank H. Johnson, Henry Eyring and Milton J. Polissar. John Wiley & Sons, Inc., 1954.

PHYSIO-CHEMICAL EFFECTS OF PRESSURE. S. D. Harmann. Academic Press, Inc., 1957.

THE PHYSIOLOGICAL EFFECTS OF PRESSURE. McKeen Cattell in *Biological Reviews of the Cambridge Philosophical Society*, Vol. 11, No. 4, pages 441–476; October, 1936.

23. Animals of the Abyss

ASPECTS OF DEEP SEA BIOLOGY. N. B. Marshall. Hutchinson's Scientific and Technical Publications, 1954.

DEEP-SEA RESEARCH. Vols. 1–4. Pergamon Press, Inc., 1953–57.

THE DEPTHS OF THE OCEAN: A GENERAL ACCOUNT OF THE MODERN SCIENCE OF OCEANOGRAPHY BASED LARGELY ON THE SCIENTIFIC RESEARCH OF THE NORWEGIAN STEAMER MICHAEL SARS IN THE NORTH ATLANTIC. Sir John Murray and Johan Hjort. Macmillan and Company, Limited, 1912.

THE GALATHEA DEEP SEA EXPEDITION 1950-1952. Edited by Anton F. Bruun, Sv. Greve, Hakon Mielche and Ragnar Spärck. The Macmillan Company, 1957.

MARINE GEOLOGY. Ph. H. Kuenen. John Wiley & Sons, Inc., 1950.

THE OCEANS: THEIR PHYSICS, CHEMISTRY AND GENERAL BIOLOGY. H. U. Sverdrup, Martin W. Johnson and Richard H. Fleming. Prentice-Hall, Inc., 1942.

PAPERS IN MARINE BIOLOGY AND OCEANOGRAPHY. *Deep-Sea Research*, Supplement to Vol. 3. Pergamon Press, 1955.

24. The Ear

THE EARLY HISTORY OF HEARING—OBSERVATIONS AND THEORIES. Georg v. Békésy and Walter A. Rosenblith in *The Journal of the Acoustical Society of America*, Vol. 20, No. 6, pages 727–748; November, 1948.

HEARING: ITS PSYCHOLOGY AND PHYSIOLOGY. Stanley Smith Stevens and Hallowell Davis. John Wiley & Sons, Inc., 1938.

PHYSIOLOGICAL ACOUSTICS. Ernest Glen Wever and Merle Lawrence. Princeton University Press, 1954.

25. The Sensitivity of Organisms to Gravity

GRAVITY AND THE ORGANISM. Edited by Solon A. Gordon and Melvin J. Cohen. University of Chicago Press, 1971.

GEOTROPIC RESPONSE PATTERN OF THE AVENA COLEOPTILE. I. DEPENDENCE ON ANGLE AND DURATION OF STIMULATION. Barbara Gillespie Pickard in *Canadian Journal of Botany*, Vol. 51, No. 5, pages 1003–1021; May, 1973.

THE TRANSPORT OF AUXIN. Mary Helen M. Goldsmith in *Annual Review of Plant Physiology*, Vol. 19, pages 347–360; 1968.

EPILOGUE

26. The Search for Extraterrestrial Intelligence

WHAT SHALL WE SAY TO MARS? H. W. Nieman and C. Wells Nieman in *Scientific American*, Vol. 122, No. 12, page 298; March 20, 1920.

PROJECT CYCLOPS. B. M. Oliver et al. National Aeronautics and Space Administration, Publication CR-114445; 1972.

COMMUNICATION WITH EXTRATERRESTRIAL INTELLIGENCE. Edited by Carl Sagan. The MIT Press, 1973.

INDEX